Gas Flows in Microsystems

Gas Flows in Microsystems

Special Issue Editors

Stéphane Colin
Lucien Baldas

MDPI • Basel • Beijing • Wuhan • Barcelona • Belgrade

MDPI

Special Issue Editors

Stéphane Colin
Université de Toulouse
France

Lucien Baldas
Université de Toulouse
France

Editorial Office
MDPI
St. Alban-Anlage 66
4052 Basel, Switzerland

This is a reprint of articles from the Special Issue published online in the open access journal *Micromachines* (ISSN 2072-666X) in 2019 (available at: https://www.mdpi.com/journal/micromachines/special_issues/Gas_Flows_in_Microsystems)

For citation purposes, cite each article independently as indicated on the article page online and as indicated below:

LastName, A.A.; LastName, B.B.; LastName, C.C. Article Title. *Journal Name* **Year**, *Article Number*, Page Range.

ISBN 978-3-03921-542-3 (Pbk)
ISBN 978-3-03921-543-0 (PDF)

Contents

About the Special Issue Editors

Stéphane Colin has been a Professor in the Mechanical Engineering Department of the National Institute of Applied Sciences, University of Toulouse, France, since 2002. He obtained his degree in Engineering in 1987 and received his PhD in Fluid Mechanics from the Polytechnic National Institute of Toulouse in 1992. He established, in 1999, the Microfluidics Group of the Hydrotechnic Society of France. He initiated and co-chaired a series of French (µFlu'02 to µFlu'06) and European (µFlu'08 to µFlu'18) Microfluidics Conferences. His current research is mainly focused on gas microflows, with a particular interest in the experimental analysis of rarefied flows. He was the coordinator of the GASMEMS European Network aimed at training young researchers in the field of gas flows in MEMS. He is the author of more than 140 scientific papers in international journals or conferences and the editor or co-author of four textbooks.

Lucien Baldas has been an Associate Professor in the mechanical engineering department of the National Institute of Applied Sciences (INSA) of Toulouse/University of Toulouse, France since 1997. He received his PhD in Fluid Mechanics in 1993 from the National Polytechnic Institute of Toulouse. His current research, at the Institute Clément Ader in Toulouse, focuses on compressible flows in pneumatic systems and on microfluidics with a special interest in rarefied gaseous flows in microsystems and in microfluidic actuators for flow control and heat transfer enhancement. From 2008 to 2012, he was assistant coordinator of the European Network GASMEMS, which contributed to the structuration of the European research community in the field of gas microflows. He is now strongly involved in the European Training Network MIGRATE, dealing with heat and mass transfer in gas-based microscale processes.

micromachines

MDPI

Editorial

Editorial for the Special Issue on Gas Flows in Microsystems

Stéphane Colin * and Lucien Baldas *

Institut Clément Ader (ICA), Université de Toulouse, CNRS-INSA-ISAE-Mines Albi-UPS, 31400 Toulouse, France
* Correspondence: stephane.colin@insa-toulouse.fr (S.C.); lucien.baldas@insa-toulouse.fr (L.B.)

Received: 22 July 2019; Accepted: 23 July 2019; Published: 25 July 2019

The last two decades have witnessed a rapid development of microelectromechanical systems (MEMS) involving gas microflows in various technical fields. Gas microflows can, for example, be observed in micro heat exchangers designed for chemical applications or for cooling of electronic components, in fluidic microactuators developed for active flow control purposes, in micronozzles used for the micropropulsion of nano- and picosatellites, in micro gas chromatographs, analyzers or separators, in vacuum generators and in Knudsen micropumps, as well as in some organs-on-a-chip such as artificial lungs. These flows are rarefied due to the small MEMS dimensions, and the rarefaction can be increased by low-pressure conditions. The flows relate to the slip flow, transition, or free molecular regimes, and can involve monatomic or polyatomic gases and gas mixtures. Hydrodynamics and heat and mass transfer are strongly impacted by rarefaction effects, and temperature-driven microflows offer new opportunities for designing original MEMS for gas pumping or separation. Accordingly, this Special Issue of *Micromachines*, entitled "Gas Flows in Microsystems" contains 14 papers (1 review and 13 research articles), which focus on novel theoretical and numerical models or data, as well as on new experimental results and techniques, for improving knowledge on heat and mass transfer in gas microflows.

A few papers of this Special Issue have addressed fundamental issues on gas microflow modeling. Many microfluidic systems involving gases operate in the slip or early transition regimes, and the bulk flow can then be modeled in these slightly rarefied regimes by continuum approaches. In the Knudsen layer close to the walls, however, local thermodynamic disequilibrium takes place and specific approaches are required. An effective mean free path model was implemented by Bhagat et al. [1] in OpenFOAM, an open source computational fluid dynamics (CFD) code based on the Navier–Stokes equations. A hybrid Langmuir–Maxwell–Smoluchowski velocity slip and temperature jump boundary condition was used with a Knudsen layer formulation and tested on the backward facing step channel. Comparison with direct simulation Monte Carlo (DSMC) demonstrated a significant improvement over existing CFD solvers. Pressure drop in microchannels is a fundamental quantity to control for many engineering problems. In a number of devices, the entrance region is not negligible and should be taken into account. Duan et al. [2] proposed a semi-analytical model based on the momentum equation coupled with first-order slip boundary conditions. A good accuracy of this model, within 5%, was demonstrated in the slip flow regime by comparison with CFD simulations, as well as with experimental and numerical data from the literature. Even in non-rarefied regimes, the determination of friction factors is not straightforward, as demonstrated by Rehman et al. [3] who determined the average friction factor in gas flows along adiabatic microchannels with rectangular cross-section. From an experimental and numerical analysis, covering a large range of the Reynolds number from 200 to 20,000, they pointed out the role of minor loss coefficients and demonstrated that they should not be considered as constant. Gas microflows can also be encountered in gas microbearings where the aerodynamic lubrication performance has a critical impact on the stability of the bearing-rotor system in micromachines. The interactive effects of gas rarefaction and surface roughness on the static and dynamic characteristics of ultra-thin film gas lubrication in journal microbearings were investigated by Wu et al. [4] under various operative conditions and structure parameters. On the basis of the fractal geometry theory and the Boltzmann slip correction factor, the authors demonstrated that high values of the eccentricity ratio and bearing number tend to significantly increase the principal stiffness coefficients, and the fractal roughness surface considerably affects the ultra-thin film damping characteristics compared to smooth surface bearing. Controlling gas damping at microscale is also of high interest for the development of new compliant resonant microsystems. Mirzazadeh and Mariani [5] developed simple analytical solutions to estimate the dissipation in the ideal case of air flow between infinite plates, at atmospheric pressure, for application to comb-drive actuators. The results of numerical simulations were also reported to assess the effect of the finite size of actual geometries on damping.

These fundamental papers underline the importance of experimental data for validating simplified or more complex models. Unfortunately, the amount of experimental data on gas microflows is very limited, compared to the high number of numerical studies. The main difficulty, as explained in the review by Brandner [6], is due to the fact that conventional measurement techniques (for temperature, pressure, etc.) cannot be adapted to gas microflows, due to their intrusiveness and/or low signal delivery, especially when timely and spatially correlated measurements are required. In that review, the potential of nuclear magnetic resonance and magnetic resonance imaging for analyzing gas microflows is discussed. Some issues linked to the intrusiveness of sensors, even highly miniaturized, are also treated in the paper by Mironov et al. [7], in which the interaction between a Pitot microtube and a supersonic microjet is investigated.

The last series of papers published in this Special Issue are devoted to specific microsystems designed for the control or the analysis of gas microflows. One specific phenomenon experienced in rarefied gas flows is thermal transpiration, which allows the design of thermally driven pumps without any moving mechanical part. These so-called Knudsen pumps are very appealing for a number of applications requiring the control of a pressure, a flow rate, or the intake of a gas sample. Lopez Quesada et al. [8] provided some guidelines for the design of Knudsen micropumps based on architectures adapted to target applications which can require a high vacuum, a high flowrate, or a compromise between these two parameters. Their work is based on kinetic modeling and simulations, but takes into account some manufacturing constraints. Zhang et al. [9] focused their numerical analysis on the behavior of N_2–O_2 gas mixtures in a more classic design of the Knudsen pump. The thermal transpiration efficiency is related to the molecular mass of the gas and, even with a molecular mass close to that of O_2, N_2 was submitted to a stronger thermal transpiration effect. In addition, the lighter gas, N_2, could effectively promote the motion of the heavier gas, O_2. If separation of gas species from a mixture is of practical interest at a microfluidic level, it is also the case of mixing. Meskos et al. [10] numerically investigated the mixing process of two pressure-driven rarefied gas flows between parallel plates and evaluated the mixing length using a DSMC approach. They proposed a simple approach to control the output mixture composition, by only adding a splitter in an appropriate location of the microsystem's mixing zone. This mixer was working in a steady state, differently from the option analyzed by Noël et al. [11] who proposed a new multi-stage design of pulsed micromixer. For example, they demonstrated that, for a 1 s pulse of pure gas (formaldehyde) followed by a 9 s pulse of pure carrier gas (air), an effective mixing up to 94–96% was obtained at the exit of the micromixer. There is currently a high demand for compact, accurate, and rapid gas detectors. Several papers in this Special Issue are focused on this subject. Khan et al. [12] developed a toluene detector based on deep ultraviolet (UV) absorption spectrophotometry. They implemented two types of hollow-core waveguides, namely, a glass capillary tube with aluminum-coated inner walls and an aluminum capillary tube, and obtained limits of detection of 8.1 ppm and 12.4 ppm, respectively. Rezende et al. [13] proposed a micro milled microfluidic photoionization detector of volatile organic compounds. The device does not require any glue, which facilitates the easy replacement of components, and the estimated detection limit is 0.6 ppm for toluene without any amplification unit. Finally, Lara-Ibeas et al. [14] developed a compact prototype of gas chromatograph equipped with a preconcentration unit, able to detect sub-ppb levels of benzene, toluene, ethylbenzene, and xylenes (BTEX) in gaseous mixtures. Detection limits of 0.20, 0.26, 0.49, 0.80, and 1.70 ppb were determined for benzene, toluene, ethylbenzene, m/p-xylenes, and o-xylene, respectively.

We wish to thank all authors who submitted their papers to this Special Issue. We would also like to acknowledge all the reviewers for dedicating their time to provide careful and timely reviews to ensure the quality of this Special Issue.

Conflicts of Interest: The authors declare no conflict of interest.

References

1. Bhagat, A.; Gijare, H.; Dongari, N. Modeling of Knudsen layer effects in the micro-scale backward-facing step in the slip flow regime. *Micromachines* **2019**, *10*, 118. [CrossRef] [PubMed]
2. Duan, Z.; Ma, H.; He, B.; Su, L.; Zhang, X. Pressure drop of microchannel plate fin heat sinks. *Micromachines* **2019**, *10*, 80. [CrossRef] [PubMed]
3. Rehman, D.; Morini, G.L.; Hong, C. A comparison of data reduction methods for average friction factor calculation of adiabatic gas flows in microchannels. *Micromachines* **2019**, *10*, 171. [CrossRef] [PubMed]
4. Wu, Y.; Yang, L.; Xu, T.; Xu, H. Interactive effects of rarefaction and surface roughness on aerodynamic lubrication of microbearings. *Micromachines* **2019**, *10*, 155. [CrossRef] [PubMed]
5. Mirzazadeh, R.; Mariani, S. Estimation of air damping in out-of-plane comb-drive actuators. *Micromachines* **2019**, *10*, 263. [CrossRef] [PubMed]
6. Brandner, J.J. In-Situ measurements in microscale gas flows—conventional sensors or something else? *Micromachines* **2019**, *10*, 292. [CrossRef] [PubMed]
7. Mironov, S.G.; Aniskin, V.M.; Korotaeva, T.A.; Tsyryulnikov, I.S. Effect of the Pitot tube on measurements in supersonic axisymmetric underexpanded microjets. *Micromachines* **2019**, *10*, 235. [CrossRef] [PubMed]
8. López Quesada, G.; Tatsios, G.; Valougeorgis, D.; Rojas-Cárdenas, M.; Baldas, L.; Barrot, C.; Colin, S. Design guidelines for thermally driven micropumps of different architectures based on target applications via kinetic modeling and simulations. *Micromachines* **2019**, *10*, 249. [CrossRef] [PubMed]
9. Zhang, Z.; Wang, X.; Zhao, L.; Zhang, S.; Zhao, F. Study of flow characteristics of gas mixtures in a rectangular Knudsen pump. *Micromachines* **2019**, *10*, 79. [CrossRef] [PubMed]
10. Meskos, S.; Stefanov, S.; Valougeorgis, D. Gas mixing and final mixture composition control in simple geometry micro-mixers via DSMC analysis. *Micromachines* **2019**, *10*, 178. [CrossRef] [PubMed]
11. Noël, F.; Serra, C.A.; Le Calvé, S. Design of a novel axial gas pulses micromixer and simulations of its mixing abilities via computational fluid dynamics. *Micromachines* **2019**, *10*, 205. [CrossRef] [PubMed]
12. Khan, S.; Newport, D.; Le Calvé, S. Development of a toluene detector based on deep UV absorption spectrophotometry using glass and aluminum capillary tube gas cells with a LED source. *Micromachines* **2019**, *10*, 193. [CrossRef] [PubMed]
13. Rezende, G.C.; Le Calvé, S.; Brandner, J.J.; Newport, D. Micro milled microfluidic photoionization detector for volatile organic compounds. *Micromachines* **2019**, *10*, 228. [CrossRef] [PubMed]
14. Lara-lbeas, I.; Rodríguez-Cuevas, A.; Andrikopoulou, C.; Person, V.; Baldas, L.; Colin, S.; Le Calvé, S. Sub-ppb level detection of BTEX gaseous mixtures with a compact prototype GC equipped with a preconcentration unit. *Micromachines* **2019**, *10*, 187. [CrossRef] [PubMed]

micromachines

MDPI

Article

Modeling of Knudsen Layer Effects in the Micro-Scale Backward-Facing Step in the Slip Flow Regime

Apurva Bhagat , Harshal Gijare and Nishanth Dongari *

Department of Mechanical and Aerospace Engineering, Indian Institute of Technology, Hyderabad, Kandi, Medak 502285, India; me13m15p000001@iith.ac.in (A.B.); me13m15p000002@iith.ac.in (H.G.)
* Correspondence: nishanth@iith.ac.in

Received: 27 December 2018; Accepted: 2 February 2019; Published: 12 February 2019

Abstract: The effect of the Knudsen layer in the thermal micro-scale gas flows has been investigated. The effective mean free path model has been implemented in the open source computational fluid dynamics (CFD) code, to extend its applicability up to slip and early transition flow regime. The conventional Navier-Stokes constitutive relations and the first-order non-equilibrium boundary conditions are modified based on the effective mean free path, which depends on the distance from the solid surface. The predictive capability of the standard 'Maxwell velocity slip—Smoluchwoski temperature jump' and hybrid boundary conditions 'Langmuir Maxwell velocity slip—Langmuir Smoluchwoski temperature jump' in conjunction with the Knudsen layer formulation has been evaluated in the present work. Simulations are carried out over a nano-/micro-scale backward facing step geometry in which flow experiences adverse pressure gradient, separation and re-attachment. Results are validated against the direct simulation Monte Carlo (DSMC) data, and have shown significant improvement over the existing CFD solvers. Non-equilibrium effects on the velocity and temperature of gas on the surface of the backward facing step channel are studied by varying the flow Knudsen number, inlet flow temperature, and wall temperature. Results show that the modified solver with hybrid Langmuir based boundary conditions gives the best predictions when the Knudsen layer is incorporated, and the standard Maxwell-Smoluchowski can accurately capture momentum and the thermal Knudsen layer when the temperature of the wall is higher than the fluid flow.

Keywords: rarefied gas flows; micro-scale flows; Knudsen layer; computational fluid dynamics (CFD); OpenFOAM; Micro-Electro-Mechanical Systems (MEMS); Nano-Electro-Mechanical Systems (NEMS); backward facing step

1. Introduction

Conventional Navier-Stokes (NS) equations are based on the assumption that the mean free path (MFP) of the particle is much smaller than the characteristic length scale of the system. However, in a few engineering applications of interest, this continuum assumption deviates, if the flow is highly rarefied (e.g., vehicles operating at high altitude conditions), or length scale of the system is of the order of MFP of the gas (e.g., micro-scale gas flows). Flow through the nano-/micro-scale devices is dominated by non-equilibrium effects such as rarefaction and gas molecule-surface interactions. Knudsen layer (KL) is one such phenomenon, where a non-equilibrium region is formed near the solid surface in rarefied/micro-scale gas flows. Molecule-surface collisions are dominated by the presence of a solid surface reducing the mean time between collisions, i.e., unconfined MFP of the gas is effectively reduced in the presence of a solid surface [1]. Molecules collide with the wall more frequently than with other molecules, leading to the formation of the Knudsen layer as demonstrated in Figure 1. Linear constitutive relations for shear stress and heat flux are no longer valid in this region [2–4].

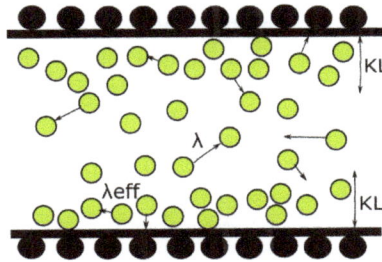

Figure 1. Schematic of inter-molecular and molecule-surface collisions leading to the formation of Knudsen layer (KL).

Behavior of the non-equilibrium gas flows and structure of KL have been extensively investigated by directly solving the Boltzmann equation [5], kinetic equations (e.g., the BGK (Bhatnagar, Gross and Krook) model, rigid-sphere model, the Williams model) [6–9] or alternative hydrodynamic models such as the Burnett equation, super-Burnett equations, Grad 13 moment equation and the regularized Grad moment equations [2,10–13]. However, obtaining solutions using these models is computationally challenging due to the complicated structure of molecular collisions term, lack of well-posed boundary conditions and inherent instability. The direct simulation Monte Carlo (DSMC) technique [14,15] is one of the most accepted and reliable methods for solving gas flows in the non-equilibrium region. Collisions of some representative particles with each other and wall boundaries, are handled in a stochastic manner [16]. As a result, computational cost becomes pretty intensive in case of micro-scale flows due to high density and low flow velocity. A few researchers [17–19] have applied DSMC method to analyze gas flow through micro-channel, and statistical scatter have been a critical issue. A huge sample size is required to reduce the statistical scatter, which makes the DSMC simulation tedious and time-consuming. These difficulties can be overcome if NS equations are extended with higher-order constitutive relations and boundary conditions so that they can accurately capture the Knudsen layer and non-linear flow physics of micro-scale gas flows.

Few researchers have attempted to include non-equilibrium effects in NS framework from different viewpoints. Myong [13] has derived the second-order macroscopic constitutive equation from the kinetic Boltzmann equation and obtained analytical solutions to the KL in Couette flow within continuum frame-work. Li et al. [20] have proposed an effective viscosity model to account for the wall effect in the wall adjacent layer. Lockerby et al. [21] have introduced the concept of wall function into a scaled stress-strain rate relation by fitting the velocity profile obtained from the linearized Boltzmann equation. This idea has been further extended to obtain Kn dependent functions [22], power-law scaling of constitutive relations [23] and discontinues correction function for near wall and far wall region [24,25]. The key disadvantage of these models is that they usually contain some empirical parameters which are specific for the geometry and the flow conditions, and it is not very convenient to extract them for various practical applications. Unlike these models, Guo et al. [26] developed a model based on the effective mean free path in which the wall bounding effect is considered with an assumption that MFP follows an exponential probability distribution. On the other hand, Dongari et al. [27] have hypothesized that the MFP of molecules follow a power-law based distribution, which is also valid in thermodynamic non-equilibrium.

Although several attempts have been made to improve constitutive relations, not much attention is given to the wall boundary conditions. Most previous studies are based on the classical velocity slip boundary conditions, as Lockerby et al. [21] and Dongari et al. [28] have used the first order velocity slip boundary condition by replacing MFP with effective MFP. Generalized second order slip boundary condition for velocity has been used by a few researchers [23,26,29]. Also, studies have been limited to low-speed isothermal gas flows over simple geometries like planar surface and cylinder. The temperature jump boundary condition with KL effects within NS framework, in thermal rarefied gas flows, has been overlooked in the literature to the best of authors knowledge. Present work aims to bridge the gap in the literature and different non-equilibrium boundary conditions, for both velocity and temperature, have been extended using effective MFP model proposed by Dongari et al. [27,28]. This model is rigorously validated against molecular dynamics (MD), DSMC, and experimental data, and also compared with other theoretical models [30–33]. The backward-facing step geometry is chosen in this manuscript as the flow experiences adverse pressure gradient and the separation.

In the present work, the effective MFP model [27,28] has been implemented in NS frame-work in open source CFD tool OpenFOAM. The mean free path is modified based on local flow density, and linear constitutive relations for shear stress and heat flux are modified to account for the effect of KL. In addition to this, first-order boundary conditions, (i) Maxwell velocity slip [34], (ii) Smoluchwoski temperature jump [35], as well as (iii) Langmuir Maxwell [36] and (iv) Langmuir Smoluchwoski [36] are modified with the effective mean free path. The simulations are carried out over a 2D backward-facing step nano- and micro-channel in the slip and early transition flow regime (0.01 < Kn < 0.1, Kn is the non-dimensional Knudsen number defined as λ/L, to indicate the degree of rarefaction, and L is the length-scale of the system). The novel contribution of the present work is that NS equations, combined with KL based constitutive relations and boundary conditions, are investigated for the flows with separation and reattachment. Results are compared with DSMC data [37], and validity of the proposed method is investigated. Effect of change in Knudsen number, inlet flow and wall temperature on the flow properties such as velocity slip and temperature jump is studied.

2. Computational Methodology

OpenFOAM (Open Field Operation and Manipulation, CFD Direct Ltd, UK) is a popular open source, parallel friendly *CFD* software, which is based on C++ library tools and a collection of various applications (created using these libraries). Implementation of tensor fields, partial differential equations, boundary conditions, etc. can be handled using these libraries [38,39].

The *rhoCentralFoam* solver is used as a base solver in the present study. It is a density-based compressible flow solver based on the central-upwind schemes of Kurganov and Tadmor [40,41]. Calculation of transport properties, formulation of KL within NS equations, governing equations with non-linear constitutive relations, and non-equilibrium various boundary conditions are explained in the Sections 2.1–2.4.

2.1. Transport Properties

Transport coefficients are obtained using kinetic theory treatment [4,42,43], and the dynamic viscosity is calculated as :

$$\mu = 2.6693 \times 10^{-5} \frac{\sqrt{MT}}{d^2 F(k_B T/\epsilon)}, \tag{1}$$

where M is the molecular weight, T is the temperature and d is the characteristic molecular diameter. $F(k_B T/\epsilon)$ is the function of $k_B T/\epsilon)$, which gives the variation of the effective collision diameter as a function of temperature (values are obtained from Bird et al. [14]), where ϵ is a characteristic energy of interaction between the molecules and k_B is the Boltzmann constant. Values of d and ϵ/k_B for different gases are associated with the Lennard-Jones potential, and are tabulated by Anderson et al. [44].

Thermal conductivity is calculated by Eucken's relation [45] :

$$\kappa = \mu \left(C_p + \frac{5}{4} R \right), \tag{2}$$

where C_p is the specific heat capacity at constant pressure and R is the specific gas constant.

Micromachines **2019**, *10*, 118

2.2. Knudsen Layer Formulation

Using kinetic theory of gases [42], Maxwellian mean free path of a gas can be expressed as,

$$\lambda = \frac{\mu}{\rho}\sqrt{\frac{\pi}{2RT}}, \tag{3}$$

where μ is obtained from Equation (1), and ρ is the gas density.

The geometry dependent effective MFP model proposed by Dongari et al. [27,28,46–48] is defined as,

$$\lambda_{eff} = \lambda\beta, \tag{4}$$

where β is the normalized MFP which is function of local MFP and normal distance from the solid surface ($\hat{y}$) defined as,

$$\beta = 1 - \frac{1}{96}\left[\left(1+\frac{\hat{y}}{\lambda}\right)^{1-n} + 2\sum_{j=1}^{7}\left(1+\frac{\hat{y}}{\lambda\cos(j\pi/16)}\right)^{1-n} + 4\sum_{j=1}^{8}\left(1+\frac{\hat{y}}{\lambda\cos((2j-1)\pi/32)}\right)^{1-n}\right] \tag{5}$$

where exponent $n = 3$. This function is based on the assumption that molecules follow a non-Brownian motion when flow is confined by a solid surface. Detailed mathematical derivation and formulation of β for planar and cylindrical surfaces can be obtained in references [28,47] (refer to Equation (12) in [28] for planar geometry and Equation (18) in [47] for non-planar geometry).

Using Equations (3) and (4), effective viscosity is calculated as:

$$\mu_{eff} = \mu\beta. \tag{6}$$

MFP for thermal cases (i.e., if temperature gradient exists in the flow) can be expressed as $\lambda_T = 1.922\lambda$ [5] for hard sphere molecules. It has been stated by Sone et al. [5,49] on the basis of solution of linearized Boltzmann equation for hard sphere molecular model. Therefore, effective MFP expression for thermal cases becomes:

$$\lambda_{eff(T)} = \lambda_T\beta_T, \tag{7}$$

where β_T is the normalized MFP for thermal cases [28].

Using Equations (2), (3) and (7), effective thermal conductivity is calculated as:

$$\kappa_{eff} = \kappa\beta_T. \tag{8}$$

One should note that the transport properties μ and κ of the fluid are initially calculated from the kinetic theory based transport model described in the Section 2.1. Their effective values, i.e., μ_{eff} and κ_{eff} are obtained to achieve the non-linear form of constitutive relations, which account for the non-equilibrium effect of KL.

2.3. Governing Equations

The *rhoCentralFoam* solver computes the following governing equations, namely conservation of total mass, momentum and energy [50]:

$$\frac{\partial\rho}{\partial t} + \nabla\cdot[\rho\mathbf{u}] = 0, \tag{9}$$

$$\frac{\partial(\rho\mathbf{u})}{\partial t} + \nabla\cdot[\mathbf{u}(\rho\mathbf{u})] + \nabla p + \nabla\cdot\mathbf{\Pi} = 0, \tag{10}$$

$$\frac{\partial(\rho E)}{\partial t} + \nabla\cdot[\mathbf{u}(\rho E)] + \nabla\cdot[\mathbf{u}p] + \nabla\cdot[\mathbf{\Pi}\cdot\mathbf{u}] + \nabla\cdot\mathbf{j} = 0, \tag{11}$$

where **u** is the velocity of the flow, p is pressure, $E = e + \frac{|u|^2}{2}$ is the total energy, e is specific internal energy, and Π is the shear stress tensor calculated as :

$$\Pi = \mu_{eff} \left(\nabla \mathbf{u} + (\nabla \mathbf{u})^T - \frac{2}{3} \mathbf{I} tr(\nabla \cdot \mathbf{u}) \right), \tag{12}$$

where μ_{eff} is the effective shear viscosity of the fluid, which accommodates non-linearity due to KL effects (see Equation (6)), and, $\mathbf{I}$ and tr denotes, identity matrix and trace, respectively. The heat flux due to conduction of energy (**j**) by temperature gradients (Fourier's law) is defined as:

$$\mathbf{j} = -\kappa_{eff} \nabla T, \tag{13}$$

where κ_{eff} is the effective thermal conductivity of the fluid based on effective thermal MFP (see Equation (8)). And temperature is calculated iteratively from the total energy as :

$$T = \frac{1}{C_v(T)} \left(E(T) - \frac{|\mathbf{u}|^2}{2} \right), \tag{14}$$

where $C_v(T)$ is the specific heat at constant volume as a function of temperature.

Perfect gas equation is solved to update the pressure as :

$$p = \rho R T. \tag{15}$$

2.4. Boundary Conditions

The first-order Maxwell velocity slip boundary condition [34], is modified to take into account the KL correction [47] as follows:

$$\mathbf{u} = \mathbf{u}_w - \left(\frac{2 - \sigma_v}{\sigma_v} \right) \lambda_{eff} \nabla_n (\mathbf{S} \cdot \mathbf{u}) - \left(\frac{2 - \sigma_v}{\sigma_v} \right) \frac{\lambda_{eff}}{\mu_{eff}} \mathbf{S} \cdot (n \cdot \Pi_{\mathbf{mc}}) - \frac{3}{4} \frac{\mu_{eff}}{\rho} \frac{\mathbf{S} \cdot \nabla T}{T}, \tag{16}$$

where $\mathbf{u}_w$ is the reference wall velocity, σ_v is tangential momentum accommodation coefficient, subscript n denotes normal direction to the surface, the tensor $\mathbf{S} = \mathbf{I} - \mathbf{nn}$ removes normal components of non-scalar field, and $\Pi_{\mathbf{mc}} = \Pi - \mu_{eff} \nabla \mathbf{u}$ is obtained from Equation (12). Here, third term on the RHS of Equation (16) accounts for the curvature effect and fourth term considers the thermal creep.

Smoluchowski temperature jump [35] is modified as follows:

$$T = T_w - \frac{2 - \sigma_T}{\sigma_T} \frac{2\gamma}{\gamma + 1} \frac{\lambda_{eff(T)}}{Pr} \nabla_n T, \tag{17}$$

where T_w is the reference wall temperature, Pr is Prandtl number, σ_T is thermal accommodation coefficient and γ is specific heat ratio.

In addition to above widely used boundary conditions, following hybrid boundary conditions, which consider the effect of adsorption of molecules on the surface, are also evaluated in the present work. These boundary conditions are developed by Le et al. [36], and have proven to give good results for rarefied hypersonic flow cases, and low-speed rarefied micro-scale gas flows [37]. These boundary conditions are based on the concept that the molecules are adsorbed by the solid surface, as a function of pressure at constant temperature. If molecules are adsorbed by the fraction α, they do not contribute to the fluid shear stress and conduction of heat due to receding molecules $(1 - \alpha)$. This fraction of coverage α is computed by the Langmuir adsorption isotherm [51,52] for mono-atomic gases,

$$\alpha = \frac{\zeta p}{1 + \zeta p}, \tag{18}$$

and for diatomic gases,

$$\alpha = \frac{\sqrt{\zeta p}}{1 + \sqrt{\zeta p}}, \qquad (19)$$

where ζ is an equilibrium constant related to surface temperature, which is represented as,

$$\zeta = \frac{A_m \lambda_{eff}}{R_u T_w} exp\left(\frac{D_e}{R_u T_w}\right), \qquad (20)$$

where A_m is approximately calculated as $N_A \pi d^2/4$ for gases [52,53], N_A is Avogadros's number, $D_e = 5255$ (J/mol) is the heat of adsorption for argon and nitrogen given in literature [52,53], and R_u is the universal gas constant.

Langmuir-Maxwell slip velocity [36] boundary condition is modified as,

$$\mathbf{u} = \mathbf{u}_w - \left(\frac{1}{1-\alpha}\right)\lambda_{eff}\nabla_n(\mathbf{S}\cdot\mathbf{u}) - \left(\frac{1}{1-\alpha}\right)\frac{\lambda_{eff}}{\mu_{eff}}\mathbf{S}\cdot(n\cdot\mathbf{\Pi_{mc}}) - \frac{3}{4}\frac{\mu_{eff}}{\rho}\frac{\mathbf{S}\cdot\nabla T}{T}, \qquad (21)$$

and Langmuir-Smoluchwoski temperature jump [36] boundary condition is modified as,

$$T = T_w - \frac{1}{1-\alpha}\frac{2\gamma}{\gamma+1}\frac{\lambda_{eff(T)}}{Pr}\nabla_n T. \qquad (22)$$

These boundary conditions consider the effect of KL as well as adsorption on the wall.

It should be noted that all equations stated above reduce to their conventional form when $\beta = \beta_T = 1$. All simulations are carried out using the conventional *rhoCentralFoam* solver without the effect of KL initially. Local MFP (λ) in Equation (3) and the geometry-dependent effective MFP (λ_{eff}) in Equation (4) are updated using the post-processing utility developed by authors within OpenFOAM framework and simulations are carried out again. Results obtained using conventional NS equations, using linear constitutive relations, along with Maxwell velocity slip and Smoluchwoski temperature jump (MS) are referred as "NS-MS", whereas, Langmuir-Maxwell velocity slip and Langmuir-Smoluchwoski temperature jump (LMS) boundary conditions are referred as "NS-LMS" throughout the manuscript. Current results, which are referred as "NS-MS-withKL" and "NS-LMS-withKL" are obtained using the modified constitutive relations and respective boundary conditions with effective MFP (β and β_T). Flow is modeled using a single gas species in chemical equilibrium in the present study.

3. Results and Discussion

A schematic of the backward-facing step is illustrated in Figure 2. Dirichlet boundary condition is imposed for pressure at inlet and outlet, whereas zero-gradient boundary condition is used for velocity (extrapolated from the interior solution), as it is a pressure driven flow. The temperature of flow is specified at the inlet boundary and zero-gradient at the outlet. Various non-equilibrium surface boundary conditions (described in Section 2.4) have been applied at the top wall, upstream wall, step, and bottom wall. Dimensions of the nano-/micro step channel vary depending on the Knudsen number and are given in Table 1. The authors have compared the results with the DSMC data obtained by Mahadavi et al. [37]. Specified inlet and outlet pressure boundary conditions have been implemented in DSMC simulations, through correcting density and velocity implicitly from the characteristics theory [54,55]. A fully diffuse reflection wall patch (perfect exchange of momentum and energy), which corresponds to $\sigma_v = \sigma_T = 1$ in CFD simulations, has been used for all simulations.

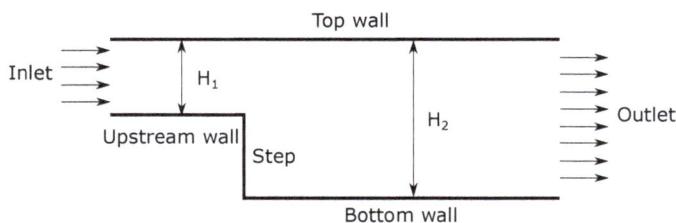

Figure 2. Schematic of backward-facing step.

A grid is created using the 'blockMesh' utility in OpenFOAM. A grid independence study has been carried out by gradually increasing cells in x and y-direction as shown Figure 3. Slip velocity on the bottom wall is plotted for nano-channel (refer Figure 3a) and micro-channel (refer Figure 3b). Results obtained are independent of the grid resolution. Final grid size have 200 cells in x-direction (minimum cell size δx = 0.427 nm) and 120 cells in y-direction (minimum cell size δy = 0.142 nm) for nano-channel, and micro-channel grid has 300 cells in x-direction (δx = 0.0187 μm) and 60 cells in y-direction (δy = 0.0166 μm).

Table 1. Dimensions of nano- and micro-step channel.

Dimensions of	Nano-Step Channel	Micro-Step Channel
Top wall	85.47 nm	5.61 μm
Upstream wall	25.641 nm	1.81 μm
Bottom wall	59.829 nm	3.8 μm
H_1	17.095 nm	1 μm
H_2	8.547 nm	0.5 μm
Step	8.547 nm	0.5 μm

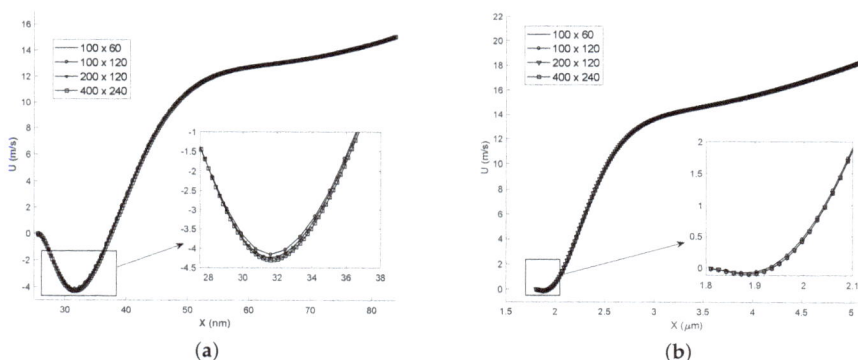

Figure 3. Grid independence study by analysing the slip velocity distribution on the bottom wall. Legends show the number of cells in x direction (along the length of channel) $\times$ y direction (along the height of channel). (**a**) Kn = 0.01; (**b**) Kn = 0.05.

3.1. Effect of Change in Knudsen Number

In this section, simulations are carried out for various Knudsen numbers in slip (Kn = 0.01, 0.05), and early transition (Kn = 0.1) flow regime. Flow parameters for all 3 cases are given in Table 2. The temperature of the inlet flow and wall is same for Kn = 0.01 case, and minimal difference of 30 K for other 2 cases. The nano-step channel is used for Kn = 0.01 case, whereas a micro-step channel is used to simulate high Kn cases [37]. Although the height of backward-facing step is less for Kn = 0.01 case, inlet pressure is very high compared to higher Kn cases. Slip velocity distribution obtained using different solvers on the bottom wall of the backward-facing step is plotted. Gradient-based local Kn ($Kn_{g_{ll}} = \frac{1}{Q}\frac{dQ}{dl}$) is plotted on the secondary y-axis for all plots. Here, $Kn_{g_{ll}}$ is calculated based on velocity gradients and Q in the denominator is taken as maximum of ($\mathbf{u}$, $\sqrt{\gamma RT}$).

Table 2. Flow parameters for various Kn number cases.

Kn (Based on H$_2$)	0.01	0.05	0.1
P$_{in}$ (MPa)	31.077	0.150735	0.075397
T$_{in}$(K)	300	330	330
P$_{in}$/P$_{out}$	2	2.32	2.32
T$_w$(K)	300	300	300
σ_v, σ_T	1	1	1
Geometry	Nano-step	Micro-step	Micro-step
Gas	N$_2$	N$_2$	N$_2$

Figure 4 shows the slip velocity distribution for Kn = 0.01 case i.e., slip flow regime. Flow accelerates through the nano-step channel along its length and undergoes Prandtl-Meyer expansion at the upstream wall-step corner. Flow is separated from the wall, and a wake is formed immediately after the step. Negative slip velocity components in Figure 4 for x < 38 nm, indicates the reverse flow and the adverse flow gradient. Flow is reattached to the wall at x = 38 nm, and slip velocity increases along the streamline. It is observed that in the separation region, solvers using LMS boundary conditions give better results than usual MS boundary conditions when compared with DSMC data. Also, the introduction of KL effects in NS-MS solver does not change results, as they exactly overlap with each other. Their deviation w.r.t. DSMC increases as flow becomes more rarefied towards the outlet, maximum deviation being 26.66%. On the other hand, results are considerably improved when KL effects are incorporated in NS-LMS solver, and they are in excellent agreement towards the outlet. Flow is more rarefied near outlet, as $Kn_{g_{ll}}$ is higher, which leads to the growth of thickness of KL. However, with and without KL results are similar in the separation zone (x < 38 nm), as local Kn < 0.03, and KL effects are minimal in this region.

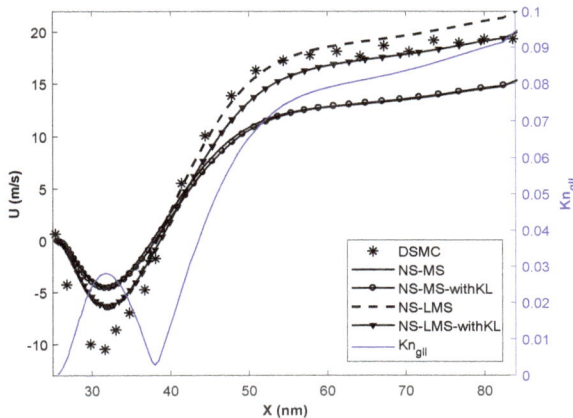

Figure 4. Velocity slip distribution on the bottom wall at Kn = 0.01.

Figures 5 and 6 demonstrate the slip velocity distribution on the bottom wall for Kn = 0.05 and 0.1 cases, respectively. As the flow enters the early transition regime, the phenomena of flow separation, recirculation, and re-attachment disappear. This is because slip velocity on the wall becomes comparable to the flow velocity. The Knudsen layer thins the shear layer, and fluid follows wall direction without separating even at the sudden step corner. It can be observed that NS-MS solver under-predicts, whereas NS-LMS solver over-predicts the slip velocity when compared to DSMC, throughout the length of the bottom wall. After the incorporation of KL effects, slip velocity values predicted by NS-MS solver, move closer to DSMC data, but the improvement is not much significant ($\sim$5.88%), and maximum deviation with DSMC is $\sim$ 31.57%. On the other hand, results are greatly improved for NS-LMS-KL solver, i.e., 10% improvement over NS-LMS solver and deviations within 10% with DSMC data. It is noticed that the introduction of KL is more effective when $Kn_{gll} > 0.05$, as the growth of KL thickness increases with increase in Kn. Visualization of KL thickness for various Kn cases is demonstrated in Figure 7, using contours of normalized MFP (β). Thickness of KL is minimal for the Kn = 0.01 case, whereas it almost covers the entire flow domain for Kn = 0.1 case.

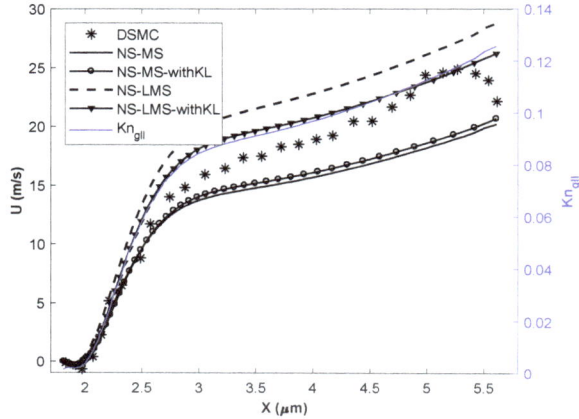

Figure 5. Velocity slip distribution on the bottom wall at Kn = 0.05.

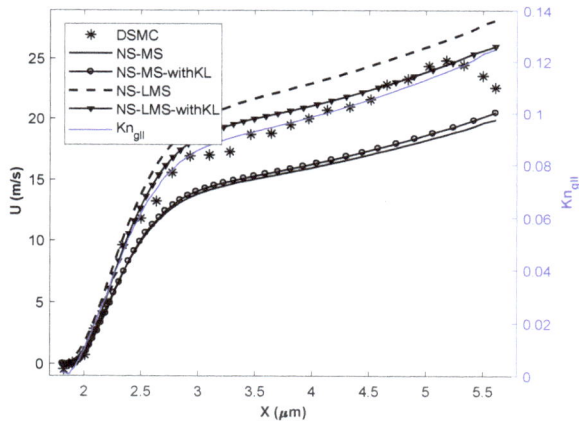

Figure 6. Velocity slip distribution on the bottom wall at Kn = 0.1.

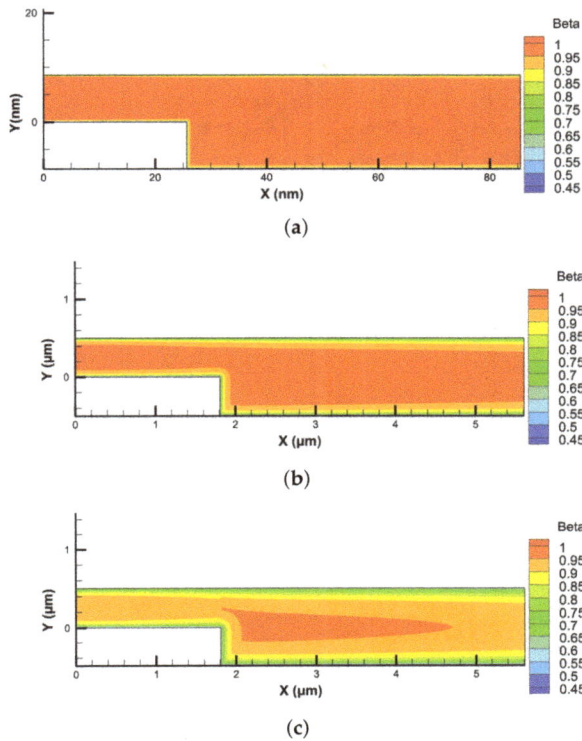

(a)

(b)

(c)

Figure 7. Knudsen layer formation in terms of the normalized MFP (β) contours for various Knudsen numbers (obtained using Equation (5)). (**a**) Kn = 0.01; (**b**) Kn = 0.05; (**c**) Kn = 0.1.

3.2. Effect of Change in Inlet Temperature

In this section, the temperature of the flow at the inlet is varied to investigate the effect of KL on temperature jump on the bottom wall. Simulations are carried out for Kn = 0.01 case with T_{in} = 500 K and T_{in} = 700 K.

Figures 8 and 9 demonstrate the temperature of the fluid on the bottom wall obtained using different solvers. Local Kn based on velocity gradients is plotted on the secondary y-axis. It is observed that NS-LMS solver results are closer to DSMC data when compared to NS-MS solver. This is due to the fact that surface boundary conditions in the DSMC method are based on gas-surface interactions, and particles are adsorbed on the surface and re-emitted. LMS boundary conditions account for the effect of molecules adsorption on the surface and are able to predict better surface properties than MS boundary conditions.

However, the introduction of KL effects has noticeably improved predictions for both the solvers, NS-MS and NS-LMS. The main reason behind this is that the thermal KL is formed near the wall, whose thickness is more than the momentum KL. It not only modifies the constitutive relation for the heat flux but also considers the thermal MFP in the calculation of surface temperature jump. The introduction of KL effects has improved NS-MS solver results even in the separation zone, with maximum relative improvement of 30% for T_{in} = 500 K case, and 41.05% for T_{in} = 700 K case w.r.t. DSMC results. NS-LMS solver with and without KL accurately captures the peak temperature value for T_{in} = 500 K case. Location of peak temperature for DSMC is slightly downstream of NS solutions, as flow gradients in DSMC are diffuse. Relative improvement of around 81% is observed for T_{in} = 700 K case solver due to the addition of KL effects over NS-LMS solver.

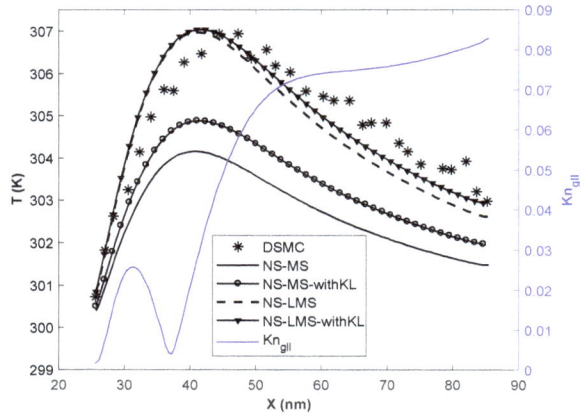

Figure 8. Temperature distribution of the fluid on the bottom wall at Kn = 0.01, T$_{in}$ = 500 K.

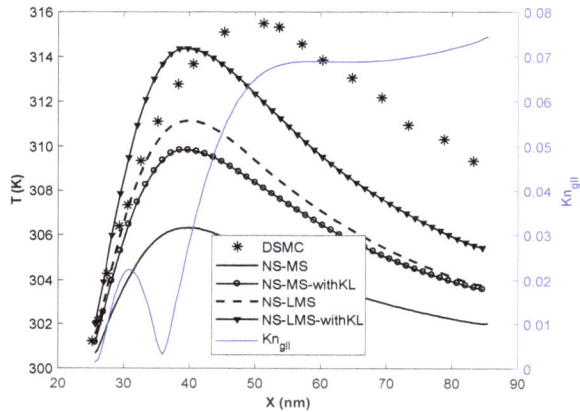

Figure 9. Temperature distribution of the fluid on the bottom wall at Kn = 0.01, T$_{in}$ = 700 K.

3.3. Effect of Change in Wall Temperature

In this section, the temperature of the bottom wall and step has been varied for Kn = 0.01 case. Inlet flow and other walls have the temperature of 300 K. Simulations are carried out by setting temperature of both the bottom wall and step to 500 K and 700 K.

Figure 10 shows the temperature distribution of fluid on the bottom wall for T$_w$ = 500 K (Figure 10a) and T$_w$ = 700 K (Figure 10b). As wall temperature is higher than the flow temperature, heat is transferred from the wall to the fluid flow, and temperature abruptly drops near the step - bottom wall region where a separation zone is formed.

It is interesting to note that for this particular case, NS-MS solvers predict better surface temperature than NS-LMS solver, unlike previous cases. This can be attributed to the fact that reference temperature T$_w$ is high, which affects the calculation of α and ζ parameters in Equations (19) and (20), respectively. In LMS boundary conditions, approaching stream of molecules has lower temperature than the surface, and high temperature surface adsorbs molecules by fraction of α. These molecules are re-emitted with T = T$_w$, which could be the reason that temperature is under-predicted by NS-LMS solvers. Therefore, LMS boundary conditions are not suitable when temperature gradients are negative or not uniform. Though NS-LMS solver has deviations w.r.t DSMC data, KL effects have led to the relative improvement of around 24% for T$_w$ = 500 K case and 29% for T$_w$ = 700 K case. Results are minutely improved for NS-MS solvers with the addition of KL effects. Peak temperature values are accurately captured by NS-MS-withKL solver.

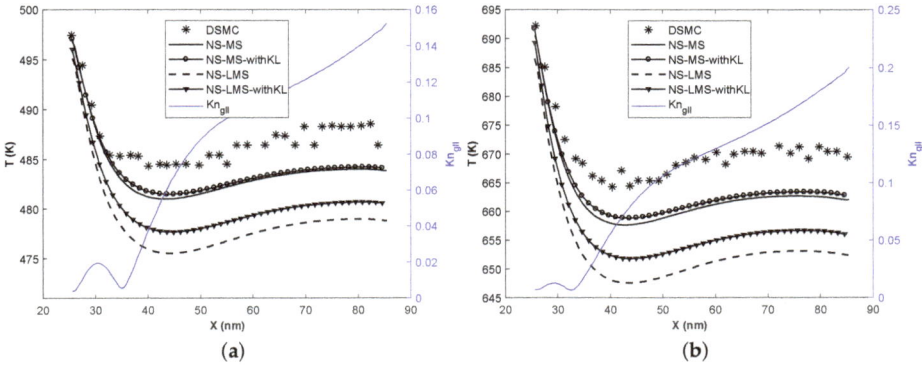

Figure 10. Temperature distribution of the fluid on the bottom wall at Kn = 0.01. (a) T_w = 500 K ; (b) T_w = 700 K.

Figure 11 demonstrates the comparison of slip velocity for various solvers for T_w = 500 K (Figure 11a) and T_w = 700 K (Figure 11b) cases. It is observed that NS-LMS solver gives better predictions in separated region, however, it overpredicts the slip velocity after reattachment i.e., x > 35 nm for T_w = 500 K and x > 32 nm for T_w = 700 K case. Slip velocity moves closer to DSMC when KL effect is taken into account in NS-LMS solver. As described earlier, KL promotes shear thinning phenomenon, which leads to less slip velocity on the wall, and higher normal gradients of velocity on the wall. There is a minor discrepancy between DSMC and NS-MS, NS-MS-withKL solvers for T_w = 500 K case, whereas there is a good agreement for T_w = 700 K case.

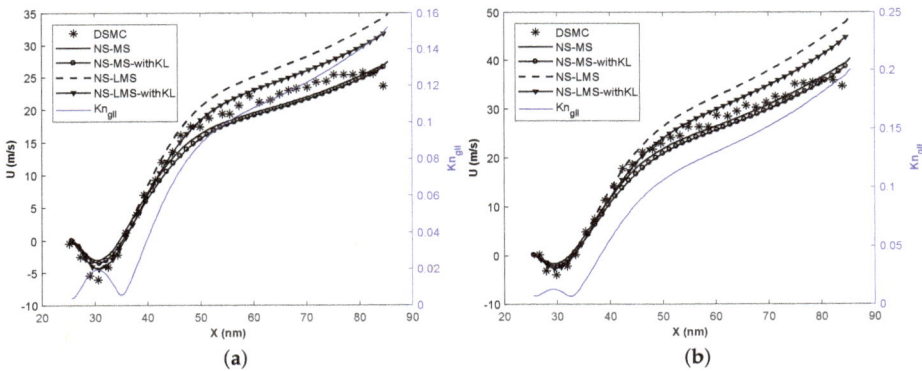

Figure 11. Velocity slip distribution on the bottom wall at Kn = 0.01. (a) T_w = 500 K ; (b) T_w = 700 K.

4. Conclusions

The effect of the Knudsen layer on the surface flow properties of a nano- and micro-scale backward facing step has been investigated in the slip and the early transition flow regime. The effective mean free path model has been implemented within the Navier-Stokes framework, and the constitutive relations for shear stress and heat flux are modified. First order non-equilibrium boundary conditions, i.e., Maxwell velocity slip and Smoluchowski temperature jump, as well as hybrid boundary conditions based on Langmuir adsorption isotherm are effectively modified to incorporate the non-equilibrium effects associated with KL.

The NS-LMS solver has proven to give better predictions in the separation zone than NS-MS solver when compared with the benchmark DSMC results. The velocity slip and temperature jump results are significantly improved for the NS-LMS method when KL effects are incorporated, implying that non-linear effects due to momentum and thermal KL are captured by the proposed method. On the other hand, for the case of negative temperature gradient near the wall, the NS-MS solver have accurately predicted the slip velocity and temperature jump and has good agreement with DSMC data, and NS-LMS method have higher deviations with DSMC.

The present results have demonstrated the potential of the effective MFP based approach in the modeling of rarefied nano- and micro-scale gas flows. Although non-linear effects associated with momentum and thermal KL are captured up to some extent, no strong conclusions can be drawn. In the future, the detailed analysis should be carried out over a wide range of Knudsen numbers, and arbitrary geometries subjected to a range of complex flow conditions.

Author Contributions: Conceptualization, A.B. and H.G.; methodology, A.B.; software, A.B. and H.G.; validation, A.B., H.G. and N.D.; formal analysis, A.B.; investigation, A.B.; writing—original draft preparation, A.B.; writing—review and editing, H.G. and N.D.; funding acquisition, N.D.

Funding: The research was supported by Department of Science and Technology (DST): SERB/F/2684/2014-15 and Ministry of Human Resource Development (MHRD) fellowship. The APC was funded by IIT Hyderabad.

Conflicts of Interest: The authors declare no conflict of interest.

References

1. Stops, D. The mean free path of gas molecules in the transition regime. *J. Phys. D Appl. Phys.* **1970**, *3*, 685. [CrossRef]
2. Burnett, D. The distribution of molecular velocities and the mean motion in a non-uniform gas. *Proc. Lond. Math. Soc.* **1936**, *2*, 382–435. [CrossRef]
3. Grad, H. Note on N-dimensional hermite polynomials. *Commun. Pure Appl. Math.* **1949**, *2*, 325–330. [CrossRef]
4. Chapman, S.; Cowling, T.G.; Burnett, D. *The Mathematical Theory of Non-Uniform Gases: An Account of The Kinetic Theory of Viscosity, Thermal Conduction and Diffusion in Gases*; Cambridge University Press: Cambridge, UK, 1970.
5. Sone, Y. *Kinetic Theory and Fluid Dynamics*; Springer Science & Business Media: Berlin, Germany, 2012.
6. Cercignani, C. *Mathematical Methods in Kinetic Theory*; Springer: New York, NY, USA, 1969; pp. 232–243.
7. Barichello, L.; Siewert, C. The temperature-jump problem in rarefied-gas dynamics. *Eur. J. Appl. Math.* **2000**, *11*, 353–364. [CrossRef]
8. Barichello, L.B.; Bartz, A.C.R.; Camargo, M.; Siewert, C. The temperature-jump problem for a variable collision frequency model. *Phys. Fluids* **2002**, *14*, 382–391. [CrossRef]
9. Su, W.; Wang, P.; Liu, H.; Wu, L. Accurate and efficient computation of the Boltzmann equation for Couette flow: Influence of intermolecular potentials on Knudsen layer function and viscous slip coefficient. *J. Comput. Phys.* **2019**, *378*, 573–590. [CrossRef]
10. Jin, S.; Slemrod, M. Regularization of the Burnett equations via relaxation. *J. Stat. Phys.* **2001**, *103*, 1009–1033. [CrossRef]
11. Al-Ghoul, M.; Eu, B.C. Generalized hydrodynamics and microflows. *Phys. Rev. E* **2004**, *70*, 016301. [CrossRef]
12. Struchtrup, H.; Torrilhon, M. Higher-order effects in rarefied channel flows. *Phys. Rev. E* **2008**, *78*, 046301. [CrossRef]
13. Myong, R. Theoretical description of the gaseous Knudsen layer in Couette flow based on the second-order constitutive and slip-jump models. *Phys. Fluids* **2016**, *28*, 012002. [CrossRef]
14. Bird, R.B.; Stewart, W.E.; Lightfoot, E.N. *Transport Phenomena*; John Wiley & Sons: Hoboken, NJ, USA, 2007.
15. Bird, G. *The DSMC Method*; CreateSpace Independent Publishing Platform: Scotts Valley, CA, USA, 2013.

16. White, C.; Borg, M.K.; Scanlon, T.J.; Longshaw, S.M.; John, B.; Emerson, D.; Reese, J.M. dsmcFoam+: An OpenFOAM based direct simulation Monte Carlo solver. *Comput. Phys. Commun.* **2018**, *224*, 22–43. [CrossRef]

17. Piekos, E.; Breuer, K. DSMC modeling of micromechanical devices. In Proceedings of the 30th Thermophysics Conference, San Diego, CA, USA, 19–22 June 1995; p. 2089.

18. Oh, C.; Oran, E.; Sinkovits, R. Computations of high-speed, high Knudsen number microchannel flows. *J. Thermophys. Heat Transf.* **1997**, *11*, 497–505. [CrossRef]

19. Nance, R.P.; Hash, D.B.; Hassan, H. Role of boundary conditions in Monte Carlo simulation of microelectromechanical systems. *J. Thermophys. Heat Transf.* **1998**, *12*, 447–449. [CrossRef]

20. Li, J.M.; Wang, B.X.; Peng, X.F. Wall-adjacent layer analysis for developed-flow laminar heat transfer of gases in microchannels. *Int. J. Heat Mass Transf.* **2000**, *43*, 839–847. [CrossRef]

21. Lockerby, D.A.; Reese, J.M.; Gallis, M.A. Capturing the Knudsen layer in continuum-fluid models of nonequilibrium gas flows. *AIAA J.* **2005**, *43*, 1391–1393. [CrossRef]

22. Cercignani, C.; Frangi, A.; Lorenzani, S.; Vigna, B. BEM approaches and simplified kinetic models for the analysis of damping in deformable MEMS. *Eng. Anal. Bound. Elem.* **2007**, *31*, 451–457. [CrossRef]

23. Lockerby, D.A.; Reese, J.M. On the modelling of isothermal gas flows at the microscale. *J. Fluid Mech.* **2008**, *604*, 235–261. [CrossRef]

24. Lilley, C.R.; Sader, J.E. Velocity gradient singularity and structure of the velocity profile in the Knudsen layer according to the Boltzmann equation. *Phys. Rev. E* **2007**, *76*, 026315. [CrossRef]

25. Lilley, C.R.; Sader, J.E. Velocity profile in the Knudsen layer according to the Boltzmann equation. *Proc. R. Soc. Lond. A Math. Phys. Eng. Sci.* **2008**, *464*, 2015–2035. [CrossRef]

26. Guo, Z.; Shi, B.; Zheng, C.G. An extended Navier-Stokes formulation for gas flows in the Knudsen layer near a wall. *EPL (Europhys. Lett.)* **2007**, *80*, 24001. [CrossRef]

27. Dongari, N.; Zhang, Y.; Reese, J.M. Molecular free path distribution in rarefied gases. *J. Phys. D Appl. Phys.* **2011**, *44*, 125502. [CrossRef]

28. Dongari, N.; Zhang, Y.; Reese, J.M. Modeling of Knudsen layer effects in micro/nanoscale gas flows. *J. Fluids Eng.* **2011**, *133*, 071101. [CrossRef]

29. Guo, Z.; Qin, J.; Zheng, C. Generalized second-order slip boundary condition for nonequilibrium gas flows. *Phys. Rev. E* **2014**, *89*, 013021. [CrossRef] [PubMed]

30. Jaiswal, S.; Dongari, N. Implementation of knudsen layer effects in open source cfd solver for effective modeling of microscale gas flows. In Proceedings of the 23rd National Heat and Mass Transfer Conference and 1st International ISHMT-ASTFE Heat and Mass Transfer Conference IHMTC2015, 17–20 December 2015; pp. 1–8.

31. To, Q.D.; Léonard, C.; Lauriat, G. Free-path distribution and Knudsen-layer modeling for gaseous flows in the transition regime. *Phys. Rev. E* **2015**, *91*, 023015. [CrossRef] [PubMed]

32. Norouzi, A.; Esfahani, J.A. Two relaxation time lattice Boltzmann equation for high Knudsen number flows using wall function approach. *Microfluid. Nanofluid.* **2015**, *18*, 323–332. [CrossRef]

33. Tu, C.; Qian, L.; Bao, F.; Yan, W. Local effective viscosity of gas in nano-scale channels. *Eur. J. Mech.-B/Fluids* **2017**, *64*, 55–59. [CrossRef]

34. Maxwell, J.C. On the dynamical theory of gases. *Philos. Trans. R. Soc. Lond.* **1867**, *157*, 49–88.

35. Smoluchowski von Smolan, M. Ueber wärmeleitung in verdünnten gasen [On heat conduction in diluted gases]. *Annalen der Physik* **1898**, *300*, 101–130. (In German) [CrossRef]

36. Le, N.T.; White, C.; Reese, J.M.; Myong, R.S. Langmuir–Maxwell and Langmuir–Smoluchowski boundary conditions for thermal gas flow simulations in hypersonic aerodynamics. *Int. J. Heat Mass Transf.* **2012**, *55*, 5032–5043. [CrossRef]

37. Mahdavi, A.M.; Le, N.T.; Roohi, E.; White, C. Thermal rarefied gas flow investigations through micro-/nano-backward-facing step: Comparison of dsmc and cfd subject to hybrid slip and jump boundary conditions. *Numer. Heat Transf. Part A Appl.* **2014**, *66*, 733–755. [CrossRef]

38. Weller, H.G.; Tabor, G.; Jasak, H.; Fureby, C. A tensorial approach to computational continuum mechanics using object-oriented techniques. *Comput. Phys.* **1998**, *12*, 620–631. [CrossRef]

39. Jasak, H. OpenFOAM: a year in review. In Proceedings of the 5th OPENFOAM Workshop, Gothenburg, Sweden, 21–24 June 2010; pp. 21–24.

Micromachines **2019**, *10*, 118

40. Kurganov, A.; Tadmor, E. New high-resolution central schemes for nonlinear conservation laws and convection–diffusion equations. *J. Comput. Phys.* **2000**, *160*, 241–282. [CrossRef]
41. Kurganov, A.; Noelle, S.; Petrova, G. Semidiscrete central-upwind schemes for hyperbolic conservation laws and Hamilton–Jacobi equations. *SIAM J. Sci. Comput.* **2001**, *23*, 707–740. [CrossRef]
42. Kennard, E.H. *Kinetic Theory of Gases, with an Introduction to Statistical Mechanics*; McGraw-Hill: New York, NY, USA, 1938.
43. Hirschfelder, J.; Bird, R.B.; Curtiss, C.F. *Molecular Theory of Gases and Liquids*; Wiley: Hoboken, NY, USA, 1964.
44. Anderson, J.D. *Hypersonic and High Temperature Gas Dynamics*; AIAA: Reston, VA, USA, 2000.
45. Vincenti, W.G.; Kruger, C.H. *Introduction to Physical Gas Dynamics*; Wiley: New York, NY, USA, 1965; Volume 246.
46. Dongari, N. Micro Gas Flows: Modelling the Dynamics of Knudsen Layers. Ph.D. Thesis, University of Strathclyde, Glasgow, UK, 2012.
47. Dongari, N.; Barber, R.W.; Emerson, D.R.; Stefanov, S.K.; Zhang, Y.; Reese, J.M. The effect of Knudsen layers on rarefied cylindrical Couette gas flows. *Microfluid. Nanofluid.* **2013**, *14*, 31–43. [CrossRef]
48. Dongari, N.; White, C.; Scanlon, T.J.; Zhang, Y.; Reese, J.M. Effects of curvature on rarefied gas flows between rotating concentric cylinders. *Phys. Fluids* **2013**, *25*, 052003. [CrossRef]
49. Sone, Y.; Ohwada, T.; Aoki, K. Evaporation and condensation on a plane condensed phase: Numerical analysis of the linearized Boltzmann equation for hard-sphere molecules. *Phys. Fluids A Fluid Dyn.* **1989**, *1*, 1398–1405. [CrossRef]
50. Blazek, J. *Computational Fluid Dynamics: Principles and Applications*; Butterworth-Heinemann: Oxford, UK, 2015.
51. Dushman, S.; Lafferty, J.M.; Pasternak, R. Scientific foundations of vacuum technique. *Phys. Today* **1962**, *15*, 53. [CrossRef]
52. Myong, R.; Reese, J.; Barber, R.W.; Emerson, D. Velocity slip in microscale cylindrical Couette flow: the Langmuir model. *Phys. Fluids* **2005**, *17*, 087105. [CrossRef]
53. Bhattacharya, D.; Eu, B.C. Nonlinear transport processes and fluid dynamics: Effects of thermoviscous coupling and nonlinear transport coefficients on plane Couette flow of Lennard-Jones fluids. *Phys. Rev. A* **1987**, *35*, 821. [CrossRef]
54. Wang, M.; Li, Z. Simulations for gas flows in microgeometries using the direct simulation Monte Carlo method. *Int. J. Heat Fluid Flow* **2004**, *25*, 975–985. [CrossRef]
55. Akhlaghi, H.; Roohi, E.; Stefanov, S. A new iterative wall heat flux specifying technique in DSMC for heating/cooling simulations of MEMS/NEMS. *Int. J. Therm. Sci.* **2012**, *59*, 111–125. [CrossRef]

micromachines

MDPI

Article

Pressure Drop of Microchannel Plate Fin Heat Sinks

Zhipeng Duan [1],*, Hao Ma [1],*, Boshu He [1], Liangbin Su [1] and Xin Zhang [1,2]

[1]   School of Mechanical, Electronic and Control Engineering, Beijing Jiaotong University, Beijing 100044, China; hebs@bjtu.edu.cn (B.H.); 16116368@bjtu.edu.cn (L.S.); zhangxin@bjtu.edu.cn (X.Z.)
[2]   Beijing Key Laboratory of Powertrain for New Energy Vehicle, Beijing Jiaotong University, Beijing 100044, China
*   Correspondence: zpduan@bjtu.edu.cn (Z.D.); 18116018@bjtu.edu.cn (H.M.); Tel.: +86-10-5168-8542 (Z.D. & H.M.); Fax: +86-10-5168-8404 (Z.D. & H.M.)

Received: 30 December 2018; Accepted: 22 January 2019; Published: 24 January 2019

Abstract: The entrance region constitutes a considerable fraction of the channel length in miniaturized devices. Laminar slip flow in microchannel plate fin heat sinks under hydrodynamically developing conditions is investigated semi-analytically and numerically in this paper. The semi-analytical model for the pressure drop of microchannel plate fin heat sinks is obtained by solving the momentum equation with the first-order velocity slip boundary conditions at the channel walls. The simple pressure drop model utilizes fundamental solutions from fluid dynamics to predict its constitutive components. The accuracy of the model is examined using computational fluid dynamics (CFD) simulations and the experimental and numerical data available in the literature. The model can be applied to either apparent liquid slip over hydrophobic and superhydrophobic surfaces or gas slip flow in microchannel heat sinks. The developed model has an accuracy of 92 percent for slip flow in microchannel plate fin heat sinks. The developed model may be used to predict the pressure drop of slip flow in microchannel plate fin heat sinks for minimizing the effort and expense of experiments, especially in the design and optimization of microchannel plate fin heat sinks.

Keywords: pressure drop; microchannels; heat sinks; slip flow; electronic cooling

1. Introduction

Fluid flow in microchannels has emerged as an important research area. This has been motivated by their various applications such as medical and biomedical use, computer chips, and chemical separations. The high level of heat dissipation requires a dramatic reduction in the channel dimensions. The high flux heat dissipation from high-speed microprocessors provided the impetus for studies on heat transfer in microchannels [1]. It is well known that the greatest challenge is overheating because of an increasing power flux and a higher thermal resistance in computer chips [2]. Due to a rapid increase in power density and miniaturization, very high heat flux chip cooling requires a large flow rate, which can lead to a significant pressure drop. The reliability, performance, and power dissipation of interconnects and transistors are heavily dependent on the operating temperature [2]. An effective cooling scheme for use in chip technologies needs to be developed to solve cooling limitations. One potential solution to the thermal management of a chip is to use microchannel heat sinks due to the small size and high heat transfer coefficient. Copper heat sinks with integrated microchannels are expected to dominate heat sink applications to handle high heat removal rates. These practical advantages of microchannel heat sinks have stimulated experimental, theoretical, and numerical research [3–26]. The use of microchannel heat sinks is becoming more common in various industrial applications. The flow and heat transfer in microchannel/nanochannel heat sinks have become subjects of growing research attention, and the analysis of microchannel heat sinks has become increasingly important.

Microchannels are the fundamental part of microfluidic systems. In addition to connecting different devices, microchannels are also utilized as biochemical reaction chambers, in physical particle separation, in inkjet print heads, in infrared detectors, in diode lasers, in miniature gas chromatographs, in aerospace technology, or as heat exchangers for cooling computer chips. Understanding the flow characteristics of microchannel flows is very important in determining the pressure drop, heat transfer, and transport properties of the flow for minimizing the effort and expense of experiments, especially in the design and optimization of microchannel plate fin heat sinks. Microchannel heat sinks have received considerable attention owing to their high surface-area-to-volume ratio, large convective heat transfer coefficient, and small mass and volume. For the effective design and optimization of microchannel heat sinks, it is significant to understand the fundamental characteristics of fluid flow and heat transfer in microchannels.

The purpose of this paper is to study the pressure drop characteristics of fluid flow through microchannel plate fin heat sinks. The developing slip flow friction factor Reynolds number parameter is determined for the laminar regime. Following this, there we analyze the pressure drop through microchannel plate fin heat sinks.

In this study, the developing laminar flow is analyzed through plate fin heat sinks with velocity slip boundary conditions, and semi-analytical closed-form solutions are obtained for the friction factor and Reynolds number product and pressure drop in terms of the channel aspect ratio, channel length, hydraulic diameter, Reynolds number, and modified Knudsen number. The primary goal of the present paper is to provide a general simple pressure drop model of fluid flow through microchannel plate fin heat sinks.

2. Literature Review

The Knudsen number (*Kn*) relates the molecular mean free path of gas to a characteristic dimension of the duct. The Knudsen number is very small for continuum flows. However, for microscale gas flows, the gas mean free path becomes comparable to the characteristic dimension of the duct. Rarefaction effects must be considered in gases in which the molecular mean free path is comparable to the channel's characteristic dimension. The continuum assumption is no longer valid, and the gas exhibits non-continuum effects such as velocity slip and temperature jump at the channel walls. Traditional examples of non-continuum gas flows in channels include low-density applications such as high-altitude aircraft or vacuum technology. The recent development of microscale fluid systems has motivated great interest in this field of study. Microfluidic systems should take into account slip effects. There is strong evidence to support the use of Navier–Stokes and energy equations to model the slip flow problem, while the boundary conditions are modified by including velocity slip and temperature jump at the channel walls. The slip length can be interpreted as the distance from the wall that the slip velocity profile extends to if extrapolated away from the boundary. In gas flows, the slip length is related to the Knudsen number, while in liquid flows, the slip length depends on the surface microstructure [27–37]. When considering liquids, the molecular mean free path may be replaced by the slip length and the Knudsen number may be replaced by the modified Knudsen number. The slip lengths reported experimentally span several orders of magnitude, from molecular lengths up to hundreds of nanometers with dependence on wetting conditions, surface roughness structure (shape and distribution), dissolved gas, surface charge, shear rate, and pressure. Hydrophobic coatings on the walls of the microchannels facilitate larger flow rates compared to hydrophilic counterparts for the same pressure drop as they offer less resistance to flow. Even though there have been considerable efforts made to study the fluid transport in hydrophobic microchannels in the fully developed regime, not much attention had been paid to the entrance effects. The entry region is not yet so deeply investigated. There is a clear need to investigate the coupled characteristics between velocity slip and entrance effects in order to understand microchannel flows.

There is extensive literature appearing on micro-scale slip flow and heat transfer. Morini et al. [38] investigated slip flow in rectangular microchannels. They presented the 2D velocity distribution of steady-state, laminar slip flow for Newtonian fluids in a hydrodynamically fully developed region. Morini et al. [39] focused on the role of the main scaling effects in adiabatic and diabatic microchannels, and analyzed the effects of viscous dissipation, conjugate heat transfer, and entrance effects on the mean value of the Nusselt number.

Wang [40] developed accurate analytical solutions for fully developed slip flow and H1 heat transfer in rectangular and equilateral triangular ducts. He pointed out that both velocity slip and temperature jump have significant influences on the Poiseuille and Nusselt numbers. Further, Wang [41] used an efficient analytical method to investigate H1 and H2 forced convection heat transfer in rectangular ducts, especially for large aspect ratios.

Some studies on the developing region in microchannels have been presented [42–54]. Steinke and Kandlikar [42] identified single-phase heat transfer enhancement techniques for use in microchannels. They speculated that this increase in heat transfer performance from these techniques could place a single-phase liquid system in competition with a two-phase system, thus simplifying the overall complexity and improving the overall reliability. However, they pointed out that the added pressure drop resulting from the techniques should be carefully evaluated. Steinke and Kandlikar [43] reviewed the available literature on single-phase liquid friction factors in microchannels and indicated that there seems to be a common thread between all of the papers that had reported some form of discrepancy between the experimental data and the predicted theoretical values. Those authors that have not discussed the developing length seem to be the same authors reporting the discrepancies. Further, the authors that have considered the added friction factor for the developing region report good agreement with the predicted traditional theory.

Ranjith et al. [44] studied numerically the hydrodynamics of a steady developing flow between two infinite parallel plates with hydrophilic and hydrophobic surfaces using dissipative particle dynamics. The hydrophobic and hydrophilic surfaces were modeled using partial-slip and no-slip boundary conditions, respectively. The simulation results of the developing flow are in good agreement with analytical solutions from Duan and Muzychka [45] for no-slip and partial-slip surfaces. Mishan et al. [46] experimentally investigated developing flow and heat transfer in rectangular microchannels. The experimental results of pressure drop confirmed that the data presented by other research can be due to entrance effects.

Kohl et al. [47] experimentally investigated entrance effects and hydrodynamic developing flow for pressure drop calculations. The experimental results suggested that it is important to include entrance effects, especially for the channels with $L/D_h < 300$.

Bayraktar and Pidugu [48] indicated that even though a tremendous effort in microfluidics research is currently underway, there is little work done to study the entry flow in microchannels. Researchers have generally assumed microchannel flows to be laminar and fully developed, ignoring entrance effects. However, in the entry region, the velocity distribution and skin friction show significant variations in the stream-wise direction, which could influence separation efficiency in processes such as electrophoresis.

Barber and Emerson [49] conducted a numerical investigation of gaseous slip flow at the entrance of circular and parallel plate microchannels using a two-dimensional Navier–Stokes solver. Barber and Emerson [49] examined the role of Reynolds and Knudsen numbers on the hydrodynamic development length at the entrance to circular and parallel plate microchannels. They carried out numerical simulations over a range of Knudsen numbers covering the continuum and slip flow regimes. They proposed expressions for the hydrodynamic entrance length but did not provide any expressions for pressure drop.

Vocale and Spiga [50] conducted a numerical study of hydrodynamically developing flow in the entrance region of rectangular microchannels with different values of aspect ratio using a solver based on the finite element method with first-order slip boundary conditions.

Hettiarachchi et al. [51] numerically studied three-dimensional laminar slip flow and heat transfer in rectangular microchannels having constant-temperature walls using the finite-volume method for thermally and simultaneously developing flows. They evaluated the effect of rarefaction on the hydrodynamically developing flow field, pressure gradient, and entrance length.

Recently, many efforts have been made applying innovative geometries [3–13], using fluids with excellent thermal features [14–17], and utilizing micro-pin-fins [15–21] in order to enhance the performance of microchannel heat sinks.

Bahiraei and Heshmatian [5] evaluated the flow, heat transfer, and second law characteristics of a hybrid nanofluid containing grapheme–silver nanoparticles inside two new microchannel heat sinks.

Khan et al. [6] numerically investigated the thermal and hydraulic performances of various geometric shapes of a microchannel heat sink with water as the cooling fluid. The performances of seven microchannel shapes were compared at the same microchannel hydraulic diameter and the same average height of the bottom silicon substrate. Their results showed that an inverse trapezoidal shape gives the lowest thermal resistance for a Reynolds number up to 300, and the values of friction factor and Reynolds number product are almost similar for all the shapes because of the constant hydraulic diameter. Ansari and Kim [7] proposed a transverse-flow arrangement concept of a double-layer microchannel heat sink. They performed a numerical analysis using three-dimensional Navier–Stokes equations for the entire heat sink domain to evaluate the thermal and hydraulic performance of the proposed heat sink.

Al Siyabi et al. [8] experimentally examined the behavior of a multilayered microchannel heat sink integrated with a concentrating photovoltaic assembly. In their work, experiments were conducted to evaluate the thermal performance of multilayered microchannel heat sinks for different numbers of layers, different heat transfer fluid flow rates, and different heating power rates. They further developed a numerical model to analyze the heat sink behavior for measurements that cannot be obtained using the experimental approach.

Shen et al. [9] proposed a modification of double-layer microchannel heat sinks into a wavy configuration with a swap of the upper and lower layer design. They demonstrated that this design could effectively improve the comparatively low cooling effectiveness at the upper layer, yet the swap of flow could promote re-developing of the flow field and heat transfer, accordingly. Some studies on flow and heat transfer characteristics in double-layered microchannel heat sinks have also been presented [10–13].

Xia et al. [14] provided an overall analysis of nanofluids flowing through microchannel heat sinks. They obtained the temperature distribution on the substrate of microchannel heat sinks. Their results indicated that the thermal conductivity and dynamic viscosity of Al_2O_3 and TiO_2 nanofluids are both improved with increasing particle volume fraction. The thermal motion of nanoparticles could promote the interruption of laminar flow and intensify the heat transfer between fluids and channel walls.

Hassani et al. [15] investigated the effects of different interruptions of fins on the transport characteristics of a nanofluid-cooled electronic heat sink with a chevron shape. Seven interruptions of fins were studied, and water-based nanofluid with Al_2O_3 nanoparticles at volume fractions of 0.5% and 1.0% were tested as the coolant in a laminar flow regime.

Zargartalebi and Azaiez [16] analyzed the effects of nanoparticle properties and pin sizes on heat removal performance using a nanofluid two-component model. Their results indicated that the nanoparticle distribution plays an important role in heat transfer.

Xu and Wu [18] experimentally investigated water flow and heat transfer characteristics in silicon micro-pin-fin heat sinks with various pin-fin configurations and a conventional microchannel. Their results indicated the better heat transfer performance of the micro-pin-fin heat sinks than of the conventional microchannel. The dominant mechanism of heat transfer enhancement caused by the micro-pin-fins is the hydrodynamic effects, including fluid disturbance as well as the breakage and re-initialization of the thermal boundary layer near the wall of the heat sinks.

Ansari and Kim [19] presented a novel hotspot-targeted cooling technique combining microchannels and pin-fins for efficient thermal management of microprocessors with heterogeneous power distributions. The performance of the proposed microchannel-pin-fin hybrid heat sink was evaluated numerically and compared with that of a simple microchannel heat sink in their study. They indicated that the hybrid heat sink exhibited a remarkable improvement in thermal performance compared to the non-hybrid heat sink with a reasonable increase in the pumping power.

Adewumi et al. [21] numerically investigated steady incompressible flow and forced convection heat transfer through a microchannel heat sink with micro-pin-fin inserts for both fixed and variable axial lengths. The effects of the micro-pin-fins on the optimized microchannel were evaluated in detail.

Ribs mounted in the microchannel heat sink generally result in a higher heat transfer coefficient but are usually accompanied by a higher pressure drop per unit length. Khan et al. [22] presented an analysis of microchannel heat sinks with ribbed channels in various configurations using the three-dimensional Navier–Stokes equations and compared them to smooth channels in a Reynolds number range of 100–500. Their results indicated that the thermal resistance of microchannels was greatly reduced by introducing ribs, and the pressure drop was increased greatly because of the ribs.

Chai et al. [23] presented a detailed numerical study on local laminar fluid flow and heat transfer characteristics in microchannel heat sinks with tandem triangular ribs for a Reynolds number of 443. Three-dimensional conjugate heat transfer models considering entrance effect, viscous heating, and temperature-dependent thermophysical properties were employed in their studies [23,24]. Their results showed that the triangular ribs could significantly reduce the temperature rise of the heat sink base and efficiently prevent the drop of the local heat transfer coefficient along the flow direction, but also result in a higher local friction factor than the straight microchannel. Further, the effects of the geometry and arrangement of triangular ribs on the thermohydraulic performance were examined by the variations in the average friction factor and Nusselt number for Reynolds numbers ranging from 187 to 715.

Chai et al. [25] performed a study of the laminar flow and heat transfer characteristics in an interrupted microchannel heat sink with ribs in the transverse microchambers. They investigated five different rib configurations, including rectangular, backward triangular, diamond, forward triangular, and ellipsoidal. The role of such ribs in the velocity contour, pressure distribution and temperature distribution, and the local pressure drop and heat transfer characteristics in such microchannel heat sinks was studied.

Zhai et al. [26] summed up the empirical correlations of laminar convective heat transfer in microchannel heat sinks in previous studies. They established an empirical model of laminar convective heat transfer in microchannel heat sinks. Further, a corresponding experiment and simulation were used to validate the accuracy of their theoretical model.

Fully developed flows have been widely investigated in different geometries, in continuum flow and slip flow conditions. In the case of developing flows with slip, only parallel plates and circular ducts are considered in the literature due to the slip boundary conditions which make this particular hydrodynamically developing flow problem even more complicated. A survey of the available literature indicates a shortage of pressure drop information for three-dimensional entrance flows with velocity slip boundary conditions, such as relatively short plate fin microchannels where the entrance region plays a very important role. There is currently no published model and data for pressure drop which can be utilized by the research community in the design and optimization of microchannel plate fin heat sinks. The entrance region in a microchannel heat sink is particularly of interest due to the presence of comparatively large pressure drop and heat transfer. Wall shear stress effects and velocity distributions vary significantly at the entrance region, and these may eventually affect the separation efficiency of the microfluidic processes. Moreover, the entrance region in hydrophobic channels is much longer than in hydrophilic channels. Reduction of the entrance length is very important for the design of some types of lab-on-a-chip devices. Given that the convective heat transfer behavior in the developing region differs from that in the fully developed region, and given that many microchannel heat sinks are short, this effect of the entrance region is significant. The apparent friction factor and Reynolds number product $f_{app}Re$ could be significantly higher than the fully developed value of friction factor and Reynolds number product *fRe*. The first approximate solution was that for the parallel plates obtained by Sparrow et al. [52] using Targ's linearization theory. Later, Quarmby [53] extended this technique to a circular tube.

Even though a tremendous effort in microchannel heat sink research is currently underway and a vast amount of literature is available for an interested researcher, there is little work performed to study the entrance flow in microchannel heat sinks. Researchers have generally assumed microchannel heat sink flows to be fully developed, ignoring entrance effects; therefore, it may be pointed out that most of the statements, formulas, and charts are valid only for long ducts. When dealing with fluid flow within microchannels, in most applications, the short length of the channels is not enabling the flow to ignore entrance effects and completely reach the fully developed regime. However, in the entrance region, skin friction shows quite significant variations. The deviation of experimental data from fully developed theoretical predictions could be misinterpreted as an early transition to turbulence or other reasons. Although numerous papers have proposed reasons for the conflicting results obtained by different researchers, it can be still concluded that no universally accepted physical interpretation has been found among studies focusing on the characterization of friction factor/pressure drop for microchannel flows. A survey of the available literature indicates a shortage of information for microchannel heat sink entrance flows in the slip regime, such as short plate fin microchannel heat sinks. There is currently no published model or tabulated data for pressure drop which can be utilized by the research community. The lack of a general pressure drop model of fluid flow through microchannel plate fin heat sinks is a major problem. This paper is concentrated on simple compact modeling methods for predicting microchannel plate fin heat sink pressure drop.

3. Theoretical Analysis

The geometry of a microchannel heat sink is shown in Figure 1a. The length of the heat sink is *L*, the width is *W*, and the height is *H*. The top surface is insulated, and the bottom surface is uniformly heated. A coolant passes through a number of microchannels along the *z* axis and takes heat away from the heat-dissipating electronic component attached below. The flow in the channels is steady, laminar, and developing. There are *N* channels, and each channel has a height 2*a* and width 2*b*. The thickness of each fin is *t*. At the channel wall, the slip flow velocity boundary condition is applied to calculate the apparent friction factor and Reynolds number product and pressure drop. One of the most fundamental problems in fluid dynamics is that of laminar flow in circular and non-circular channels under a constant pressure gradient. The starting point in the theoretical discussion will be the definition of the friction factor and Reynolds number parameter. Upon obtaining the velocity distribution $u(x, y)$ and mean velocity $\bar{u}$, the friction factor and Reynolds number parameter may be defined using the simple expression denoted in some texts by the Poiseuille number:

$$Po_{D_h} = \frac{\bar{\tau} D_h}{\mu \bar{u}} = \frac{\left(-\frac{A}{P}\frac{dp}{dz}\right)D_h}{\mu \bar{u}} = \frac{fRe_{D_h}}{2} \tag{1}$$

The above grouping *Po* is interpreted as the fully developed dimensionless average wall shear. The fully developed mean wall shear stress may also be related to the pressure gradient by means of the force balance $\bar{\tau} = -A/P \, dp/dz$.

We examine the momentum equation and consider the various force balances using the method of scale analysis. Comparing the force scale between friction and inertial forces, we obtain the following relation:

$$\frac{\mu\frac{\partial^2 u}{\partial y^2}}{\rho u \frac{\partial u}{\partial x}} \sim \frac{\mu \frac{U}{D_h^2}}{\frac{\rho U^2}{L}} = \frac{L}{D_h Re_{D_h}} = \xi \tag{2}$$

where ξ is the non-dimensional channel length. This analysis demonstrates that inertial forces are quite important for short ducts where $\xi \ll 1$. The fluid flow behavior in the developing region differs from that in the fully developed region. The parameter $L/(D_h Re)$ is always a significant parameter in internal fluid flows. The flow behaves differently and is dominated by different mechanisms as the parameter $L/(D_h Re)$ changes.

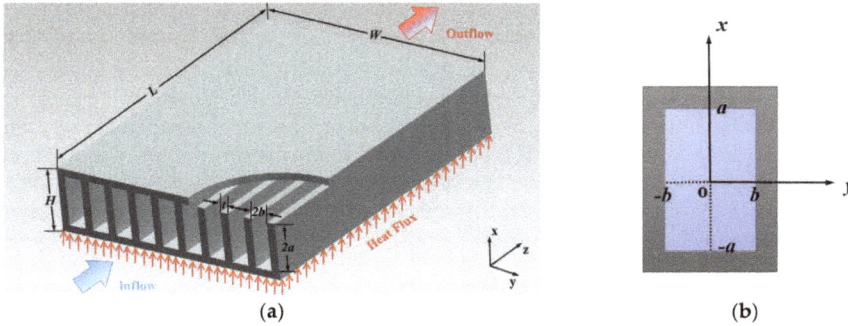

Figure 1. Schematics of the microchannel plate fin heat sink. (**a**) Microchannel plate fin heat sink; (**b**) Computational unit.

In many practical applications, the length of the channel in the developing region therefore forms a major portion of the flow length through a microchannel. To account for the developing region, the pressure drop equations are presented in terms of an apparent friction factor. The apparent friction factor accounts for the actual pressure drop due to friction and the developing region effects. It represents an actual value of the friction factor over the flow length between the inlet and the location under consideration. Therefore, the apparent friction factor, f_{app}, must be utilized to calculate the factual pressure drop. The apparent flow friction factor is used in this paper since it incorporates the pressure drops caused both by the wall shear stress due to the significant velocity gradient normal to the wall and by the momentum flux variation due to the change of velocity field from a uniform profile at the inlet to a specific profile downstream in the channel. Researchers have generally assumed microchannel heat sink flows to be fully developed, ignoring entrance effects; therefore, care should be exercised as most of the statements, formulas, and charts are valid only for long ducts. The long duct criterion will be discussed later in this paper.

Generally, there are three main components that contribute to the overall pressure drop. The inlet and exit losses need to be quantified for the microchannel. The hydrodynamic frictional pressure loss needs to be carefully evaluated. Summing all of the frictional and inlet and exit dynamic losses, the total pressure drop model function is given in terms of Bernoulli's equation,

$$\Delta P = \left[K_c + 4\left(f_{app} Re_{D_h} \right) \frac{L}{D_h Re_{D_h}} + K_e \right] \frac{1}{2} \rho \bar{u}^2 = \left[K_c + 4\left(f_{app} Re_{D_h} \right) \zeta + K_e \right] \frac{1}{2} \rho \bar{u}^2 \tag{3}$$

where K_c and K_e represent the contraction and expansion loss coefficients due to area changes. The friction factor used in Equation (3) is the apparent friction factor and accounts for the developing region. Fluid flow modelling in a plate fin heat sink is essentially a simultaneously developing hydraulic and thermal boundary layer problem in rectangular ducts. The flow may become fully developed if the heat sink channel is sufficiently long in the flow direction or with relatively small fin spacing; however, this is very unlikely for microchannel heat sinks for electronic cooling applications. The hydrodynamically developing flow can become quite important in microchannels. Due to the often-short lengths, the developing flow could dominate the entire flow length of the microchannel. When considering the developing flows, the pressure drop is now related to an apparent friction factor. The apparent friction factor, f_{app}, for a rectangular channel may be computed using a form of the model developed by Duan and Muzychka [45] for developing laminar flow. Duan and Muzychka [45] demonstrated that the boundary layer behavior in a circular tube entry is substantially identical to that on a flat plate. As the boundary layer develops further downstream, the effects of geometry become gradually more pronounced. We will take advantage of the asymptotic limit for developing an approximate model for predicting pressure drop for plate fin heat sinks. The apparent friction factor consists of two components just for convenience. Actually, the apparent friction factor should be utilized to calculate the pressure drop. The first is the friction factor from the familiar theory for the fully developed flow, and the second is the pressure defect. The first term is the frictional loss resulting from the fully developed flow. The second term represents the added pressure drop due to the developing flow region. It is convenient to report the pressure drop in a developing flow as equal to that for a fully developed flow plus a correction term $G(\xi)$ representing additional pressure drop which exceeds the fully developed pressure drop. Thus, the difference between the apparent friction factor and Reynolds number product over a length and the fully developed friction factor and Reynolds number product is expressed in terms of an incremental pressure defect $G(\xi)$:

$$f_{\text{app}} Re = f Re + G(\xi) \tag{4}$$

The commonly used customary incremental pressure drop number (also denoted in some texts by Hagenbach's factor) is given by

$$k = \left(f_{\text{app}} - f \right) \frac{4L}{D_h} = \left(f_{\text{app}} Re - f Re \right) 4\xi = 4\xi G(\xi) \tag{5}$$

The Navier–Stokes equations are assumed to be valid in their traditional form, and wall slip is merely modeled through a modification of the boundary condition. Rectangular geometries are of particular interest in microfluidics applications. We may now examine the solution for rectangular ducts for slip flow. A schematic diagram of the rectangular cross section is showed in Figure 1b. When $\xi \gg 1$, the fully developed flow momentum equation in Cartesian coordinates reduces to the form

$$\frac{\partial^2 u}{\partial x^2} + \frac{\partial^2 u}{\partial y^2} = \frac{1}{\mu} \frac{dp}{dz} \tag{6}$$

The velocity distribution must satisfy the slip boundary condition at the walls. The slip boundary condition that is applied in the analysis both of slip gas flows and of liquid flows over superhydrophobic surfaces takes the following form. The local slip velocity is proportional to the local velocity gradient normal to the wall. Due to symmetry, the boundary conditions are

$$u = -\lambda \frac{\partial u}{\partial y} \qquad at \quad y = b, \qquad 0 \leq x < a \tag{7}$$

$$u = -\lambda \frac{\partial u}{\partial x} \qquad at \quad x = a, \qquad 0 \leq y < b \tag{8}$$

$$\frac{\partial u}{\partial y} = 0 \qquad at \quad y = 0, \qquad 0 \leq x \leq a \tag{9}$$

$$\frac{\partial u}{\partial x} = 0 \qquad at \quad x = 0, \qquad 0 \leq y \leq b \tag{10}$$

where λ generally denotes the slip length for either gases or liquids, which is defined based on the physics of the fluid flow. The slip length can be used to characterize the type of flow in channels: if $\lambda = 0$, the flow is no-slip flow; if $\lambda = \infty$, the flow is plug flow; and intermediate values of λ represent partial slip flow. For gas slip flow, Maxwell's first order correction gives

$$u = -\lambda_p \frac{2 - \sigma}{\sigma} \frac{\partial u}{\partial y} \qquad at \quad y = b, \qquad 0 \leq x < a \tag{11}$$

$$u = -\lambda_p \frac{2-\sigma}{\sigma} \frac{\partial u}{\partial x} \qquad at \quad x = a, \qquad 0 \le y < b \tag{12}$$

where λ_p is the molecular mean free path. The constant σ denotes the tangential momentum accommodation coefficient. Equations (11) and (12) are mathematically equivalent to the first-order correction commonly employed in the analysis of rarefied gas flow in the slip flow regime. It is convenient to define a modified Knudsen number as $Kn^* = Kn(2-\sigma)/\sigma$. A nondimensional number similar to the modified Knudsen number (Kn^*) can be defined for liquid slip flow, i.e., the ratio of slip length to a characteristic dimension of the flow field, and the presently developed model can then be utilized for liquid flows over superhydrophobic surfaces to predict pressure drop. To look at it from a slightly more general mathematical point of view, when the no-slip condition on the solid surfaces is partially relaxed, the molecular mean free path and the term involving the accommodation coefficient ($\lambda_p(2-\sigma)/\sigma$) and the slip length ($\lambda$) have identical mathematical meaning.

Following Ebert and Sparrow [55], using the Method of Eigenfunction Expansions, a solution of the velocity may be assumed as follows:

$$u = \frac{b^2}{\mu} \frac{dp}{dz} \sum_{n=1}^{\infty} X_n\left(\frac{x}{a}\right) \cos\left(\delta_n \frac{y}{b}\right) \tag{13}$$

in which the δ_n are a set of eigenvalues, the X_n are a set of functions of x/a, and the $\cos(\delta_n.y/b)$ are a set of eigenfunctions. This solution satisfies the boundary condition, Equation (9). Furthermore, substituting the velocity solution into the boundary condition, Equation (7), we obtain

$$\delta_n \tan \delta_n = \frac{b}{\lambda} \tag{14}$$

The characteristic length scale in the present analysis is defined as the hydraulic diameter.

$$Kn_{D_h} = \frac{\lambda_p}{D_h} = \frac{\lambda_p}{\frac{4b}{1+\varepsilon}} \tag{15}$$

Thus,

$$\delta_n \tan \delta_n = \frac{1}{\frac{4}{1+\varepsilon} Kn^*_{D_h}} \tag{16}$$

The eigenvalues δ_n can be obtained from Equation (16). Finally, the velocity distribution is obtained as follows:

$$u = \frac{b^2}{\mu} \frac{dp}{dz} \sum_{n=1}^{\infty} \frac{2 \sin \delta_n \cos\left(\delta_n \frac{y}{b}\right)}{\delta_n^2(\delta_n + \sin \delta_n \cos \delta_n)} \left[\frac{\cosh\left(\frac{\delta_n}{\varepsilon}\frac{x}{a}\right)}{\cosh\left(\frac{\delta_n}{\varepsilon}\right) + \frac{4}{1+\varepsilon} Kn^*_{D_h} \delta_n \sinh\left(\frac{\delta_n}{\varepsilon}\right)} - 1 \right] \tag{17}$$

The mean velocity is found by integration of Equation (17) across the section of the duct.

$$\begin{aligned}\bar{u} &= \frac{1}{A} \int u \, dA = \int_0^1 \int_0^1 u \, d\frac{x}{a} \, d\frac{y}{b} \\ &= \frac{b^2}{\mu} \frac{dp}{dz} \sum_{n=1}^{\infty} \frac{2\varepsilon \sin^2 \delta_n}{\delta_n^4(\delta_n + \sin \delta_n \cos \delta_n)} \left[\frac{\sinh\left(\frac{\delta_n}{\varepsilon}\right)}{\cosh\left(\frac{\delta_n}{\varepsilon}\right) + \frac{4}{1+\varepsilon} Kn^*_{D_h} \delta_n \sinh\left(\frac{\delta_n}{\varepsilon}\right)} - \frac{\delta_n}{\varepsilon} \right]\end{aligned} \tag{18}$$

We can obtain the friction factor and Reynolds number product from the above equations in terms of the aspect ratio and the slip coefficient.

$$\begin{aligned}f Re_{D_h} &= \frac{2\left(-\frac{A}{P}\frac{dp}{dz}\right) D_h}{\mu \bar{u}} \\ &= \frac{4}{(1+\varepsilon)^2 \sum_{n=1}^{\infty} \frac{\varepsilon \sin^2 \delta_n}{\delta_n^4(\delta_n + \sin \delta_n \cos \delta_n)} \left[\frac{\delta_n}{\varepsilon} - \frac{\sinh\left(\frac{\delta_n}{\varepsilon}\right)}{\cosh\left(\frac{\delta_n}{\varepsilon}\right) + \frac{4}{1+\varepsilon} Kn^*_{D_h} \delta_n \sinh\left(\frac{\delta_n}{\varepsilon}\right)} \right]}\end{aligned} \tag{19}$$

The limit of Equation (19) corresponds to a parallel-plate channel for $\varepsilon \to 0$:

$$fRe_{D_h} = \frac{24}{1 + 12Kn^*_{D_h}} \tag{20}$$

It can also be demonstrated that Equation (19) reduces to its no-slip flow limits as $Kn^* \to 0$. The relationship between the flow friction coefficient, f, and Reynolds number, Re, for a fully developed laminar flow regime in a rectangular channel is only a function of the aspect ratio and may be calculated as follows:

$$\left(fRe_{D_h}\right)_{ns} = \frac{24}{(1+\varepsilon)^2 \left[1 - 6\sum\limits_{n=1}^{\infty} \frac{\varepsilon}{\delta_n^5} \tanh\left(\frac{\delta_n}{\varepsilon}\right)\right]} \tag{21}$$

Examination of the single-term solution reveals that the single-term approximation is accurate enough for engineering applications. The largest difference occurs when $\varepsilon = 1$, which gives an fRe value 0.7 percent below the exact value. When greater accuracy is desired, two terms are absolutely enough due to very rapid convergence, and the largest error is less than 0.05%. Considering only the two terms of the exact series, Equation (21) gives

$$\left(fRe_{D_h}\right)_{ns} = \frac{24}{(1+\varepsilon)^2 \left[1 - \frac{192\varepsilon}{\pi^5}\left(\tanh\left(\frac{\pi}{2\varepsilon}\right) + \frac{1}{243}\tanh\left(\frac{3\pi}{2\varepsilon}\right)\right)\right]} \tag{22}$$

This equation is founded on theory and is more accurate compared to those obtained by curve fitting [56].

The slip flow friction factor results can be presented conveniently in terms of the normalized Poiseuille number. The Poiseuille number reduction depends on the geometry of the cross section. It is convenient that the Poiseuille number results are expressible to good accuracy by the relation

$$\frac{Po}{Po_{ns}} = \frac{fRe}{(fRe)_{ns}} = \frac{1}{1 + \alpha Kn^*_{D_h}} \tag{23}$$

in which α depends on the duct geometry.

For rectangular ducts, the constants α are derived from a least-square fit of the Poiseuille number results. The constants α are a weak function of the aspect ratio, and the data points are fitted to a simple correlation [57]:

$$\alpha = 11.97 - 10.59\varepsilon + 8.49\varepsilon^2 - 2.11\varepsilon^3 \tag{24}$$

Then, Equation (19) can be simplified to facilitate practical application as follows:

$$fRe = \frac{24}{(1+\varepsilon)^2 \left[1 - \frac{192\varepsilon}{\pi^5}\left(\tanh\left(\frac{\pi}{2\varepsilon}\right) + \frac{1}{243}\tanh\left(\frac{3\pi}{2\varepsilon}\right)\right)\right]} \frac{1}{1 + \alpha Kn^*} \tag{25}$$

It is found that the entrance friction factor and Reynolds number product is of finite value and dependent on Kn^* but independent of the cross-sectional geometry [45]. Duan and Muzychka [45] also demonstrated that very near the inlet of circular and parallel plate ducts ($\xi \leq 0.001$), $f_{app}Re$ is nearly equivalent and independent of duct shape. The boundary layer behavior in the tube entry is substantially identical to that on a flat plate. At the entrance of the duct, the velocity boundary layer starts developing at each wall under the imposed flow acceleration. As long as the thickness of the boundary layer is small compared to the duct dimensions, the boundary layers from different walls do not affect each other appreciably. This explains why very near the inlet of ducts, $f_{app}Re$ is nearly equivalent and independent of duct shape. As the boundary layer develops further downstream ($\xi > 0.001$), the effects of geometry become gradually more pronounced. The incremental pressure defect $G(\xi)$ for a rectangular channel may be computed using a form of the model developed by Duan and Muzychka [45] for developing laminar flow:

$$G(\xi) = \frac{1}{3\xi(1+8Kn^*)^2} - 2\sum_{i=1}^{\infty} \frac{(3-\exp(-4\alpha_i^2\xi))\exp(-4\alpha_i^2\xi)}{\alpha_i^2\xi\left(1+8Kn^*+4(\alpha_iKn^*)^2\right)} \tag{26}$$

Equation (26) is nearly independent of the duct shape and may be used to calculate the friction factor and Reynolds number product for the short asymptote of rectangular ducts for slip flow. We will take advantage of the asymptotic limit for developing an approximate model for predicting pressure drop for plate fin heat sinks. The eigenvalue α_i satisfies the following equation [45]:

$$\alpha_i J_0(\alpha_i) - 2\left(1+Kn^*\alpha_i^2\right)J_1(\alpha_i) = 0 \tag{27}$$

where $J_v(x)$ is the Bessel function of the first kind of order v. While typical microflows are characterized by $Re < 100$, in a few microfluidic applications such as micro heat exchangers and micromixers, the Reynolds number can reach the order of a few hundreds. It is emphasized that several eigenvalues are sufficiently accurate for all values of ξ of engineering interest. The negative exponentials cause rapid convergence, especially if ξ is not too small. As an illustration, for a practical engineering application limit of $\xi \geq 0.01$, only two terms in the summation are really required. It can be demonstrated [45] that the proposed model of Equation (28) strictly correctly approaches the $\xi \to 0$ asymptote $2/Kn^*$ and approaches the $\xi \to \infty$ asymptote of Equation (19). Therefore, using the simple expression Equation (28), the apparent friction factor and Reynolds number product results can be easily obtained for practical engineering applications without sacrificing much in accuracy.

$$\begin{aligned}
f_{app}Re = &\frac{24}{(1+\varepsilon)^2\left[1-\frac{192\varepsilon}{\pi^5}\left(\tanh\left(\frac{\pi}{2\varepsilon}\right)+\frac{1}{243}\tanh\left(\frac{3\pi}{2\varepsilon}\right)\right)\right]}\frac{1}{1+\alpha Kn^*} \\
&+ \frac{1}{3\xi(1+8Kn^*)^2} - 2\sum_{i=1}^{2}\frac{(3-\exp(-4\alpha_i^2\xi))\exp(-4\alpha_i^2\xi)}{\alpha_i^2\xi(1+8Kn^*+4(\alpha_iKn^*)^2)}
\end{aligned} \tag{28}$$

For the inlet and exit pressure losses for a heat sink, Kays and London [58] provide loss coefficients in the form $\Delta P = K(\rho\bar{u}^2/2)$ as a function of the ratio of free-flow area to frontal area $\varphi = 2b/(2b+t)$. The graphs of experimental data for laminar flow in Reference [58] have been curve fitted here for laminar flow:

$$K_c = 0.4\left(1-\varphi^2\right) + 0.4 \tag{29}$$

$$K_e = (1-\varphi)^2 - 0.4\varphi \tag{30}$$

4. Results and Discussion

The developing apparent friction factor and Reynolds number parameter as a function of aspect ratio and modified Knudsen number is illustrated clearly in some graphs. Figures 2–4 present the proposed model Equation (28) $f_{app}Re$ in slip flow for various aspect ratios of the rectangular cross section. From an inspection of the graphs, it is seen that $f_{app}Re$ monotonically decreases in the entrance region, and $f_{app}Re$ decreases as the modified Knudsen number increases for the same aspect ratio. The effect of increasing Kn^* is to decrease the apparent friction and pressure drop over the channel length very significantly. In addition, the $f_{app}Re$ values decrease with increasing ε for the same Kn^*. Moreover, it is obvious that the pressure gradient for a slip flow is less than that for the corresponding no-slip flow. This effect of the developing region is significant for microchannel plate fin heat sinks.

The values in the entrance region are larger than those in the fully developed region, which demonstrates the critical importance of the entrance region in determining the pressure drop characteristics in microchannel heat sinks. To take into account the entrance effects on the overall pressure drop in the microchannel region, clearly, the dimensionless developing length ξ is the proper parameter. Figures 2–4 indicate that the local apparent friction factor and Reynolds number product decreases and approaches the fully developed constant value with increasing dimensionless developing length ξ. The effects of the aspect ratios on the pressure drop in the slip flow region are investigated in Figures 2–4. It is noted that the pressure drop is higher for lower aspect ratios. The pressure drop in microchannel plate fin heat sinks decreases with an increase in the aspect ratios.

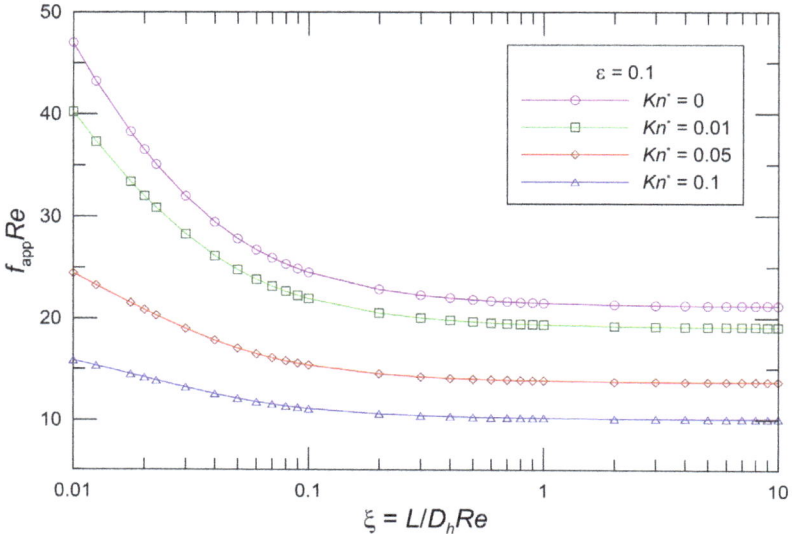

Figure 2. Effect of Kn^* on $f_{app}Re$ for rectangular ducts ($\varepsilon = 0.1$).

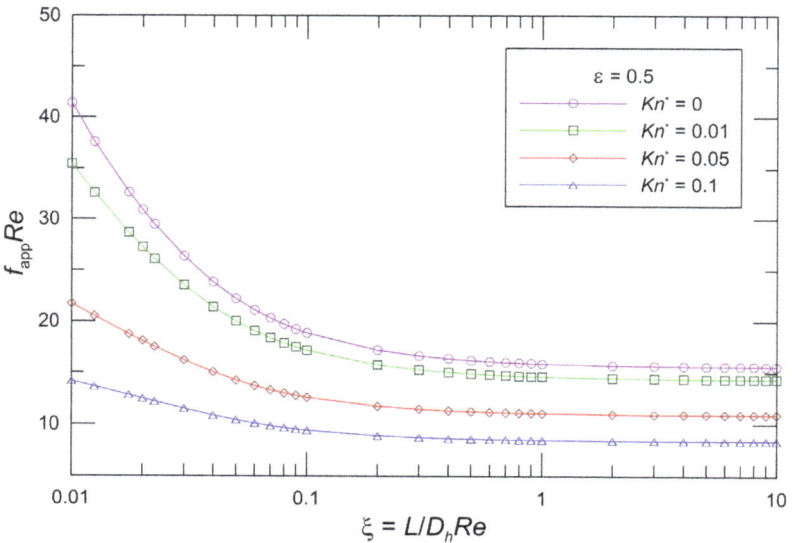

Figure 3. Effect of Kn^* on $f_{app}Re$ for rectangular ducts ($\varepsilon = 0.5$).

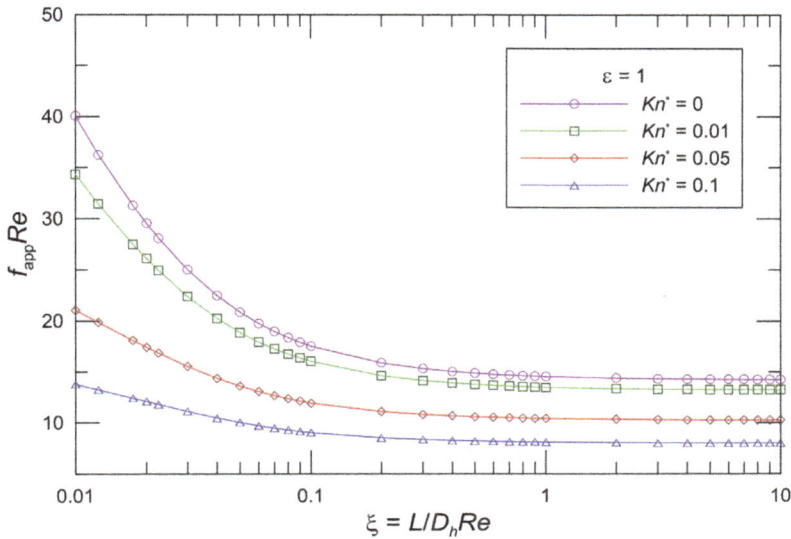

Figure 4. Effect of Kn^* on $f_{app}Re$ for rectangular ducts ($\varepsilon = 1$).

Numerical simulations were conducted using the commercial solver ANSYS Fluent 14.0 (Ansys Inc., Canonsburg, PA, USA) with user-defined functions. A structured mesh composed of rectilinear elements was constructed in a preprocessor to define the flow domains. The governing equations were solved with a commercial implementation of the finite volume method using a pressure-based solver and the SIMPLE algorithm. The slip boundary condition was implemented as a user-defined function in FLUENT. The solution algorithms were considered converged when the convergence criterion was satisfied. Local mesh refinement was performed using FLUENT's grid adaption utility at the flow inlet.

Shah and London [56] presented the results of Curr et al. [59] for hydrodynamically developing laminar flow in rectangular ducts. We validated the model with most of the available developing flow data for the rectangular channel. Figure 5 demonstrates the comparison between the proposed model Equation (28) and the available numerical data from Curr et al. [59] and our new numerical data. The model predictions are in agreement with the numerical solution. It is found that the difference between the model and available data from Curr et al. [59] is less than 7.6%. The maximum difference between our numerical data and the proposed model is less than 3.0%. This result does support the validity of the model. From the figure it is seen that the difference increases with a decrease in the aspect ratio. The difference is smaller and negligible for large aspect ratios. The apparent friction factor and Reynolds number product is significantly higher than the fully developed value. The distribution of $f_{app}Re$ does not depend on the dimensionless hydrodynamic length when the dimensionless hydrodynamic length is approximately unity, which should be the long duct criterion. In the range of $\xi < 0.06$, the $f_{app}Re$ decreases rapidly with increasing ξ due to the finite thickness of the boundary layer at the microchannel entrance region. With an increase of ξ, the $f_{app}Re$ has a comparatively gently downward tendency.

Figure 6 demonstrates the corresponding varying local $f_{app}Re$ with the non-dimensional flow distance and the comparison between the proposed model Equation (28) and slip flow numerical results when the aspect ratio $\varepsilon = 1$. It is found that the model predictions agree with our numerical results. The maximum deviation between the numerical results and the proposed model is less than 2.6%.

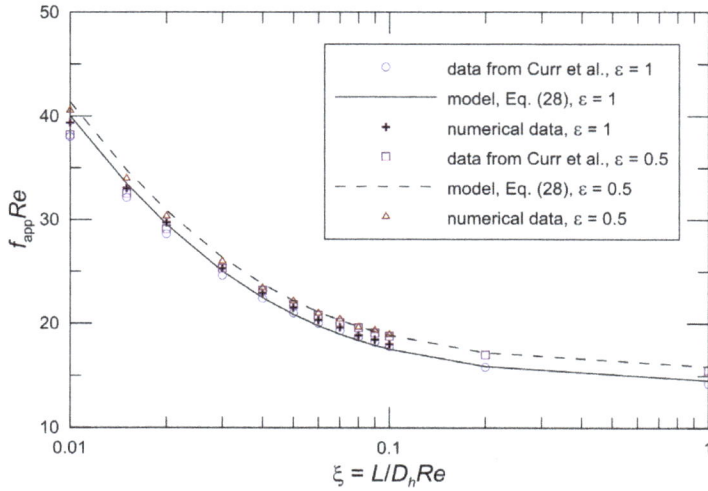

Figure 5. Comparison of $f_{app}Re$ for Curr et al. [59] and new numerical data.

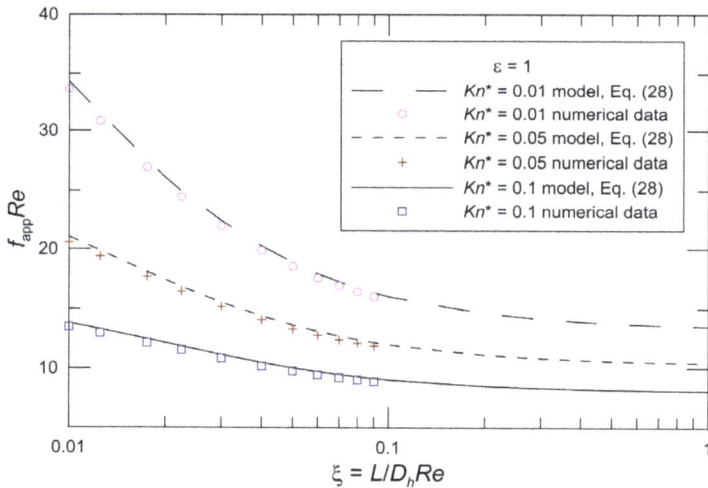

Figure 6. Comparison of $f_{app}Re$ for our numerical data.

It is clear that Equation (28) characterizes the pressure drop in microchannel plate fin heat sinks. The maximum deviation of exact values is less than 8 percent. The pressure drop may be predicted from Equation (28), unless greater accuracy is desired. The developed pressure drop model may be suitable for the parametric design of and optimization studies on microchannel plate fin heat sinks.

The fluid flow behavior in the developing region differs from that in the fully developed region. The parameter $L/(D_h Re)$ is always a significant parameter in internal fluid flows. The flow behaves differently and is dominated by different mechanisms as the parameter $L/(D_h Re)$ changes. This effect of the developing region is significant if the microchannels are short. It is shown that the data presented by some researchers may display entrance effects.

It can be seen that fully developed flow is attained at different ξ values, with low-aspect-ratio ducts reaching it a little earlier. The fully developed flow is attained when ξ is approximately 1 for rectangular ducts, as seen from these figures. The dimensionless developing length ξ is the proper parameter to take into account the entrance effects on the overall friction factor and pressure drop. The friction factor is higher in the developing flow region. Finally, beyond the entry region, the conventional theory for fully developed flow applies. Generally, the entrance region effects are less than 2% for the friction factor and Reynolds number product and can be neglected when $L/(D_h Re) \geq 1$. The fully developed flow (or long duct) criterion for pressure drop is $L/(D_h Re) \geq 1$.

5. Conclusions

This paper investigated pressure drop in the fluid flow in microchannel plate fin heat sinks. The paper deals with issues of hydrodynamic flow development in microchannel plate fin heat sinks. A model was proposed for predicting the pressure drop in microchannel plate fin heat sinks for developing slip flow and continuum flow. The accuracy of the developed model was found to be within 5 percent for practical configurations. Errors of this magnitude are acceptable for most engineering purposes. The model can be considered adequate and sufficiently reliable to analyze the pressure drop in microchannel plate fin heat sinks. As for slip flow, no solutions or tabulated data exist for microchannel plate fin heat sinks, so this developed model may be used to predict pressure drop in slip flow in microchannel plate fin heat sinks. The developed model is simple and founded on theory, and it may be used by the research community for the practical engineering design and optimization of microchannel plate fin heat sinks.

The goal of this investigation was to predict the pressure drop of the flow in microchannel plate fin heat sinks. The fully developed flow or long duct criterion for pressure drop is given. It is clearly shown that the entrance effects could be the source of the often-conflicting results previously reported in the literature.

Author Contributions: Conceptualization, Z.D. and H.M.; Methodology, Z.D. and H.M.; Software, B.H. and L.S.; Validation, Z.D. and X.Z.; Writing-Review & Editing, Z.D. and H.M.

Funding: This research was funded by the National Natural Science Foundation of China under No. 51576013 and the National Key R&D Program of China under No. 2017YFB0102101.

Acknowledgments: The authors acknowledge Professors Michael Yovanovich and Yuri Muzychka, who meant a lot to the work.

Conflicts of Interest: The authors declare no conflict of interest.

Nomenclature

A	flow area, m^2
a	major semi-axis of rectangle, m
b	minor semi-axis of rectangle, m
D_h	hydraulic diameter, $= 4A/P$
f	fanning friction factor, $= \bar{\tau}/\left(\frac{1}{2}\rho\bar{u}^2\right)$
Kn	Knudsen number, $= \lambda_p/D_h$
Kn^*	modified Knudsen number, $= Kn(2-\sigma)/\sigma$
L	channel length, m
P	perimeter, m
Po	Poiseuille number, $= \bar{\tau}D_h/\mu\bar{u}$
p	pressure, N/m^2
Re	Reynolds number, $= \bar{u}D_h/\nu$
T	temperature, K
t	fin thickness, m
U	velocity scale, m/s
u	velocity, m/s
$\bar{u}$	average velocity, m/s
X_n	function of x/a
x, y	Cartesian coordinates, m
z	coordinate in flow direction, m

Greek symbols

α	constants
γ	ratio of specific heats
δ_n	eigenvalues
ε	aspect ratio, $= b/a$
λ	hydrodynamic slip length, m
λ_p	molecular mean free path, m
μ	dynamic viscosity, $N \cdot s/m^2$
ν	kinematic viscosity, m^2/s
ξ	dimensionless hydrodynamic channel length, $= L/(D_h Re)$
σ	tangential momentum accommodation coefficient
$\overline{\tau}$	wall shear stress, N/m^2

Subscripts

D_h	based upon the hydraulic diameter
ns	no-slip

References

1. Kandlikar, S.; Garimella, S.; Li, D.; Colin, S.; King, M.R. *Heat Transfer and Fluid Flow in Minichannels and Microchannels*, 1st ed.; Elsevier: Amsterdam, Netherlands, 2006.
2. Wang, S.X.; Yin, Y.; Hu, C.X.; Rezai, P. 3D integrated circuit cooling with microfluidics. *Micromachines* **2018**, *9*, 287. [CrossRef] [PubMed]
3. Hajmohammadi, M.R.; Alipour, P.; Parsa, H. Microfluidic effects on the heat transfer enhancement and optimal design of microchannels heat sinks. *Int. J. Heat Mass Transf.* **2018**, *126*, 808–815. [CrossRef]
4. Jing, D.L.; Song, S.Y.; Pan, Y.L.; Wang, X.M. Size dependences of hydraulic resistance and heat transfer of fluid flow in elliptical microchannel heat sinks with boundary slip. *Int. J. Heat Mass Transf.* **2018**, *119*, 647–653. [CrossRef]
5. Bahiraei, M.; Heshmatian, S. Thermal performance and second law characteristics of two new microchannel heat sinks operated with hybrid nanofluid containing grapheme-silver nanoparticles. *Energy Convers. Manag.* **2018**, *168*, 357–370. [CrossRef]
6. Khan, A.A.; Kim, S.M.; Kim, K.-Y. Evaluation of various channel shapes of a microchannel heat sink. *Int. J. Air-Cond. Refrig.* **2016**, *24*, 1650018. [CrossRef]
7. Ansari, D.; Kim, K.-Y. Double-layer microchannel heat sinks with transverse-flow configurations. *Trans. ASME J. Electron. Packag.* **2016**, *138*, 031005. [CrossRef]
8. Al Siyabi, I.; Khanna, S.; Sundaram, S.; Mallick, T. Experimental and numerical thermal analysis of multi-layered microchannel heat sink for concentrating photovoltaic application. *Energies* **2019**, *12*, 122. [CrossRef]
9. Shen, H.; Zhang, Y.C.; Wang, C.-C.; Xie, G.N. Comparative study for convective heat transfer of counter-flow wavy double-layer microchannel heat sinks in staggered arrangement. *Appl. Therm. Eng.* **2018**. [CrossRef]
10. Ansari, D.; Kim, K.-Y. Performance analysis of double-layer microchannel heat sinks under non-uniform heating conditions with random hotspots. *Micromachines* **2017**, *8*, 54. [CrossRef]
11. Radwan, A.; Ookawara, S.; Mori, S.; Ahmed, M. Uniform cooling for concentrator photovoltaic cells and electronic chips by forced convective boiling in 3D-printed monolithic double-layer microchannel heat sink. *Energy Convers. Manag.* **2018**, *166*, 356–371. [CrossRef]
12. Jing, D.L.; He, L. Thermal characteristics of staggered double-layer microchannel heat sink. *Entropy* **2018**, *20*, 537. [CrossRef]
13. Wang, S.L.; Li, X.Y.; Wang, X.D.; Lu, G. Flow and heat transfer characteristics in double-layered microchannel heat sinks with porous fins. *Int. Commun. Heat Mass Transf.* **2018**, *93*, 41–47. [CrossRef]
14. Xia, G.D.; Liu, R.; Wang, J.; Du, M. The characteristics of convective heat transfer in microchannel heat sinks using Al_2O_3 and TiO_2 nanofluids. *Int. Commun. Heat Mass Transf.* **2016**, *76*, 256–264. [CrossRef]
15. Hassani, S.M.; Khoshvaght-Aliabadi, M.; Mazloumi, S.H. Influence of chevron fin interruption on thermo-fluidic transport characteristics of nanofluid-cooled electronic heat sink. *Chem. Eng. Sci.* **2018**, *191*, 436–447. [CrossRef]

16. Zargartalebi, M.; Azaiez, J. Heat transfer analysis of nanofluid based microchannel heat sink. *Int. J. Heat Mass Transf.* **2018**, *127*, 1233–1242. [CrossRef]

17. Duangthongsuk, W.; Wongwises, S. A comparison of the thermal and hydraulic performances between miniature pin fin heat sink and microchannel heat sink with zigzag flow channel together with using nanofluids. *Heat Mass Transf.* **2018**, *54*, 3265–3274. [CrossRef]

18. Xu, F.; Wu, H.Y. Experimental study of water flow and heat transfer in silicon micro-pin-fin heat sinks. *Trans. ASME J. Heat Transf.* **2018**, *140*, 122401. [CrossRef]

19. Ansari, D.; Kim, K.-Y. Hotspot thermal management using a microchannel-pinfin hybrid heat sink. *Int. J. Therm. Sci.* **2018**, *134*, 27–39. [CrossRef]

20. Vinoth, R.; Kumar, D.S. Experimental investigation on heat transfer characteristics of an oblique finned microchannel heat sink with different channel cross sections. *Heat Mass Transf.* **2018**, *54*, 3809–3817. [CrossRef]

21. Adewumi, O.O.; Bello-Ochende, T.; Meyer, J.P. Constructal design of combined microchannel and micro pin fins for electronic cooling. *Int. J. Heat Mass Transf.* **2013**, *66*, 315–323. [CrossRef]

22. Khan, A.A.; Kim, S.M.; Kim, K.-Y. Performance analysis of a microchannel heat sink with various rib configurations. *J. Thermophys. Heat Transf.* **2015**, *30*, 782–790. [CrossRef]

23. Chai, L.; Wang, L.; Bai, X. Thermohydraulic performance of microchannel heat sinks with triangular ribs on sidewalls–Part 1: Local fluid flow and heat transfer characteristics. *Int. J. Heat Mass Transf.* **2018**, *127*, 1124–1137. [CrossRef]

24. Chai, L.; Wang, L.; Bai, X. Thermohydraulic performance of microchannel heat sinks with triangular ribs on sidewalls–Part 2: Average fluid flow and heat transfer characteristics. *Int. J. Heat Mass Transf.* **2019**, *128*, 634–648. [CrossRef]

25. Chai, L.; Xia, G.D.; Wang, H.S. Laminar flow and heat transfer characteristics of interrupted microchannel heat sink with ribs in the transverse microchambers. *Int. J. Therm. Sci.* **2016**, *110*, 1–11. [CrossRef]

26. Zhai, Y.L.; Xia, G.D.; Li, Z.H.; Wang, H. Experimental investigation and empirical correlations of single and laminar convective heat transfer in microchannel heat sinks. *Exp. Therm. Fluid Sci.* **2017**, *83*, 207–214. [CrossRef]

27. Ou, J.; Perot, B.; Rothstein, J.P. Laminar drag reduction in microchannels using ultrahydrophobic surfaces. *Phys. Fluids* **2004**, *16*, 4635–4643. [CrossRef]

28. Sbragaglia, M.; Prosperetti, A. A note on the effective slip properties for microchannel flows with ultrahydrophobic surfaces. *Phys. Fluids* **2007**, *19*, 043603. [CrossRef]

29. Maynes, D.; Jeffs, K.; Woolford, B.; Webb, B.W. Laminar flow in a microchannel with hydrophobic surface patterned microribs oriented parallel to the flow direction. *Phys. Fluids* **2007**, *19*, 093603. [CrossRef]

30. Ybert, C.; Barentin, C.; Cottin-Bizonne, C.; Joseph, P.; Bocquet, L. Achieving large slip with superhydrophobic surfaces: Scaling laws for generic geometries. *Phys. Fluids* **2007**, *19*, 123601. [CrossRef]

31. Ng, C.-O.; Wang, C.Y. Stokes shear flow over a grating: Implications for superhydrophobic slip. *Phys. Fluids* **2009**, *21*, 087105. [CrossRef]

32. Feuillebois, F.; Bazant, M.Z.; Vinogradova, O.I. Effective slip over superhydrophobic surfaces in thin channels. *Phys. Rev. Lett.* **2009**, *102*, 026001. [CrossRef] [PubMed]

33. Zhang, J.X.; Yao, Z.H.; Hao, P.F. Drag reductions and the air-water interface stability of superhydrophobic surfaces in rectangular channel flow. *Phys. Rev. E* **2016**, *94*, 053117. [CrossRef]

34. Lee, C.; Choi, C.-H.; Kim, C.-J. Superhydrophobic drag reduction in laminar flows: A critical review. *Exp. Fluids* **2016**, *57*, 176. [CrossRef]

35. Patlazhan, S.; Vagner, S. Apparent slip of shear thinning fluid in a microchannel with a superhydrophobic wall. *Phys. Rev. E* **2017**, *96*, 013104. [CrossRef] [PubMed]

36. Antuono, M.; Durante, D. Analytic solutions for unsteady flows over a superhydrophobic surface. *Appl. Math. Model.* **2018**, *57*, 85–104. [CrossRef]

37. Rastegari, A.; Akhavan, R. The common mechanism of turbulent skin-friction drag reduction with superhydrophobic longitudinal microgrooves and riblets. *J. Fluid Mech.* **2018**, *838*, 68–104. [CrossRef]

38. Morini, G.L.; Spiga, M. Slip flow in rectangular microtubes. *Microscale Thermophys. Eng.* **1998**, *2*, 273–282.

39. Morini, G.L. Scaling effects for liquid flows in microchannels. *Heat Transf. Eng.* **2006**, *27*, 64–73. [CrossRef]

40. Wang, C.Y. Benchmark solutions for slip flow and H1 heat transfer in rectangular and equilateral triangular ducts. *Trans. ASME J. Heat Transf.* **2013**, *135*, 021703. [CrossRef]

41. Wang, C.Y. On the Nusselt number for H$_2$ heat transfer in rectangular ducts of large aspect ratios. *Trans. ASME J. Heat Transf.* **2014**, *136*, 074501. [CrossRef]

42. Steinke, M.E.; Kandlikar, S.G. Review of single-phase heat transfer enhancement techniques for application in microchannels, minichannels and microdevices. *Int. J. Heat Technol.* **2004**, *22*, 3–11.

43. Steinke, M.E.; Kandlikar, S.G. Single-phase liquid friction factors in microchannels. *Int. J. Therm. Sci.* **2006**, *45*, 1073–1083. [CrossRef]

44. Ranjith, S.K.; Patnaik, B.S.V.; Vedantam, S. Hydrodynamics of the developing region in hydrophobic microchannels: A dissipative particle dynamics study. *Phys. Rev. E* **2013**, *87*, 033303. [CrossRef]

45. Duan, Z.P.; Muzychka, Y.S. Slip flow in the hydrodynamic entrance region of circular and noncircular microchannels. *Trans. ASME J. Fluids Eng.* **2010**, *132*, 011201. [CrossRef]

46. Mishan, Y.; Mosyak, A.; Pogrebnyak, E.; Hetsroni, G. Effect of developing flow and thermal regime on momentum and heat transfer in micro-scale heat sink. *Int. J. Heat Mass Transf.* **2007**, *50*, 3100–3114. [CrossRef]

47. Kohl, M.J.; Abdel-Khalik, S.I.; Jeter, S.M.; Sadowski, D.I. An experimental investigation of microchannel flow with internal pressure measurements. *Int. J. Heat Mass Transf.* **2005**, *48*, 1518–1533. [CrossRef]

48. Bayraktar, T.; Pidugu, S.B. Characterization of liquid flows in microfluidic systems. *Int. J. Heat Fluid Flow* **2006**, *49*, 815–824. [CrossRef]

49. Barber, R.W.; Emerson, D.R. A numerical investigation of low Reynolds number gaseous slip flow at the entrance of circular and parallel plate microchannels. In Proceedings of the ECCOMAS Computational Fluid Dynamics Conference, Wales, UK, 4–7 September 2001.

50. Vocale, P.; Spiga, M. Slip flow in the hydrodynamic entrance region of microchannels. *Int. J. Microscale Nanoscale Therm. Fluid Transp. Phenom.* **2013**, *4*, 175–191.

51. Hettiarachchi, H.D.M.; Golubovic, M.; Worek, W.M.; Minkowycz, W.J. Three-dimensional laminar slip-flow and heat transfer in a rectangular microchannel with constant wall temperature. *Int. J. Heat Mass Transf.* **2008**, *51*, 5088–5096. [CrossRef]

52. Sparrow, E.M.; Lundgren, T.S.; Lin, S.H. Slip flow in the entrance region of a parallel plate channel. In Proceedings of the Heat Transfer and Fluid Mechanics Institute; Stanford University Press: Palo Alto, CA, USA, 1962; pp. 223–238.

53. Quarmby, A. Slip flow in the hydrodynamic entrance region of a tube and parallel plate channel. *Appl. Sci. Rev.* **1965**, *15*, 411–428. [CrossRef]

54. Ma, N.Y.; Duan, Z.P.; Ma, H.; Su, L.B.; Liang, P.; Ning, X.R.; Zhang, X. Lattice Boltzmann simulation of the hydrodynamic entrance region of rectangular microchannels in the slip regime. *Micromachines* **2018**, *9*, 87. [CrossRef] [PubMed]

55. Ebert, W.A.; Sparrow, E.M. Slip flow in Rectangular and Annular Ducts. *J. Basic Eng.* **1965**, *87*, 1018–1024. [CrossRef]

56. Shah, R.K.; London, A.L. *Laminar Flow Forced Convection in Ducts*; Academic Press: New York, NY, USA, 1978.

57. Duan, Z.P.; Muzychka, Y.S. Slip flow in non-circular microchannels. *Microfluid. Nanofluid.* **2007**, *3*, 473–484. [CrossRef]

58. Kays, W.M.; London, A.L. *Compact Heat Exchangers*, 3rd ed.; McGraw Hill: New York, NY, USA, 1984.

59. Curr, R.M.; Sharma, D.; Tatchell, D.G. Numerical predictions of some three dimensional boundary layers in ducts. *Comput. Methods Appl. Eng.* **1972**, *1*, 143–158. [CrossRef]

micromachines

MDPI

Article

A Comparison of Data Reduction Methods for Average Friction Factor Calculation of Adiabatic Gas Flows in Microchannels

Danish Rehman [1,*], Gian Luca Morini [1] and Chungpyo Hong [2]

[1] Microfluidics Laboratory, Department of Industrial Engineering (DIN), University of Bologna, Via del Lazzaretto 15/5, 40131 Bologna BO, Italy; gianluca.morini3@unibo.it
[2] Microscale Heat Transfer Laboratory, Department of Mechanical Engineering, Kagoshima University, Kagoshima Prefecture 890-8580, Japan; hong@mech.kagoshima-u.ac.jp
* Correspondence: danish.rehman2@unibo.it

Received: 9 February 2019; Accepted: 23 February 2019; Published: 28 February 2019

Abstract: In this paper, a combined numerical and experimental approach for the estimation of the average friction factor along adiabatic microchannels with compressible gas flows is presented. Pressure-drop experiments are performed for a rectangular microchannel with a hydraulic diameter of 295 μm by varying Reynolds number up to 17,000. In parallel, the calculation of friction factor has been repeated numerically and results are compared with the experimental work. The validated numerical model was also used to gain an insight of flow physics by varying the aspect ratio and hydraulic diameter of rectangular microchannels with respect to the channel tested experimentally. This was done with an aim of verifying the role of minor loss coefficients for the estimation of the average friction factor. To have laminar, transitional, and turbulent regimes captured, numerical analysis has been performed by varying Reynolds number from 200 to 20,000. Comparison of numerically and experimentally calculated gas flow characteristics has shown that adiabatic wall treatment (Fanno flow) results in better agreement of average friction factor values with conventional theory than the isothermal treatment of gas along the microchannel. The use of a constant value for minor loss coefficients available in the literature is not recommended for microflows as they change from one assembly to the other and their accurate estimation for compressible flows requires a coupling of numerical analysis with experimental data reduction. Results presented in this work demonstrate how an adiabatic wall treatment along the length of the channel coupled with the assumption of an isentropic flow from manifold to microchannel inlet results in a self-sustained experimental data reduction method for the accurate estimation of friction factor values even in presence of significant compressibility effects. Results also demonstrate that both the assumption of perfect expansion and consequently wrong estimation of average temperature between inlet and outlet of a microchannel can be responsible for an apparent increase in experimental average friction factor in choked flow regime.

Keywords: underexpansion; Fanno flow; flow choking; compressibility

1. Introduction

With the pioneering work of Tuckerman and Pease [1], flows in microchannels (MCs) have become of prime importance for heat-exchanging applications. Such systems offer high surface-to-volume ratio and record high heat transfer coefficients with low to moderate pressure drop compared to their macro counterparts [2]. Frictional characteristics of the MC flow are therefore important as they directly translate into pressure drop and hence associated fiscal penalty due to increased required power. Many of earlier experimental pressure-drop studies reported a deviation in the friction factor of MCs compared to macroflows [3–7] while only a few advocated for an agreement [8,9]. A detailed discussion of possible reasons of disagreement with conventional theory has been presented by Morini et al. [10]. The results obtained by testing both liquid and gas flows have confirmed that there exists a good agreement between the correlations used for the prediction of the pressure drop in conventional channels and the experimental data obtained for MCs if no significant scaling effects are present [11]. Due to compressibility effects, gas microflows tend to deviate slightly from macroflow laws which are developed based upon incompressible flows. Numerical results by Hong et al. [12] resulted in a correlation for Poiseuille Number with Mach number which catered for an increased pressure drop due to compressibility compared to the classical law. Experimental results of Yang [13] also showed that the laminar friction factors for microtubes with gas flow are slightly higher than Poiseuille law and follow closely the aforementioned correlation if the Mach number at exit of the microtube is employed for data reduction than its average value along the length. Moreover, most of the earlier experimental studies assumed isothermal flow [14–17] which holds true for incompressible fluids but not always for compressible gases. Literature has been divided into two main approaches for establishing experimental average frictional characteristics in MCs. When a total pressure drop of MC assembly and inlet temperature are measured, a classical methodology is to invoke minor loss coefficients and subtract minor pressure losses associated with the inlet/outlet manifold from the total measured pressure drop. The resulting pressure difference is then used along with measured temperature at manifold inlet to calculate the average isothermal friction factor. Such a treatment is quite realistic when an incompressible liquid working fluid is tested but the use of this method with compressible flows is questionable. In reality, a gas microflow does not stay isothermal and shows a strong temperature decrease close to MC outlet even for adiabatic walls. In a high-speed gas microflow, placing a thermocouple in the outlet jet will measure a value between static and total temperature [18], and therefore direct measurement is still challenging. Fortunately for an adiabatic flow, static temperature estimation at MC outlet can be done using a quadratic equation proposed by Kawashima and Asako [18]. By incorporating temperature gradient along the length of MC, it was shown numerically that local friction factor for nitrogen gas in microtubes (MTs) is in good agreement with conventional theory not only in laminar but also in the high-speed turbulent flow regime. Hong et al. [19] later measured local pressure at five axial positions of rectangular MCs and considered gas temperature change for the experimental calculation of semi-local friction factors (i.e., between two closely placed pressure ports). Semi-local turbulent friction factors considering Fanno flow were in good agreement with Blasius law whereas they were lower when gas was considered isothermal. In this study; however, authors did not calculate average friction factor between inlet and outlet of microchannel as the case with most of the previous experimental studies. Recently a new data reduction methodology for the average friction factor calculation between inlet and outlet has been developed by Hong et al. [20]. Experimental results for 867 μm MT vented to atmosphere showed better agreement with the Blasius law in the turbulent flow regime using an improved equation for adiabatic friction factor. Applicability of this new data reduction methodology on a rectangular channel with higher degree of flow choking is discussed in the present work.

From the recent literature, it can be established that most of the earlier studies used minor loss coefficients to calculate experimental average friction factors between inlet and outlet of a MC/MT. Furthermore, almost all such studies assumed an isothermal flow inside MC. On the contrary, a few studies that dealt with temperature change for compressible gas flows avoid the use of minor loss coefficients and present the evaluation of semi-local friction factors instead. With MC/MT, it is not trivial to obtain local pressure and temperature values, but it has been demonstrated how in the presence of significant compressibility effects, an accurate estimation of the friction factor is feasible only by assuming non-isothermal gas flow along the channel. If the local temperature trend cannot be measured experimentally, data reduction method for the accurate estimation of the adiabatic friction factor for compressible gases must be based on a series of assumptions that are able to predict this variation along the MC. Current study aims to demonstrate which assumptions must be adopted for a better evaluation of experimental friction factor by showing the difference in terms of average friction factors that are obtained by adopting different data reduction methods proposed in literature. Moreover, by coupling numerical modeling with experimental results, the effect of gas flow choking on evaluation of experimental average friction factor is also elucidated. Finally, a unique methodology for the analysis of pressure drop in presence of compressible gases in MCs is suggested to the reader.

2. Experimental Methodology

2.1. Setup

Schematic of test bench and MC assembly used in this work are shown in Figure 1. Nitrogen gas is stored in a high-pressure flask (①, 200 bar) and brought to approximately 11 bars and ambient temperature before entering into a 7 μm particle filter (②, Hamlet®), used to prevent possible impurities from clogging the MCs or the flow controller. The facility is equipped with three volume flow rate controllers (Bronkhorst EL-Flow E7000) operating in the 0–50 NmL/min (4a), 0–500 NmL/min (4b) and 0–5000 NmL/min (4c) ranges respectively. A three-way valve (③) then directs the working fluid to the proper flow controller by means of a computer-steered valve. This allows to impose a certain volume flow rate through the MC to achieve desired Reynolds number. Gas is then allowed to enter the MC test assembly (⑤). The total pressure drop between the inlet and the outlet of the MC assembly is measured by means of a differential pressure transducer (⑥, Validyne DP15) with an interchangeable sensing element that allows accurate measurements over the whole range of encountered pressures. Atmospheric pressure is measured using an absolute pressure sensor (⑦, Validyne AP42). To measure the temperature at the entrance of the MC a K-type, calibrated thermocouple is used (⑧). Thermocouple voltage and an amplified voltage of pressure sensors are fed to internal multiplexer board of Agilent 39470A and are read by means of a PC using a Labview® program.

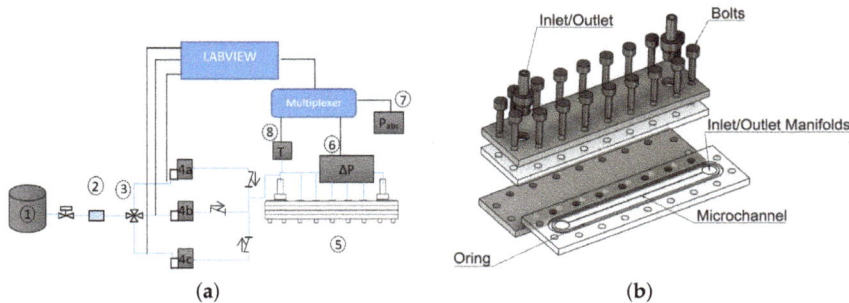

Figure 1. Experimental setup (**a**), and an exploded view of MC assembly (**b**).

The MC is fabricated by milling a PMMA plastic sheet with a nominal thickness of 5 mm. CNC milling is performed using Roland® MDX40A with a 100 W spindle motor. Modeling and CNC toolpath are generated in Autodesk® Fusion360. A flat end mill of 300 μm is used to make MC slot. Dry milling is performed using spindle rpm and speed of 10,000 rpm and 300 mm/min, respectively. A constant depth of cut of 100 μm is used. Dimensions and inner surface roughness of resulting channel are measured using an optical profilometer. Cross section and roughness measurements are performed at three locations along the length of the MC and an average of 3 readings is taken at each cross section. The average width (w), height (h) and surface roughness (ϵ) of realized MC are 360 μm, 250 μm and 1.05 μm, respectively. PMMA chip containing MC and a microscopic view of bottom and top of the MC is shown in Figure 2. Lower PMMA sheet containing milled MC is sealed by means of a top cover of same material and an O-ring as shown in Figure 1b. Both top and bottom plastics are sandwiched between two 5 mm thick Aluminum plates. Whole assembly is then bolted to ensure leak tightness. Test section is checked for leakage by applying a pressure of almost 10 bars between the inlet and outlet connectors and closing the outlet manifold. Pressure inside the assembly is monitored for at least 5 h to spot any major leakage. Such a test is repeated before initiating experimental test campaign. Small holes of 200–250 µm are drilled in top plastic at 3 axial locations $x = 0.58L, 0.72L$ and $0.87L$. Local static pressure is read through these holes using a solenoid valve-switching assembly. At each pressure tap, differential pressure between the port and atmosphere is measured using the differential pressure gauge. The atmospheric pressure is also measured at every data point and is finally added to differential reading to obtain absolute pressure of the specific cross section. Typical uncertainties associated with instruments used in current experimental work are reported in Table 1.

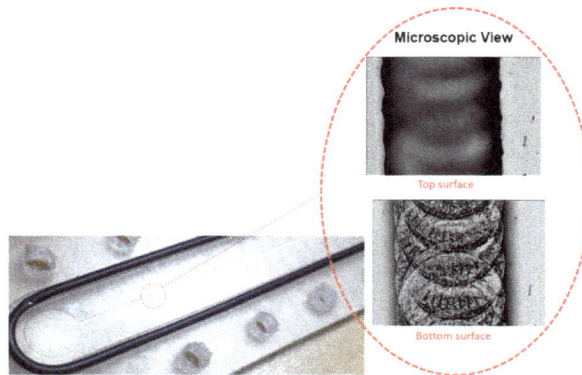

Figure 2. Zoomed part of chip containing MC and its top and bottom surfaces.

Table 1. Typical uncertainties of instruments used.

Instrument	Range (0–Full Scale (FS))	Uncertainty
Volume flow rate controllers	0–500 & 0–5000 nmL/min	0.5% FS
Pressure sensors	0–256, 0–860 & 0–1460 kPa	0.5% FS
K-type thermocouple	0–100 °C	0.25% FS

2.2. Data Reduction

Local Fanning friction factor can be defined by the following expression for a compressible flow [18]:

$$f_{f,local} = \frac{4\tau_w}{\frac{1}{2}\rho u^2} = \frac{2D_h}{p} - \frac{2D_h p}{\rho^2 u^2 RT}\frac{dp}{dx} - \frac{2D_h}{T}\frac{dT}{dx} \tag{1}$$

where hydraulic diameter of a rectangular MC is defined as:

$$D_h = \frac{4A}{Per} = \frac{2wh}{w+h} \tag{2}$$

Reynolds number at the inlet of MC can then be calculated using measured mass flow rate and calculated viscosity at inlet temperature with the following equation:

$$Re = \frac{\dot{m}D_h}{\mu A} \tag{3}$$

Considering one dimensional flow of ideal gas, Equation (1) can be integrated between two points a and b along the length (L), to calculate average friction factor between those points as follows:

$$f_f = \frac{D_h}{x_b - x_a} \left[\frac{p_a^2 - p_b^2}{R T_{av} G^2} - 2 \ln \left(\frac{p_a}{p_b} \right) + 2 \ln \left(\frac{T_a}{T_b} \right) \right] \tag{4}$$

For the rest of the text, when Equation (4) is applied between two closely spaced pressure ports (e.g., between p_4–p_5 in Figure 3 for $\Delta x = x_5 - x_4$), resulting friction factor will be referred to as semi-local friction factor ($\tilde{f}_f$) while when it is applied between inlet and outlet of MC ($\Delta x = L$), it will be known as average adiabatic friction factor (f_f). In addition, under the hypothesis of adiabatic compressible flow, the energy balance for one dimensional Fanno flow, between inlet '*in*' and any other cross section at a distance 'x' from the inlet of the MC yields the following quadratic equation for the estimation of average cross-sectional temperature [18]:

$$\left(\frac{\rho_{in}^2 u_{in}^2 R^2}{2 c_p p_x^2} \right) T_x^2 + T_x - \left(T_{in} + \frac{u_{in}^2}{2 c_p} \right) = 0 \tag{5}$$

Finally, knowing the average pressure and temperature of a specific cross section, average density and velocity of compressible gas can be obtained using gas and continuity equations, respectively. The local value assumed by Mach number, defined as the ratio of velocity and the local speed of sound, can be calculated as follows:

$$Ma_x = \frac{u_x}{\sqrt{\gamma R T_x}} \tag{6}$$

In all the published experimental results, MC/MT is attached to a conventional piping system using an entrance manifold. The geometry of the manifold may vary case by case but a pressure drop between the manifold and MC inlet exists there. Similarly, when gas exits the MC/MT, the expansion of the gas to the exit manifold or atmosphere causes an additional pressure drop. This is shown schematically in Figure 3. The current experimental MC assembly has a reducer that connects the entrance manifold to the gas supply piping. Similarly, another reducer towards the exit of assembly vents the gas to the atmosphere coming from exit manifold. Therefore, minor losses ($\Delta P_{in/out}$) in reducers and manifolds need to be accounted for to estimate the pressure drop of the MC/MT, alone (ΔP_{ch} see Figure 3) from the total measured pressure drop.

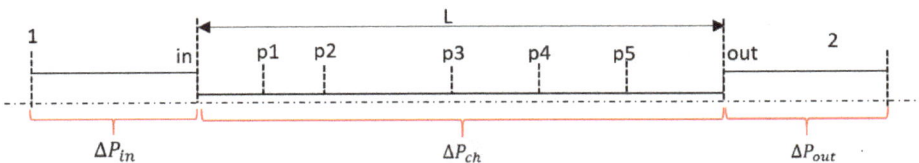

Figure 3. Schematic of minor losses.

The most used method for estimating these minor losses is to use loss coefficients ($K_{in/out}$) available in fluid mechanics texts, which are generally validated for liquids and weakly compressible gases. Minor pressure losses are defined as [21]:

$$\Delta P_{in/out} = K_{in/out} \frac{1}{2} \rho u_{in/out}^2 \tag{7}$$

Data reduction methodology where numerical inlet minor loss coefficients are used along with the temperature estimation at MC outlet using Equation (5), is referred to as M1 in the subsequent text. An alternative methodology (M2), followed by the group of Prof Asako is to estimate MC inlet flow properties by assuming isentropic flow between the manifold and MC inlet. This automatically caters for a reduction in MC inlet pressure and hence the use of K_{in} is no more required. An initial estimate of the gas velocity at MC inlet is made as in M1 (i.e., using the measured mass flow rate and density of the gas at the inlet of assembly). Inlet properties are then calculated iteratively using the following set of equations (see Figure 3):

$$T_{in} = T_1 - \frac{u_{in}^2}{2C_p} \tag{8}$$

$$p_{in} = \frac{p_1}{\left(1 + \frac{u_{in}^2}{2C_p T_{in}}\right)^{\frac{\gamma}{\gamma-1}}} \tag{9}$$

$$\rho_{in} = \frac{p_{in}}{RT_{in}} \tag{10}$$

$$u_{in} = \frac{\dot{m}}{\rho_{in} A} \tag{11}$$

Main differences between the two experimental data reduction methods (M1 and M2) described before are summarized in Table 2. In the next sections, experimental results for average friction factor are deduced by using both M1 and M2 approaches. These results are then compared with numerical predictions to put in evidence discrepancies among experiment and theory. This comparison is finally used to establish the most accurate data reduction procedure for the estimation of average friction factor for MC/MT in presence of compressible gases.

Table 2. Data Reduction Methods Used in Current Study.

Location	M1	M2
Inlet	- Numerical minor loss coefficient to estimate entrance manifold pressure drop - $T_{in} = T_1$ (see Figure 3)	- Isentropic expansion between entrance manifold (1) and MC inlet (in)
Outlet	- Fully expanded flow ($p_{out} = p_{atm}$) - T_{out} estimated using Equation (5)	

3. Numerical Methodology

Due to small dimensions of MC assemblies, it is not possible to insert as many sensors as one desires along a MC. To overcome this lack, a validated CFD model can be used to gain an insight of flow physics [11]. Therefore, a numerical model of the experimental test assembly is developed in ANSYS Workbench framework. Three MC dimensions are simulated in the current study as tabulated in Table 3. MC1 is used to replicate the channel tested experimentally and gain insight of flow physics by comparing with experiments whereas MC2 is used for the validation of adapted numerical scheme. Simulation results of aforementioned MCs along with MC3 are also used for discussing the role of loss coefficients in the calculation of friction factor. An exhaustive parametric study to individuate effects of D_h and aspect ratio ($\alpha = \frac{h}{w}$) on minor losses evaluation, is out of scope for current analysis. Therefore, a limited set of simulations with only 3 MCs with different hydraulic diameters and aspect ratios are chosen to emphasize minor losses dependence on the MC dimensions.

Table 3. Channel geometry used for simulations.

Channel	w (µm)	h (µm)	D_h (µm)	α
MC1	360	250	~295	0.7
MC2	1020	112.7	~203	0.11
MC3	550	110	~184	0.2

Geometry and meshing is done using Design Modeler and ANSYS Meshing software, respectively. A mesh of $45 \times 30 \times 200$ is used in the MCs. A structured mesh locally refined at the walls of the channel and manifolds is employed as shown in Figure 4. The mesh expansion factor is kept as 1.1 and first node point is placed such that $y+$, which is non-dimensional distance between first mesh node and MC wall, is in between 1 and 4 for the highest Re simulated. Orthogonality of mesh elements inside MC is between 0.95–1 in all the simulated cases. A commercial solver CFX based on finite volume methods is used for the flow simulations. Reducers and manifolds are also simulated along with the MCs. Height of the reducer (H_r, see Figure 4) is 30 mm with an internal diameter of 4 mm. Whereas diameter of the circular manifolds is 9 mm and height (H_{man}) is kept same as the height of the simulated MC ($H_{man} = h$). Dimensions of these parts are chosen based upon the experimentally tested MC assembly. Ideal nitrogen gas enters the entrance manifold that is orthogonal to MC and leaves again orthogonally through the exit manifold. For the MC that is also experimentally tested (MC1), measured mass flow rate is used while for other cases it is calculated from Equation (3).

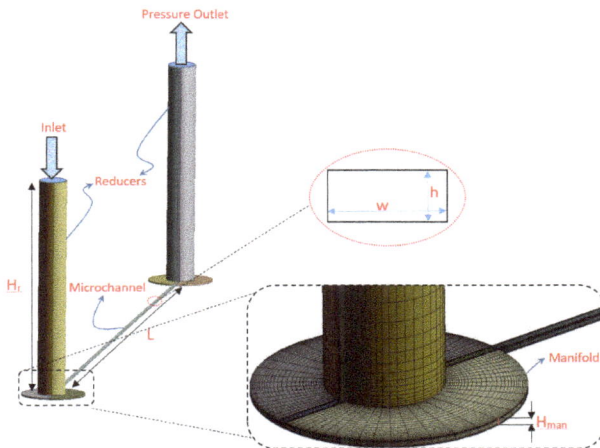

Figure 4. Details of numerical model and meshing.

Steady state RANS simulations are performed for all turbulent cases. Laminar flow solver is used for the cases where $Re \leq 1000$ and for $Re > 1000$, SST k-ω transitional turbulence model is used. A modified formulation of γ-Re_θ transition turbulence model for internal flows is applied [22]. High-resolution turbulence numerics are employed with a higher order advection scheme available in CFX. Pseudo time marching is done using a physical timestep of 0.1s. A convergence criteria of 10^{-6} for RMS residuals of governing equations is chosen while monitor points for pressure and velocity at the MC inlet and outlet are also observed during successive iterations. In case where residuals stayed higher than supplied criteria, the solution is deemed converged if monitor points did not show any variation for 200 consecutive iterations. Reference pressure of 101 kPa was used for the simulation and all the other pressures are defined with respect to this reference pressure. Energy equation was activated using total energy option available in CFX which adopts energy equation without any simplifications in governing equations solution. Kinematic viscosity dependence on gas temperature is defined using Sutherland's law. Further details of boundary conditions can be seen in Table 4.

To estimate friction factor and minor loss coefficients, five different cross-sectional planes are defined at x/L of 0.005, 0.25, 0.5, 0.75 and 0.98, respectively. In addition, two planes defined at x/L of 0.0005 and 0.9995 are treated as the inlet and outlet of MC, respectively. Results from these planes are further post processed in MATLAB to deduce required flow quantities. Numerical friction factors are then evaluated simply by using Equation (4).

Table 4. Boundary Conditions.

Boundary	Value
Inlet	- mass flow rate: experimental or from Equation (3)
	- Turbulence Intensity, TI = 5%
	- Temperature T_{in} = 23 °C
Walls	- No slip
	- Adiabatic
Outlet	Pressure outlet, Relative p = 0 Pa

Validation

To validate the adopted numerical scheme, a MC with hydraulic diameter of 203 μm (MC2) is simulated and numerical friction factors are compared with experiments of Hong et al. [19]. The width and height of the MC are 1020 μm and 112.7 μm respectively giving it an aspect ratio of 0.11. Length of the channel in numerical model is taken as 100 mm while it is 26.9 mm in experiments performed by [19]. Moreover, dimensions of the inlet manifold are slightly different in experimental settings than those adapted in numerical model. Since a comparison of semi-local friction factor ($\tilde{f}_f$) towards the last half of the MC is made, these differences should not induce a significant effect on $\tilde{f}_f$ in that region.

Hong et al. [19] reports semi-local friction factor measured between two closely placed pressure ports at the dimensionless length (x/L) of 0.67–0.8. Obtained numerical results from x/L = 0.5–0.75 (o) and x/L = 0–1 (Δ) are compared to the experimentally reported values in Figure 5. There exists an excellent agreement between the current numerical results and experimental results in the laminar flow regime where f_f follows the Shah & London correlation (S&L):

$$f_{f_{SL}} = \frac{96}{Re} \left(1 - 1.3553\alpha + 1.9467\alpha^2 - 1.7012\alpha^3 + 0.9564\alpha^4 - 0.2537\alpha^5 \right) \tag{12}$$

In the turbulent flow regime, both experimental and numerical results are slightly above the Blasius law. It is to be noted that even by assuming smooth walls in the numerical simulation, the turbulent friction factor can be slightly higher than Blasius law which is also the case of experimental results of semi-local friction factor reported by [19]. Therefore turbulent f_f can be higher than Blasius law even with smooth walls, if compressible effects are significant.

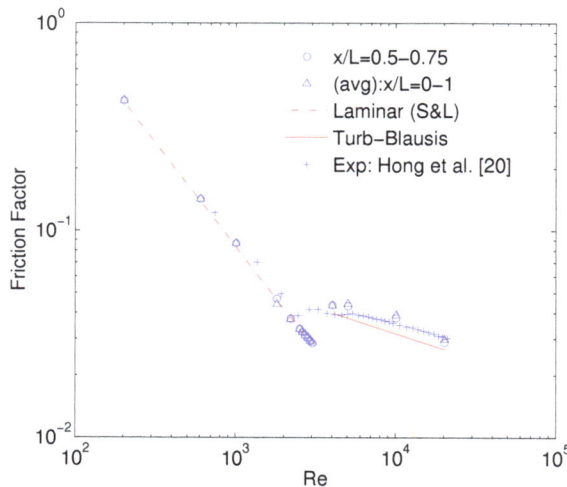

Figure 5. Numerical validation of friction factor calculation.

4. Results and Discussion

4.1. Numerical Calculation of Minor Loss Coefficients

The minor loss associated with the inlet from numerical results is calculated using Equation (7) between the inlet of assembly and MC inlet and therefore includes 90° bend loss due to orthogonal flow direction change at the manifold inlet. Results of K_{in} and K_{out} for the three MCs are compared in Figure 6. K_{in} decreases steeply in laminar and early turbulent flow regimes, whereas it becomes almost a constant in high turbulent flow regime as shown in Figure 6a. At the lower Re, MC assembly pressure drop in experimental campaign is also usually lower and hence assuming a smaller and/or constant K_{in} would certainly cause f_f to be higher than macro theory. For the smallest α simulated, K_{in} is as high as 5.19 which is significantly higher than values available in general fluid mechanics text [21]. For a rectangular MC, K_{in} is a function of α and D_h simultaneously and hence it must be evaluated numerically in advance to help in experimental data reduction. K_{out} on the other hand stays close to zero in laminar and early turbulent flow regime and shoots rapidly in higher turbulent flow regimes as seen in Figure 6b. From the investigated MC assembly, it is evident that K_{out} is highest for the smallest D_h simulated for high turbulent flow regime and decreases with an increase of D_h. Numerical results show that K_{out} for compressible flows, can also go higher than its limiting value of 1 as calculated using the following theoretical relation:

$$K_{out,th} = \left(1 - \frac{A_{MC}}{A_{man}}\right)^2 \tag{13}$$

where A_{MC} and A_{man} denote cross-sectional area of MC and manifold, respectively.

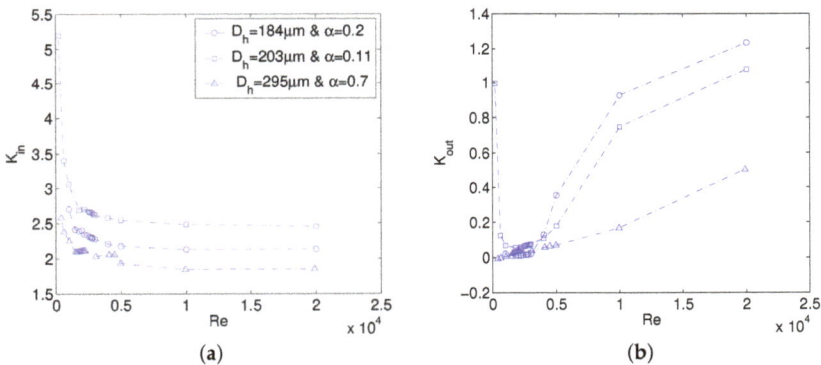

Figure 6. Numerical minor loss coefficients: K_{in} (**a**), and K_{out} (**b**).

A MC assembly similar to that considered in the present work has been tested by Mirmanto [23]. The author used a conventional value of 1.2 as loss coefficient for 90° bend which is not observed in the present numerical results. Similar assembly has also been investigated numerically by Sahar et al. [24] where inlet losses in laminar incompressible flow increased for increasing D_h for a constant value of aspect ratio α. A need for systematic investigation of minor losses using numerical modeling as a priori is also emphasized in [25]. The three MCs modeled in this paper demonstrated that minor loss coefficients are dependent on assembly as well as MC geometry and in all cases are not equal to general values used in literature for macro flows and hence a numerical model is required for an accurate estimation friction factors especially in laminar flow regime. This is due to the fact that minor losses carry a significant portion of the total pressure drop in laminar flow regime but as the Re is increased, the contribution of minor losses diminishes, and they do not have significant effect on calculation of f_f in turbulent flow regime.

4.2. Experimental Average Friction Factor

Pressure-drop experiments are performed for MC1 using nitrogen gas as working fluid. Results in terms of f_f in laminar flow regime are compared with Shah & London correlation (Equation (12)) for rectangular channels while Blasius law in turbulent flow regime is used for comparison. Laminar f_f is also compared with the correlation of Hong et al. [12] that caters for an increase in laminar f_f due to compressibility. Numerically obtained values of K_{in} are used in data reduction M1. When outlet is open to atmosphere, a convenient practice is to estimate the experimental f_f by assuming a fully expanded flow at the MC outlet. This assumption essentially makes K_{out} to be zero and is quite realistic in the laminar and early turbulent flow regimes as shown numerically in Figure 6b. Therefore, due to the lack of MC outlet pressure measurement in the current experimental campaign, $K_{out} = 0$ is used in further data reduction. Limitations to this assumption will be discussed later in the text.

Previous section already put in evidence that minor losses differ for different MC geometries. Therefore, for an accurate evaluation of inlet minor losses a curve fit on the numerical results is used to extract the variation of K_{in} with Re. This is used to model the inlet minor loss to calculate f_f with data reduction M1. The adiabatic f_f calculated using Equation (4) for both methodologies M1 and M2 is shown in Figure 7. There is an excellent agreement between experimental and numerical f_f in laminar flow regime where both follow the Hong et al. [12] correlation within experimental uncertainty. Uncertainty bars are omitted in Figure 7 for reasons of clarity. Isothermal friction factor ($f_{f,iso}$) obtained by assuming $T_{in} = T_{out}$ and hence neglecting the last term of Equation (4), stays lower than Blasius law in turbulent regime. Results of numerical f_f are slightly higher than Blasius in high turbulent regime. Experimental f_f in early turbulent flow regime is lower than Blasius law with both M1 and M2 methods and starts not only increasing again in high turbulent flow regime but with a slope that diverges from Blasius at $Re > 10,000$. On the contrary, slope for numerical f_f does not diverge significantly from Blasius law in turbulent flow regime. Since the relative roughness ($\frac{\varepsilon}{D_h}$) of the tested channel is equal to 0.5%, such an increase in experimental f_f compared to Blasius law using both M1 and M2 methods can potentially be associated with the rough channel walls. To better investigate this aspect, experimentally deduced flow properties are further compared with numerical ones.

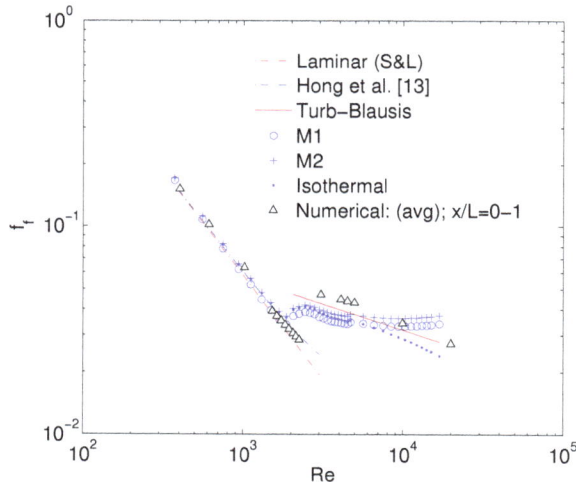

Figure 7. Comparison of numerical and experimental f_f using M1 and M2.

The temperature at the inlet and outlet of MC is compared in Figure 8a. Discrepancy in the inlet temperature for M1 method is a direct consequence of the assumption of constant density and temperature at MC inlet, both assumed equal to the measured values at the manifold inlet. In absolute sense, there is a maximum difference of 7K in turbulent flow regime. Gas stays almost isothermal at low Reynolds numbers ($Re < 1000$) but as Re is increased, relative pressure difference between the manifold inlet and MC inlet increases causing a decrease in temperature at MC inlet that is not captured in M1 method. Temperature at the outlet estimated using Equation (5) follows the numerical trend where the gas temperature is decreasing with increasing Re. For such an estimation, an absolute temperature difference of approximately 26K compared to 92K in case of isothermal assumption at Re of around 13,000 is observed. This difference furthers as Re is increased. On the other hand, when the MC inlet conditions are defined using M2 method, inlet expansion and hence temperature decrease from the entrance manifold to MC inlet is correctly modeled as shown in Figure 8a. This result suggests using M2 method for correctly predicting the MC inlet flow properties without using an additional K_{in}. Outlet flow properties, however, differ from the numerical values for both M1 and M2 methods. It is worth mentioning that outlet pressure is assumed to be atmospheric as outlined in Table 2 ($K_{out} = 0$) for both methods to estimate outlet temperature. Numerical value assumed by the Mach number at the MC outlet is shown in Figure 8b where it keeps on increasing with Re and finally gets to an almost constant value after Re =10,000. At this point, flow starts to choke and Ma at the outlet reaches a constant value of close to 1 (i.e., in this case $M_a = 1.19$ after $Re = 15,000$). An explanation of supersonic jet at the exit of constant area ducts has been presented by Lijo et al. [26]. Numerical works of Kawashima et al. [27] and Hong et al. [28] showed that Ma at the outlet can go higher than its maximum limit of 1(Fanno flow). This happens due to shear thinning of the boundary layer close to the outlet of MC/MT that serves as de-Laval nozzle for incoming high subsonic jet of gas flow. Flow choking at the outlet is not captured by experimental data reduction methodologies M1 and M2 and therefore Ma keeps on increasing with Re in both methods. This result disagrees with what is seen in numerical results. At maximum Re values, outlet Ma reaches as high as 2.46 for isothermal treatment of gas while it reduces to 1.88 in cases of M1 and M2 methods. These very high values of Ma signify that both methods lack the flow physics at the outlet. This is due to the wrong estimation of outlet jet temperature using Equation (5) at very high Re as can be seen in Figure 8a. Experimental data shows flow choking at the inlet where Ma becomes constant for $Re > 6000$. A constant temperature and Ma ensure a constant velocity at MC inlet which explains the reason for a constant value of inlet loss coefficient in the turbulent flow regime after an initial decrease in laminar and early the turbulent flow regime, as observed in Figure 6a. However, it is evident that flow choking at inlet starts much earlier than outlet and numerical results of MC inlet temperature are lower than assumed in M1. Effect of flow choking on calculation of average f_f is discussed in the next section.

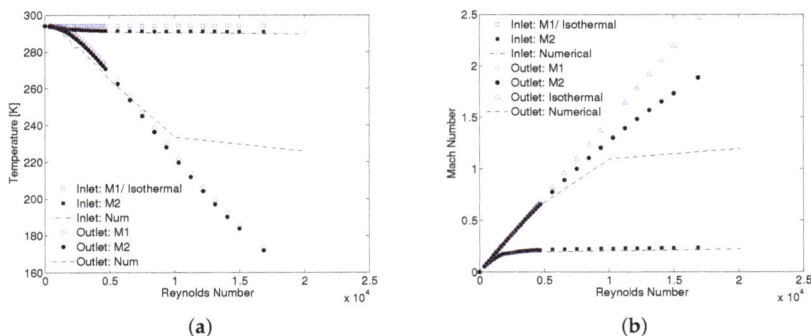

Figure 8. Comparison of flow properties using M1 & M2: local temperature (**a**), and Mach number (**b**).

4.3. Flow Choking and under Expansion at Outlet

Usually in fluid mechanics text, flow choking is defined as the flow condition in which working fluid at outlet of a pipe attains a sonic velocity and therefore mass flow rate cannot be increased any further. However, as shown by the current experiments and by the work of Hong et al. [28], mass flow rate for a compressible working fluid keeps on increasing even when the flow is choked. Ma and hence velocity, however, attains a maximum at all cross sections of the pipe in choked gas flows. This is simply because of an increase in upstream density of the gas by increasing upstream pressure causes mass flow rate to increase even if velocity is choked downstream. Definition of flow choking shall be given therefore by highlighting the presence of maximum in Ma (velocity) rather than in mass flow rate for compressible flows. Variation of numerical outlet Ma with ratio of MC outlet pressure and atmospheric pressure is shown in Figure 9a for all three simulated MCs. It can be seen that as soon as outlet Ma starts to choke, MC outlet pressure starts increasing higher than atmospheric pressure which means that the flow at MC outlet becomes underexpanded. The measure of underexpansion is not much pronounced during the experimental tests on MC1 where flow choking begins around $Re = 9934$. On the contrary, flow choking is observed earlier in MC2 and MC3 in correspondence of $Re = 5000$ and $Re = 3999$ respectively. After these Re, outlet pressure shoots above the atmospheric pressure and the flow cannot be considered as fully expanded as assumed in both M1 and M2 methods. As a result, a sudden increase in outlet numerical K_{out} after flow starts to choke can be noted in Figure 9b. As discussed earlier, a single value of the outlet loss coefficient is therefore not enough, especially for the MCs with smaller hydraulic diameter where flow choking can start even in laminar flow regime.

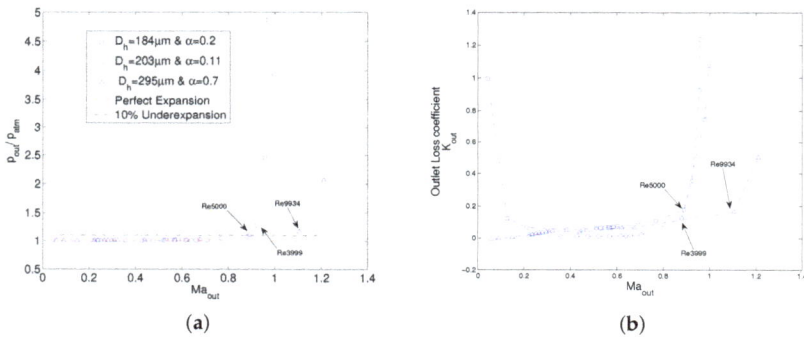

Figure 9. flow choking and underexpansion for simulated MCs: back pressure (**a**), and K_{out} (**b**).

From experimental point of view, it is difficult to measure static pressure exactly at the MC outlet and therefore an assumption of fully expanded flow serves the purpose. However, as evidenced in all three simulated cases, this holds true for unchoked gas flows only. Moreover, the measure of underexpansion becomes stronger with smaller dimensions of the channel and hence should be of concern for deducing friction factors using Equation (4) in any experimental campaign. As Equation (5) requires cross-sectional average pressure to estimate the average temperature of that cross section, full expanded flow assumption therefore always overestimates the decrease in outlet temperature in choked gas flows. This in turn causes an artificial increase in the calculation of friction factor in the choked flow regime from experimental pressure-drop data. Therefore experimental f_f shows a deviation from Blasius law in Figure 7 while numerical results agree with Blasius.

Flow choking that was observed in numerical data can experimentally be established only using local flow properties. For an adiabatic flow in a duct, fluid temperature and density at a specific cross section can be estimated using only a static pressure measurement. Thus, local pressure measurement will result in a better understanding of choked compressible flow along the length of MC. A comparison of local measured static pressure with numerical results is shown in Figure 10a for MC1 and a zoomed portion of laminar to early turbulent flow regime is shown in Figure 10b. Numerically estimated values are generally in good agreement with experimental local pressure values in laminar and early turbulent regimes. In highly turbulent flow regime ($Re > 10{,}000$) prescribed boundary conditions result in an overall lower pressure drop than what is experimentally observed. However, such difference is not too significant and therefore numerical model can be used as a priori tool to have first estimation of pressure drop and local flow physics (especially flow choking).

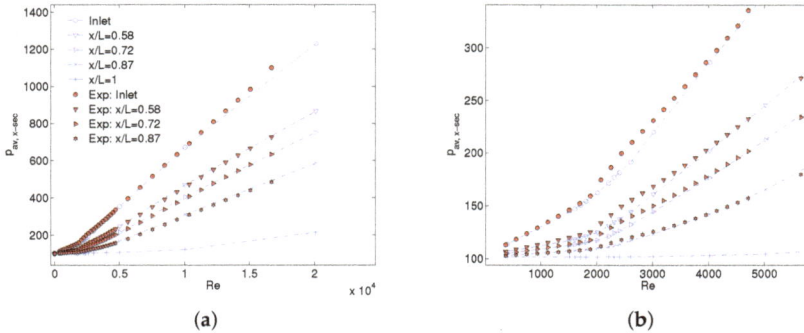

Figure 10. Comparison of measured and numerical static pressure inside MC1 (**a**), and zoomed low *Re* region (**b**).

Numerical local temperature is also in good agreement with the estimated cross-sectional temperature obtained by using Equation (5) as can be seen in Figure 11a. Experimental flow choking of the gas flow is also evident in Figure 11b where numerical and experimentally deduced *Ma* are in excellent agreement. It is, therefore, not erroneous to assume that numerical flow choking at outlet where *Ma* goes higher than 1 (as in the case of MC1), would also be encountered in experiments. It is interesting to note that all through the data reduction for the results presented in this section, no numerical input has been considered as a priori and MC inlet properties are calculated by considering an isentropic expansion between the entrance manifold and MC inlet.

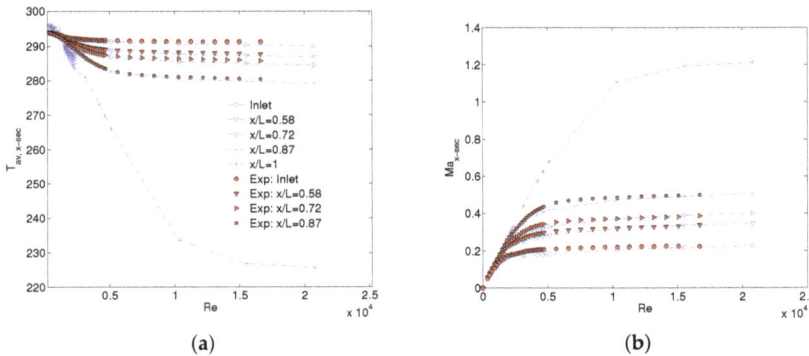

Figure 11. Comparison of measured and numerical static T (**a**), and *Ma* (**b**) inside MC1.

Another interesting observation from Figure 7 is that in choked flow regime, isothermal treatment of gas results in $f_{f,iso}$ that although is lower than Blasius law but follows its slope. The same is true for semi-local $\tilde{f}_f$ between last two pressure taps at $x/L = 0.72$–0.87 which again follows the slope but stays slightly higher than Blasius law. On the contrary f_f, using both M1 and M2 methods shows a false increase when the flow chokes. This again points towards a possibility of a wrong estimation of average gas temperature assumed in M1 and M2 methods. Because such a rapid increase in slope of f_f is not observed when difference in gas temperature is relatively small as is the case for semi-local values (see Figure 13a) or when it is zero (for isothermal gas treatment). In fact, if the numerical static temperature is analyzed along the length of MC as shown in Figure 12a, it can be established that even for the highest simulated Re, temperature almost stays isothermal along most of the length of the MC with a sudden decrease towards the end of MC. Therefore, to calculate average f_f between inlet and outlet, an equal weighted average of measured inlet temperature and estimated at the outlet using Equation (5) would underpredict the real average gas temperature. Average temperature of the gas between two pressure ports 'a' and 'b' along the length of MC can be defined by:

$$T_{av} = c_1 T_a + c_2 T_b \tag{14}$$

The effect of values of c_1 and c_2 on temperature average between inlet and outlet of MC and hence evaluation of f_f is shown in Figure 12b. It is evident that friction factor is more in agreement with Blasius law if the average is obtained with $c_1 > c_2$.

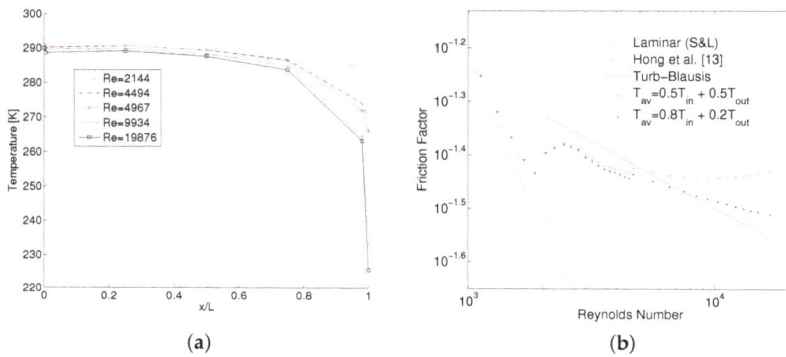

Figure 12. Numerical Temperature decrease along the length of MC1 at various Re (**a**), and Effect of T_{av} on calculation of experimental f_f (**b**).

Therefore, in absence of measured outlet pressure it is appropriate to estimate the average temperature by assuming a higher weight of the inlet temperature with respect to the outlet temperature (Equation (5)), obtained by considering the fully expanded flow assumption. For a case where pressure taps are located close to each other such that there is no significant temperature change from one tap to the other, average temperature in between can be approximated well with an equal weighted average ($c_1 = c_2 = 0.5$) and therefore semi-local $\tilde{f}_f$ follows Blasius law even in the choked turbulent flow regime. However, if the distance between these ports is large (this is the case with inlet and outlet of MC) an integral average of temperature between these ports should be considered. A detailed derivation of the average friction factor equation using such temperature average is given in [20] and therefore it is skipped in this text. The calculation of average friction factor for Fanno flow (adiabatic walls) between two pressure ports a and b can be obtained as follows:

$$f_{f,av} = \frac{D_h}{x_b - x_a} \left[\left(-2\ln\frac{p_a}{p_b} + 2\ln\frac{T_a}{T_b} \right) - \left(\frac{1}{\dot{G}^2 R \left(T_{in} + \frac{u_{in}^2}{2c_p} \right)} \right) \times \right.$$
$$\left. \left(\frac{p_b^2 - p_a^2}{2} + \frac{B^2}{2}\ln\frac{p_b + \sqrt{p_b^2 + B^2}}{p_a + \sqrt{p_a^2 + B^2}} + \frac{1}{2}\left(p_b\sqrt{p_b^2 + B^2} - p_a\sqrt{p_a^2 + B^2} \right) \right) \right] \tag{15}$$

where $B^2 = 4\beta \times \frac{\dot{G}^2 R^2}{2c_p} \times \left(T_{in} + \frac{u_{in}^2}{2c_p} \right)$ and β, kinetic energy recovery coefficient, is taken as 2 for laminar and 1 for turbulent flow.

Semi-local $\tilde{f}_f$ curves between $x/L = 0.58 - 0.72$ and $x/L = 0.72 - 0.87$ are shown in Figure 13a for MC1 experiments. Average f_f between the inlet and outlet using Equations (4) and (15) is also plotted for comparison. There is an excellent agreement between $f_{f,av}$ calculated using Equation (15) and Blasius law even in choked turbulent flow regime. This is due to the fact that an integral average of the temperature is used instead of an equal weighted average between the inlet and outlet of MC. Whereas average friction factor calculated using Equation (4) suffers a deviation from Blasius law only due to erroneous estimation of the bulk fluid temperature between the inlet and outlet of MC. It is reminded to the reader that in Figure 13a, MC outlet pressure is still assumed to be the atmospheric as was done with M1 and M2 methods to evaluate $f_{f,av}$. However, as Equation (15) does not require for an explicit approximation of T_{av} between inlet and outlet of MC, resulting $f_{f,av}$ values are therefore in a better agreement with Blasius law. To emphasize it further, results of average friction factor by Equations (4) and (15) with numerical estimated pressure (num. p_{out}) instead of atmospheric pressure at outlet, are also shown in Figure 13b. A significant improvement in the slope of turbulent f_f calculated using Equation (4) can be seen when numerical outlet pressure is used due to better estimation of outlet temperature. Results however, are still higher than values calculated using Equation (15) because an equal weighted T_{av} between inlet and outlet ($c1 = c2 = 0.5$) is assumed. On the contrary, $f_{f,av}$ with numerical outlet pressure calculated using Equation (15) is in closer agreement with Blasius law mainly due to better estimation of average temperature of gas. Therefore Equation (15) is recommended for calculating average experimental f_f between two pressure ports for adiabatic MCs. Equation (4) should be considered as an approximation of Equation (15) for two closely placed pressure taps, where temperature change between these taps can be well represented with an equal weighted average of respective temperatures, as is the case for semi-local friction factor calculations. An experimental/numerical campaign to analyze the applicability of Equation (15) on MCs of smaller D_h with even higher degree of flow choking is therefore recommended to complete this analysis.

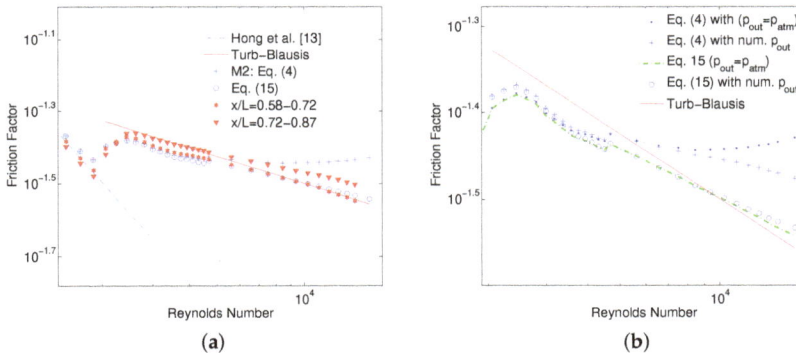

Figure 13. Comparison of f_f calculation by using: M2, Equation (15) and semi-local values (**a**), Equations (4) and (15) with different outlet pressure treatment (**b**).

To summarize the results and discussion of current work following recommendations are being made:

- In laminar flow regime, due to their significant relative contribution towards pressure drop, minor losses can only be estimated using validated numerical results for compressible gas microflows. To avoid numerical input while deducing experimental friction factor results, M2 method can be used to correctly model pressure and temperature at the inlet of MC.
- In choked turbulent regime, MC outlet pressure should be measured to better estimate the cross-sectional average temperature using Equation (5). For an experimental campaign where it is not possible, adiabatic friction factor with an assumption of perfect expansion ($p_{out} = p_{atm}$) should be evaluated using Equation (15).

5. Conclusions

From the above-mentioned discussion followings points can be inferred:

1. For the correct experimental estimation of f_f from total pressure drop and inlet temperature measurements employing minor losses methodology proposed for incompressible flows (M1), it is essential to have inlet and outlet loss coefficients calculated numerically as a priori else results will be misleading in the presence of strong compressibility effects.

2. Gas flow properties at MC inlet can be obtained by considering an isentropic expansion of gas between the entrance manifold and MC (M2). This, when coupled with the assumption of perfect expansion at MC outlet, results in a self-sustained experimental data reduction method with no numerical input as a priori. Furthermore, this avoids possibility of inducing errors in inlet minor loss estimation due to poor numerical modeling.

3. While deducing experimental results for f_f between the inlet and outlet, perfect expansion ($K_{out} = 0$) can be used as a first approximation till the flow starts to choke at outlet. After this limit, gas flow is underexpanded and a fully expanded treatment will result in a significant overestimation of MC outlet temperature using Equation (5) and hence an artificial increase of f_f in choked flow regime is observed.

4. Detailed experimental and numerical analysis shows that the gas flow can be assumed isothermal only in the laminar flow regime for the evaluation of friction factors in MCs.

5. Experiments also show that the isothermal treatment of the gas results in friction factors that are usually lower than adiabatic ones in the choked turbulent flow regime with $f_{f,iso}$ following the slope of Blasius law and adiabatic f_f diverging from it at higher Re. The reason for this diversion is the inappropriate data reduction at MC outlet during flow choking regime.

6. As gas flow accelerates towards the end of the MC causing a steep decrease in temperature near the outlet, average friction factors between inlet and outlet should be calculated using Equation (15).

Author Contributions: Formal analysis, D.R.; Funding acquisition, G.L.M.; Investigation, D.R.; Methodology, D.R., G.L.M. and C.H.; Supervision, G.L.M.; Validation, C.H., D.R.; Writing—original draft, D.R.; Writing—review & editing, D.R., G.L.M. and C.H.

Funding: This research received funding from the European Union's Framework Programme for Research and Innovation Horizon 2020 (2014–2020) under the Marie Sklodowska-Curie Grant Agreement No. 643095 (MIGRATE Project).

Acknowledgments: Technical support of Franceso Vai for fabrication of MC test section assembly is acknowledged. Also technical support of Adrian H. Lutey from University of Parma, Italy is acknowledged for surface roughness measurements.

Conflicts of Interest: The authors declare no conflict of interest.

Abbreviations

The following abbreviations are used in this manuscript:

MC	Microchannel
MT	Microtube
man	manifold
f_f	Average fanno friction factor
$f_{f,local}$	Local Fanno/Adiabatic friction factor
$\tilde{f}_f$	Semi local Fanno friction factor
f_{iso}	Isothermal friction factor
τ_w	Wall shear stress
ρ	Density of the gas
u	Gas average axial velocity

D_h	Hydraluic diamter
α	Aspect ratio of MC
ϵ	Roughness of MC
β	Kinetic energy recovery coefficient
Per	Wetted perimeter
p	Cross sectional average pressure
ΔP	Minor pressure loss
R	Universal gas constant
T	Cross sectional average temperature
T_{av}	Average temperature between two axial locations
$\dot{G}$	mass flow rate per unit area
c_p	Specific heat at constant pressure
A	MC cross sectional area
$\dot{m}$	mass flow rate
μ	Kinematic viscosity
Ma	Mach number
γ	Ratio of specific heat
ΔP	Pressure drop
K	Minor loss coefficient
L	MC length
w	MC width
h	MC height
H_{man}	Manifold Height
H_r	Height of reducer
Subscripts	
in/out	MC inlet and outlet
1	Entrance of inlet manifold
2	Exit of outlet manifold
th	Theoretical
atm	Atmospheric
man	Manifold

References

1. Tuckerman, D.B.; Pease, R.F.W. High-performance heat sinking for VLSI. *IEEE Electron Device Lett.* **1981**, *5*, 126–129. [CrossRef]
2. Asadi, M.; Gongnan, X. A review of heat transfer and pressure drop characteristics of single and two-phase microchannels. *Int. J. Heat Mass Transf.* **2014**, *79*, 34–53. [CrossRef]
3. Wu, P.; Little, W.A. Measurement of friction factors for the flow of gases in very fine channels used for microminiature Joule–Thomson refrigerators. *Cryogenics* **1983**, *23*, 273–277. [CrossRef]
4. Choi, S.B.; Barron, R.F.; Warrington, R.O. Fluid Flow and Heat Transfer in Microtubes. In Proceedings of the Winter Annual Meeting of the American Society of Mechanical Engineers on Micro-Mechanical Sensors, Actuators and System, Atlanta, GA, USA, 1–6 December 1991; pp. 123–134.
5. Peng, X.F.; Peterson, G.P. The effect of thermofluid and geometrical parameters on convection of liquids through rectangular microchannels. *Int. J. Heat Mass Transf.* **1995**, *38*, 755–758. [CrossRef]
6. Xu, B.; Ooi, K.T.; Wong, NT.; Choi, W.K. Experimental investigation of flow friction for liquid flow in microchannels. *Int. Commun. Heat Mass Transf.* **2000**, *27*, 1165–1176. [CrossRef]
7. Pfund, D.; Rector, D.; Shekarriz, A.; Popescu, A.; Welty, J. Pressure drop measurements in a microchannel. *AIChE* **2000**, *46*, 1496–1507. [CrossRef]
8. Harms, T.M.; Kazmierczak, M.J. Developing convective heat transfer in deep rectangular microchannels. *Int. J. Heat Fluid Flow* **1999**, *20*, 149–157. [CrossRef]
9. Adams, T.; Dowling, M.; Abdel-Khalik, S.; Jeter, S. Applicability of traditional turbulent single-phase forced convection correlations to non-circular microchannels. *Int. J. Heat Mass Transf.* **1999**, *42*, 4411–4415. [CrossRef]

10. Morini, G.L.; Yang, Y.; Chalabi, H.; Lorenzini, M. A critical review of measurement techniques for the analysis of gas microflows through microchannels. *Exp. Therm. Fluid Sci.* **2011**, *35*, 849–865. [CrossRef]

11. Morini, G.L.; Yang, Y. Guidelines for the analysis of single-phase forced convection in microchannels. *ASME J. Heat Transf.* **2013**, *135*, 101004. [CrossRef]

12. Hong, C.; Asako, Y.; Lee, J.H. Poiseuille Number Correlation for high speed microflows. *J. Phys. D Appl. Phys.* **2008**, *41*, 105111. [CrossRef]

13. Yang, Y. Experimental and Numerical Analysis of Gas Forced Convection through Microtubes and Micro Heat Exchangers. Ph.D. Thesis, University of Bologna, Bologna, Italy, 2013. [CrossRef]

14. Morini, G.L.; Lorenzini, M.; Colin, S.; Geoffroy, S. Experimental analysis of pressure drop and laminar to turbulent transition for gas flows in smooth microtubes. *Heat Transf. Eng.* **2007**, *8–9*, 670–679. [CrossRef]

15. Morini, G.L.; Lorenzini, M.; Colin, S.; Geoffroy, S. Laminar, Transitional and Turbulent Friction Factors For Gas Flows in Smooth and Rough Microtubes. In Proceedings of the 6th International Conference on Microchannels and Minichannels, Darmstadt, Germany, 23–25 June 2008.

16. Tang, G.H.; Li, Z.; He, Y.L.; Tao, W.Q. Experimental study of compressibility, roughness and rarefaction influences on microchannel flow. *Int. J. Heat Mass Transf.* **2007**, *50*, 2282–2295. [CrossRef]

17. Turner, S.E.; Lam, L.C.; Faghri, M.; Gregory, O.J. Experimental investigation of gas flow in microchannel. *J. Heat Transf.* **2004**, *127*, 753–763. [CrossRef]

18. Kawashima, D.; Asako, Y. Data reduction of friction factor for compressible flow in micro-channels. *Int. J. Heat Mass Transf.* **2014**, *77*, 257–261. [CrossRef]

19. Hong, C.; Nakamura, T.; Asako, Y.; Ueno, I. Semi-local friction factor of turbulent gas flow through rectangular microchannels. *Int. J. Heat Mass Transf.* **2016**, *98*, 643–649. [CrossRef]

20. Hong, C.; Asako, Y.; Morini, G.L.; Rehman, D. Data reduction of average friction factor of gas flow through adiabatic micro-channels. *Int. J. Heat Mass Transf.* **2019**, *129*, 427–431. [CrossRef]

21. Munson, B.R.; Young, D.F.; Okiishi, T.H.; Huebsch, W.W. *Fundamentals of Fluid Mehanics*, 6th ed.; Jhon Willey & Sons Inc.: Hoboken, NJ, USA, 2007; ISBN ES8-0-470-39881-4.

22. Abraham, J.P.; Sparrow, E.M.; Tong, J.C.K.; Bettenhausen, D.W. Internal flows which transist from turbulent through intermittent to laminar. *Int. J. Therm. Sci.* **2010**, *49*, 256–263. [CrossRef]

23. Mirmanto, M. Developing flow pressure drop and friction factor of water in copper microchannels. *J. Mech. Eng. Autom.* **2013**, *3*, 641–649.

24. Sahar, A.M.; Wissink, J.; Mahmoud, M.M.; Karayiannis, T.G.; Ishak, M.S.A. Effect of hydraulic diameter and aspect ratio on single phase flow and heat transfer in a rectangular microchannel. *Appl. Therm. Eng.* **2017**, *115*, 793–814. [CrossRef]

25. Vocale, P.; Rehman, D.; Morini, G.L. Inlet and outlet minor losses in presence of compressibile gas flows in a microtube. In Proceedings of the 9th World Conference on Experimental Heat Transfer, Fluid Mechanics and Thermodynamics, Iguazu Falls, Brazil, 11–15 June 2017.

26. Lijo, V.; Kin, H.D.; Setoguchi, T. Analysis of choked viscous flows through a constant area duct. *IMechE Part G J. Aerosp. Eng.* **2010**, *224*, 1151–1162. [CrossRef]

27. Kawashima, D.; Yamada, T.; Hong, C.; Asako, Y. Mach number at outlet plane of a straight micro-tube. *IMechE Part C J. Mech. Eng. Sci.* **2016**, *19*, 3420–3430. [CrossRef]

28. Hong, C.; Tanaka, G.; Asako, Y. Katanoda, Flow characteristics of gaseous flow through microtube discharged into atmosphere. *Int. J. Heat Mass Transf.* **2018**, *121*, 187–195. [CrossRef]

micromachines

MDPI

Article

Interactive Effects of Rarefaction and Surface Roughness on Aerodynamic Lubrication of Microbearings

Yao Wu [1,2,3], Lihua Yang [1,2,3,*], Tengfei Xu [1,2,3] and Haoliang Xu [1,2,3]

[1] State Key Laboratory for Strength and Vibration of Mechanical Structures, Xi'an Jiaotong University, Xi'an 710049, China; nealjackman@stu.xjtu.edu.cn (Y.W.); xtf1992@stu.xjtu.edu.cn (T.X.); xuhaoliang1988@stu.xjtu.edu.cn (H.X.)
[2] Shaanxi Key Laboratory of Environment and Control for Flight Vehicle, Xi'an 710049, China
[3] School of Aerospace Engineering, Xi'an Jiaotong University, Xi'an 710049, China
*   Correspondence: lihuayang@mail.xjtu.edu.cn

Received: 30 January 2019; Accepted: 22 February 2019; Published: 25 February 2019

Abstract: The aerodynamic lubrication performance of gas microbearing has a particularly critical impact on the stability of the bearing-rotor system in micromachines. Based on the Duwensee's slip correction model and the fractal geometry theory, the interactive effects of gas rarefaction and surface roughness on the static and dynamic characteristics were investigated under various operation conditions and structure parameters. The modified Reynolds equation, which governs the gas film pressure distribution in rough bearing, is solved by employing the partial derivative method. The results show that high values of the eccentricity ratio and bearing number tend to increase the principal stiffness coefficients significantly, and the fractal roughness surface considerably affects the ultra-thin film damping characteristics compared to smooth surface bearing.

Keywords: gaseous rarefaction effects; fractal surface topography; modified Reynolds equation; aerodynamic effect; bearing characteristics

---

1. Introduction

The microfluidic devices are widely used in many applications such as ultra-precision machine tool spindles, inertial navigation system (INS), medical devices, and hard disk drives (HDDs) in micro-electro-mechanical systems (MEMS). Since the development of microsystems engineering technology, gas journal microbearing has been generally preferred over the electromagnetic bearing and rolling bearing owing to its advantages of simple structure, high rotary accuracy, high running speed, low friction power loss, and wide working temperature range [1–3]. The real surface of a mechanical part produced by various machining and finish operations is composed of a large number of distributed peaks and valleys. The lubricating film thickness between the surfaces of shaft and bearing has continually decreased, which results in the increase of the roughness heights that are of the same order of magnitude as the minimum clearance gap. Thus, the assumptions that the gas flow is typically treated as a continuum flow with no slip boundary condition and the bearing surface roughness is considered negligible in classical fluid mechanics have to be revised at microscales [4–7]. The interaction between rarefaction and surface roughness in microbearing can influence the reliability and operational efficiency of micro rotating machinery obviously, so the gas journal microbearing performance should be comprehensively analyzed.

In order to accurately predict the effects of surface roughness and rarefaction on the bearing characteristics, many researchers devoted numerous research efforts to study the complicated flow behaviors at very small clearances over the last few decades. In consideration of rarefaction effects in ultra-thin gas film lubrication, the Knudsen number K_n is utilized to describe the rarefied gas flow, which is defined as the ratio of molecular mean free path λ_0 to the characteristic length of gas film thickness. Burgdorfer [8], Hsia et al. [9], and Mitsuya [10], based on the slip velocity boundary condition, derived the classical first order, second order, and 1.5 order slip models in the slider/disk interface for HDDs to take into account the effect of gas rarefaction. Fukui and Kaneko [11,12] developed a Poiseuille flow rate database for a wide Knudsen number range to modify the compressible Reynolds type equation including thermal creep flow and accommodation coefficient from linearized Boltzmann equation. In order to account for the effect of surface roughness, Christensen and Tønder [13,14] presented a stochastic model of hydrodynamic lubrication for finite width journal bearing in which they considered the lubricant film thickness as a stochastic process. The operating characteristics of bearing was theoretically analyzed with roughness pattern, nominal geometric features, and statistical properties by surface averaging techniques. Via linear transformation of random matrices, the Gaussian or non-Gaussian distribution of surface heights were generated by Patir [15] using the prescribed autocorrelation functions and frequency density functions. Patir and Cheng [16] further derived the average Reynolds equation suitable for various roughness structure and discussed the effect of roughness on mean hydrodynamic pressure, mean viscous friction, and mean bearing inflow in finite slider bearings. The average flow model of Patir and Cheng was extended by Tripp [17], in which the statistical expectation of flow factors were calculated with a perturbation expansion of the film pressure. The results showed that the flow factors are closely correlated with roughness parameters. White et al. [18] introduced the transverse sinusoidal roughness pattern to study the influence of surface roughness on steady-state pressure profiles of wedge bearing by variable grid implicit finite difference method and found that the load capacity could be decreased to a limiting value at higher bearing numbers. For the applications of perturbation technique and mapping function, Li et al. [19] studied the effects of roughness orientations and rarefaction on static performance of magnetic recording systems. The results demonstrated that the flow factors changed with the orientation angle and Peklenik number, and the effect of moving surface on surface characteristics is more significant than that of the stationary surface. Turaga et al. [20] proposed the stochastic finite method to solve Reynolds equation and obtained the static and dynamic performance of hydrodynamic journal bearings with the longitudinal, transverse and isotropic roughness pattern. Naduvinamani et al. [21] established the surface roughness by a stochastic random variable with nonzero mean, variance and skewness, and the average Reynolds equations were adopted to analyze the performance of porous step-slider bearings with Stokes couple stress fluid. Zhang et al. [22,23] presented the modified Reynolds equation by including fractal roughness effect and velocity slip boundary condition and concluded that the flow behaviors in gas-lubricated journal microbearings was appreciably affected by Knudsen number, bearing number and fractal dimension. The coupling effects of non-Newtonian micropolar fluids and roughness on the dynamic characteristics of plane slider bearings were investigated by Lin et al. [24] on the basis of the microcontinuum theory and Christensen stochastic roughness model. They indicated that the transverse roughness serves to somewhat increase bearing dynamic property, whereas the longitudinal roughness would tend to decrease the dynamic coefficients. Jao et al. [25] examined the influences of surface roughness and anisotropic slips on hydrodynamic lubrication of journal bearings. They described the lubricant flow in rough bearing surface by the product of flow factors and flow in nominal film thickness, and also identified that boundary slip reduced the effect of surface roughness. Kalavathi et al. [26] reported a generalized Reynolds equation for finite porous slider bearing with both longitudinal and transverse roughness. The authors showed the surface roughness enhanced the pressure distribution and load carrying capacity while the permeability parameter diluted the load. Quiñonez [27] utilized the linear superposition of perturbation method and Flourier transformation to provide a solution for the flow characteristics of wide exponential land slider bearings with rough surfaces. The results were in good agreement with the cases of sinusoidal and single Gaussian dent. The linear perturbation method was used by Wang et al. [28] to solve the unsteady Reynolds equation for rough aerostatic journal bearings during the iterative process, and the dynamic performance was obtained by taking into account the interactions of journal rotation and surface waviness. However, likely due to the nonlinear and complexity of dynamic flow behavior, previous papers were mainly focusing on steady-state characteristics in rough journal bearings, and the dynamic characteristics of hydrodynamic gas-lubricated microbearings were seldom reported in the research literature. Moreover, the statistical parameters (such as root mean square of asperity heights, surface slope, curvature, skewness, and kurtosis), which are conventionally applied to characterize surface roughness, vary with the sampling length and resolution of measuring equipment. A scale-invariant surface characterization should be considered. Hence, the analytical studies of surface roughness effect on dynamic characteristics of gas slider bearings with rarefaction coefficients in microfluidic engineering devices is motivated.

In this paper, the Weierstrass-Mandelbrot (W-M) fractal function is used to characterize the homogeneous surface roughness, and the Boltzmann slip correction model is applied to represent the rheological behavior of compressible rarefied gas film. The generalized Reynolds-type equation considering gas rarefaction, as well as roughness effect, is mathematically derived and solved by the partial derivative method and relaxation iteration algorithm. Bearing performances (including the load-carrying capacity, friction coefficient and corresponding attitude angle, dynamic stiffness, and damping coefficients) are presented and discussed in comparison with smooth surface bearings. The work is expected to elucidate the performance characteristics of gas microbearings with Poiseuille flow and random asperities, which is conducive to understand the fluid mechanisms of very low clearance gas films for microfluidics devices.

2. Characterization of Fractal Rough Surface

The distribution of asperity heights on the bearing surface consisting of long narrow ridges and valleys in engineering practice. A rough surface is a random system and the fractal geometry is introduced to characterize the random and multiscale topographies. Mandelbrot [29] initially developed the fractal theory by researching the coastal geomorphology in 1967. He found that the most machined surfaces can be constructed using Weierstrass-Mandelbrot (W-M) function with randomness, multiscale nature, self-similarity, and self-affine property. Unlike the traditional characterization parameters of surface roughness, fractal roughness parameters are independent of scan lengths and provide all the surface topography information of rough profiles. The W-M function that explicitly expressed the homogeneous rough surface in self-acting gas-lubricated journal bearings is given by

$$
\begin{aligned}
h_r(x,y) = L\left(\frac{G}{L}\right)^{D_f-2} \cdot \sqrt{\frac{\ln\gamma}{M}} \cdot \sum_{m=1}^{M}\sum_{n=0}^{n_{max}}\gamma^{(D_f-3)n} \times \\
\left(\cos\phi_{m,n} - \cos\left\{\frac{2\pi\gamma^n\sqrt{x^2+y^2}}{L} \cdot \cos\left[\tan^{-1}\left(\frac{y}{x}\right) - \frac{\pi m}{M}\right] + \phi_{m,n}\right\}\right)
\end{aligned}
\tag{1}
$$

where $h_r(x,y)$ is the height of rough surface, x and y are the measure distances in the vertical and horizontal position, respectively. L is the sampling length of the profile of surface. D_f is the fractal dimension, varying from 2 to 3 in three-dimensional surface topography. G is the scaling constant that relates to the roughness profile. γ is the scaling parameter ($\gamma > 1$), which determines the spectral density, γ is equal to 1.5 for a Gaussian and isotropic surface. M is the number of overlapped ridges on the surface, m and n are the frequency index, $n_{max} = \text{int}[\log(L/L_s)/\log\gamma]$, $\varphi_{m,n}$ is the random phase, L_s is the cut-off length that depends on cut-off wavelength of resolution in measuring machines.

The values of asperity heights can be changed by the fractal dimension D_f, comparisons of the distributions of asperity heights for different D_f are illustrated in Figure 1. It is seen that the heights of rough surfaces increase as the fractal dimension decreases.

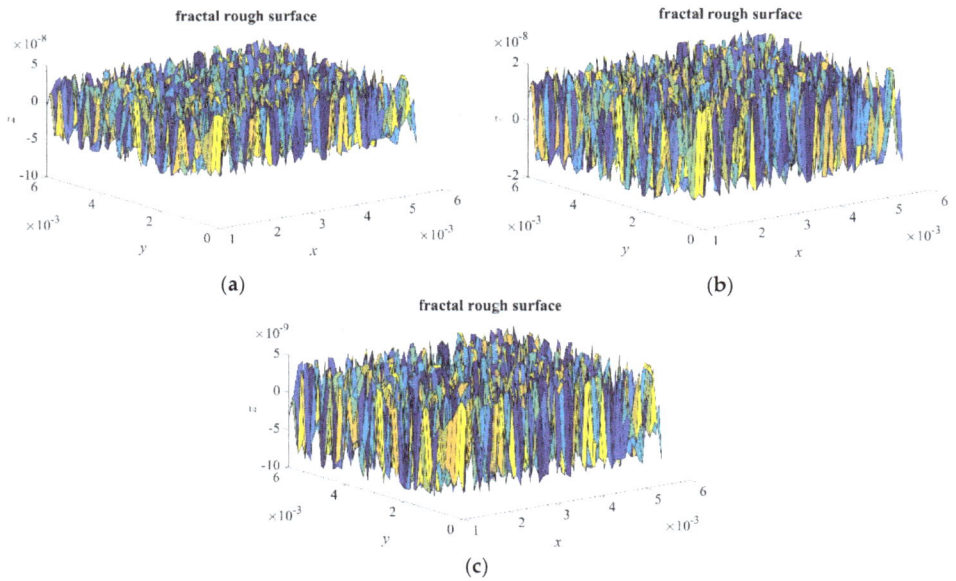

Figure 1. Simulation of a three-dimensional fractal surface topography for different fractal dimensions. (**a**) D_f = 2.2, and $G = 1 \times 10^{-10}$ m; (**b**) D_f = 2.3, and $G = 1 \times 10^{-10}$ m; (**c**) D_f = 2.4, and $G = 1 \times 10^{-10}$ m.

In the three-dimensional isotropic roughness type, the schematic presentation of rough gas microbearing is plotted in Figure 2.

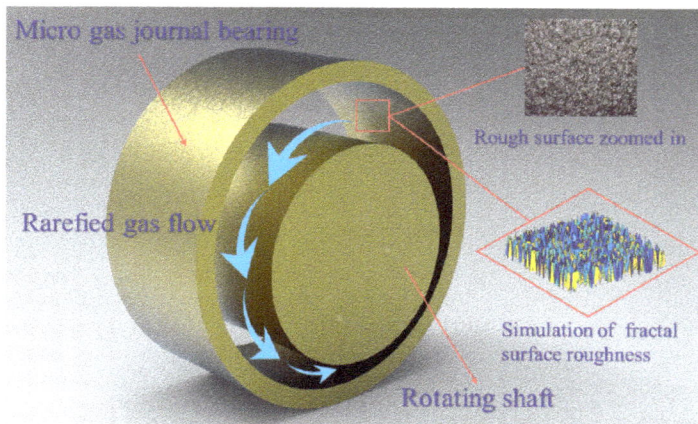

Figure 2. Schematic of a rough-surface gas-lubricated journal microbearings.

3. Numerical Model and Solution Method

In the current analysis, rarefied gas flow between the surfaces of bearing and journal is treated as isothermal, laminar flow with uniform viscosity. and elastic deformation of bearing surface is not considered. The Boltzmann slip correction is used to model the gas rarefaction effect at the gas-solid interface for arbitrary Knudsen numbers. The physical configuration of micro gas bearing with rough surface is shown in Figure 3. The modified Reynolds equation incorporating gas rarefaction and surface roughness effects in non-dimensional form appears as

$$\frac{\partial}{\partial \varphi}\left(QPH^3\frac{\partial P}{\partial \varphi}\right) + \frac{\partial}{\partial \lambda}\left(QPH^3\frac{\partial P}{\partial \lambda}\right) = \Lambda\frac{\partial(PH)}{\partial \varphi} + 2\Lambda\frac{\partial(PH)}{\partial T} \tag{2}$$

where $P = p/p_a$, $H = h/c_b$, $\varphi = x/R$, $\lambda = z/R$ are the dimensionless gas film pressure, the dimensionless gas film thickness, and the coordinates in the circumferential and axial direction, p_a is the ambient pressure, c_b is the radius clearance, R is the radius of journal, p is the local gas pressure, h is the clearance spacing of ultra-thin gas film, ε is the eccentricity ratio and $\varepsilon = e/c_b$, e is the eccentricity. $\Lambda = 6\mu\omega R^2/(p_a c_b^2)$ is the gas bearing number, μ is the viscosity coefficient, ω is the rotating angular velocity of journal, T is the dimensionless time.

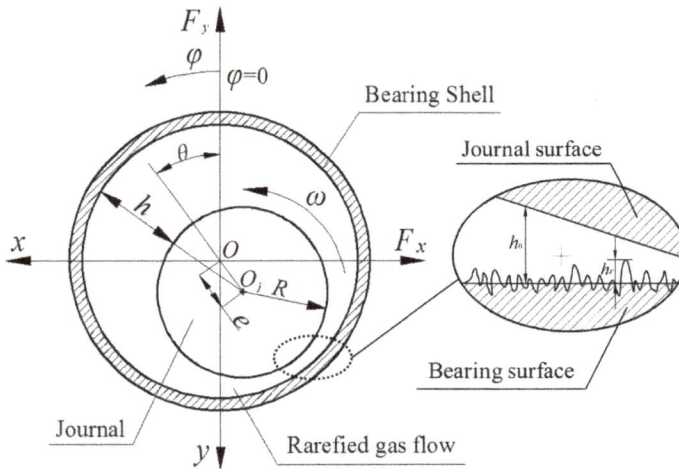

Figure 3. Geometrical configuration of the micro gas-lubricated journal bearing with rough surface.

For the slip correction factors in rarefied gas flows under ultra-low spacing, the dimensionless Poiseuille flow rate between rotor and bearing surfaces with the gas thickness h is given as [30]

$$Q_P = \frac{\frac{1}{2\mu}\frac{\partial p}{\partial x}h^3\left(b \cdot D^c + \frac{a\sqrt{\pi}}{2D} + \frac{1}{6}\right)}{\frac{-h^3}{2\mu D} \cdot \frac{\partial p}{\partial x}} = b \cdot D^{c+1} + \frac{a\sqrt{\pi}}{2} + \frac{D}{6} \tag{3}$$

The Poiseuille flow rate ratio Q appears as

$$Q = \frac{Q_P}{Q_{continuum}} = \frac{b \cdot D^{c+1} + \frac{a\sqrt{\pi}}{2} + \frac{D}{6}}{\frac{D}{6}} = 1 + \frac{3a\sqrt{\pi}}{D} + 6b \cdot D^c \tag{4}$$

where the inverse Knudsen number $D = \frac{\sqrt{\pi}}{2K_n}$, the three adjustable coefficients $a = 0.01807$, $b = 1.35355$ and $c = -1.17468$. The Boltzmann Poiseuille flow rate ratio [31] Q is expressed as

$$Q = 1 + 0.10842K_n + 9.3593/K_n^{-1.17468} \tag{5}$$

Here the lubricant film thickness h is the sum of nominal smooth film thickness h_0 and random roughness h_r measured from the nominal smooth height.

$$h = h_0 + h_r = c_b(1 + \varepsilon \cos \varphi) + L\left(\frac{G}{L}\right)^{D_f - 2} \cdot \sqrt{\frac{\ln\gamma}{M}} \cdot \sum_{m=1}^{M} \sum_{n=0}^{n_{max}} \gamma^{(D_f - 3)n} \times$$
$$\left(\cos \phi_{m,n} - \cos\left\{ \frac{2\pi\gamma^n \sqrt{x^2+y^2}}{L} \cdot \cos\left[\tan^{-1}\left(\frac{y}{x}\right) - \frac{\pi m}{M}\right] + \phi_{m,n} \right\} \right) \quad (6)$$

As in the steady state, the transient term $\partial(PH)/\partial T$ of Equation (2) can be ignored, the static dimensionless modified Reynolds equation can be gained as

$$\frac{\partial}{\partial \varphi}\left(QPH^3 \frac{\partial P}{\partial \varphi} \right) + \frac{\partial}{\partial \lambda}\left(QPH^3 \frac{\partial P}{\partial \lambda} \right) = \Lambda \frac{\partial(PH)}{\partial \varphi} \quad (7)$$

In order to obtain the aerodynamic performance of gas microbearings, the modified Reynolds equation for homogeneous surface roughness and gas rarefaction should be solved numerically. Equation (2), governing the gas pressure distribution of the bearing, is a nonlinear two-dimensional partial differential equation (PDE). It is difficult to get its analytical solution. So, the partial derivative method [32,33], with a relaxation iteration algorithm, is employed to ensure a reasonable and efficient solution. By introducing the mathematical transformation $PH = S$, $(PH)^2 = S^2 = \Pi$, Equation (2) can be converted to the ellipse-type partial differential equation $-\nabla \cdot (c\nabla u) + au = f$ in the following form:

$$-\left(\frac{\partial^2 \Pi}{\partial \varphi^2} + \frac{\partial^2 \Pi}{\partial \lambda^2}\right) + \frac{2\Pi}{H}\left(\frac{\partial^2 H}{\partial \varphi^2} + \frac{\partial^2 H}{\partial \lambda^2}\right) + \frac{2\Pi}{QH}\left(\frac{\partial Q}{\partial \varphi}\frac{\partial H}{\partial \varphi} + \frac{\partial Q}{\partial \lambda}\frac{\partial H}{\partial \lambda}\right) =$$
$$-\frac{1}{H}\left(\frac{\partial H}{\partial \varphi}\frac{\partial \Pi}{\partial \varphi} + \frac{\partial H}{\partial \lambda}\frac{\partial \Pi}{\partial \lambda}\right) + \frac{1}{Q}\left(\frac{\partial Q}{\partial \varphi}\frac{\partial \Pi}{\partial \varphi} + \frac{\partial Q}{\partial \lambda}\frac{\partial \Pi}{\partial \lambda}\right) - \frac{2\Lambda}{QH}\frac{\partial S}{\partial \varphi} - \frac{4\Lambda}{QH}\frac{\partial S}{\partial T} \quad (8)$$

The pressure boundary conditions for the Reynolds equation are:

$$\begin{cases} P\big|_{\varphi,\lambda = \pm\frac{B}{2R}} = 1, \\ P\big|_{\varphi=0,\lambda} = P\big|_{\varphi=2\pi,\lambda}, \\ \frac{\partial P}{\partial \varphi}\big|_{\varphi=0,\lambda} = \frac{\partial P}{\partial \varphi}\big|_{\varphi=2\pi,\lambda} \end{cases} \quad (9)$$

where B is the bearing width.

The non-dimensional hydrodynamic gas film forces are obtained by integrating the film pressure acting on the microbearing along the both horizontal and vertical directions.

$$\begin{cases} \overline{F}_x = p_a R^2 \int_{-\frac{B}{2R}}^{\frac{B}{2R}} \int_0^{2\pi} (P-1)\sin\varphi d\varphi d\lambda \\ \overline{F}_y = p_a R^2 \int_{-\frac{B}{2R}}^{\frac{B}{2R}} \int_0^{2\pi} (P-1)\cos\varphi d\varphi d\lambda \end{cases} \quad (10)$$

The attitude angle θ of the gas-lubricated journal microbearing is calculated by

$$\theta = \arctan\left(\frac{\overline{F}_x}{\overline{F}_y}\right) \quad (11)$$

The total non-dimensional load-carrying capacity is written as

$$C_L = \frac{W}{p_a RB} = \frac{R}{B}\int_{-\frac{B}{2R}}^{\frac{B}{2R}} \int_0^{2\pi} (P-1)\cos\varphi d\varphi d\lambda \quad (12)$$

The non-dimensional skin friction coefficient on the journal surface can be computed by

$$F_b = -\int_{-\frac{B}{2R}}^{\frac{B}{2R}} \int_0^{2\pi} \left(\frac{\Lambda}{6}\frac{1}{H} + \frac{H}{2}\frac{\partial P}{\partial \varphi}\right) d\varphi d\lambda \quad (13)$$

Suppose that the journal center whirls around its static equilibrium position with a small amplitude periodic motion under the perturbation frequency ratio Ω, the linear perturbation method is adopted for calculating the dynamic stiffness and damping coefficients. The steady-state position is indicated as $(\varepsilon_0, \theta_0)$, and its dynamic disturbance about $(\varepsilon_0, \theta_0)$ are denoted as E and Θ. The positions of eccentricity ratio ε and attitude angle θ of journal at a random position are represented by the static and dynamic components as follows:

$$\begin{cases} \varepsilon = \varepsilon_0 + E = \varepsilon_0 + E_0 e^{i\Omega T} \\ \theta = \theta_0 + \Theta = \theta_0 + \Theta_0 e^{i\Omega T} \end{cases} \tag{14}$$

where E_0 and Θ_0 are the small perturbation amplitude of journal eccentricity ratio and attitude angle in the complex field. The dimensionless perturbation frequency Ω, which is defined as the ratio of journal disturbance frequency ν to rotating angular velocity ω of journal, $i = \sqrt{-1}$.

Thus, the non-dimensional gas-film pressure and gas-film thickness of the gas microbearing can be expressed as

$$\begin{cases} P = P_0 + Q_{gd} = P_0 + \widetilde{P}_0 e^{i\Omega T} \\ H = H_0 + H_{gd} = H_0 + \widetilde{H}_0 e^{i\Omega T} \end{cases} \tag{15}$$

where $\widetilde{H}_0 = E_0 \cos(\varphi - \theta_0) + \varepsilon_0 \Theta_0 \sin(\varphi - \theta_0)$, P_0 is the static gas-film pressure and H_0 is the static gas-film thickness. Q_{gd}, H_{gd} are the dynamic gas film pressure and gas film thickness, respectively. $\widetilde{P}_0$ and $\widetilde{H}_0$ are the perturbation magnitudes in terms of complex numbers for dynamic gas film pressure and gas film thickness.

Substituting Equation (15) into Equation (2), the generalized dynamic lubrication equation for molecular model and surface roughness can be derived as:

$$\begin{aligned} &\frac{\partial}{\partial \varphi}(QP_0 H_0{}^3 \frac{\partial \widetilde{P}_0}{\partial \varphi}) + \frac{\partial}{\partial \lambda}(QP_0 H_0{}^3 \frac{\partial \widetilde{P}_0}{\partial \lambda}) + \frac{\partial}{\partial \varphi}(Q\widetilde{P}_0 H_0{}^3 \frac{\partial P_0}{\partial \varphi}) + \frac{\partial}{\partial \lambda}(Q\widetilde{P}_0 H_0{}^3 \frac{\partial P_0}{\partial \lambda}) + \\ &\frac{\partial}{\partial \varphi}(3QP_0 H_0{}^2 \widetilde{H}_0 \frac{\partial P_0}{\partial \varphi}) + \frac{\partial}{\partial \lambda}(3QP_0 H_0{}^2 \widetilde{H}_0 \frac{\partial P_0}{\partial \lambda}) = \Lambda \frac{\partial}{\partial \varphi}(P_0 \widetilde{H}_0 + \widetilde{P}_0 H_0) + i2\Lambda\Omega(P_0 \widetilde{H}_0 + \widetilde{P}_0 H_0) \end{aligned} \tag{16}$$

As mentioned in Reference [34], some variables are defined by

$$\begin{cases} P_E = \frac{\partial \widetilde{P}_0}{\partial E_0}, \\ P_\theta = \frac{1}{\varepsilon_0} \frac{\partial \widetilde{P}_0}{\partial \Theta_0}, \\ H_E = \frac{\partial \widetilde{H}_0}{\partial E_0}, \\ H_\theta = \frac{1}{\varepsilon_0} \frac{\partial \widetilde{H}_0}{\partial \Theta_0} \end{cases} \tag{17}$$

After differentiating $\widetilde{P}_0$ and $\widetilde{H}_0$ in Equation (16) with respect to E_0 and Θ_0, and combining with some mathematical transformation, the resulting dynamic PDE equations are obtained for rough surface microbearings concerning the variables P_E, P_θ, H_E, and H_θ.

$$\begin{aligned} &\frac{\partial}{\partial \varphi}(QP_0 H_0{}^3 \frac{\partial P_E}{\partial \varphi}) + \frac{\partial}{\partial \lambda}(QP_0 H_0{}^3 \frac{\partial P_E}{\partial \lambda}) + \frac{\partial}{\partial \varphi}(QP_E H_0{}^3 \frac{\partial P_0}{\partial \varphi}) + \frac{\partial}{\partial \lambda}(QP_E H_0{}^3 \frac{\partial P_0}{\partial \lambda}) + \\ &\frac{\partial}{\partial \varphi}(\frac{\partial Q}{\partial E_0} P_0 H_0{}^3 \frac{\partial \widetilde{P}_0}{\partial \varphi}) + \frac{\partial}{\partial \lambda}(\frac{\partial Q}{\partial E_0} P_0 H_0{}^3 \frac{\partial \widetilde{P}_0}{\partial \lambda}) + \frac{\partial}{\partial \varphi}(\frac{\partial Q}{\partial E_0} \widetilde{P}_0 H_0{}^3 \frac{\partial P_0}{\partial \varphi}) + \\ &\frac{\partial}{\partial \lambda}(\frac{\partial Q}{\partial E_0} \widetilde{P}_0 H_0{}^3 \frac{\partial P_0}{\partial \lambda}) + 3QP_0 H_0{}^3 \frac{\partial P_0}{\partial \varphi} \frac{\partial}{\partial \varphi}(\frac{H_E}{H_0}) + 3QP_0 H_0{}^3 \frac{\partial P_0}{\partial \lambda} \frac{\partial}{\partial \lambda}(\frac{H_E}{H_0}) + \\ &3\frac{\partial}{\partial \varphi}(\frac{\partial Q}{\partial E_0} P_0 H_0{}^2 \widetilde{H}_0 \frac{\partial P_0}{\partial \varphi}) + 3\frac{\partial}{\partial \lambda}(\frac{\partial Q}{\partial E_0} P_0 H_0{}^2 \widetilde{H}_0 \frac{\partial P_0}{\partial \lambda}) + 3\Lambda \frac{H_E}{H_0} \frac{\partial(P_0 H_0)}{\partial \varphi} = \\ &\Lambda \frac{\partial}{\partial \varphi}(P_0 H_E + P_E H_0) + i2\Lambda\Omega(P_0 H_E + P_E H_0) \end{aligned} \tag{18}$$

$$H_E = \cos(\varphi - \theta_0) \tag{19}$$

$$\frac{\partial}{\partial\varphi}(QP_0H_0{}^3\frac{\partial P_\theta}{\partial\varphi}) + \frac{\partial}{\partial\lambda}(QP_0H_0{}^3\frac{\partial P_\theta}{\partial\lambda}) + \frac{\partial}{\partial\varphi}(QP_\theta H_0{}^3\frac{\partial P_0}{\partial\varphi}) + \frac{\partial}{\partial\lambda}(QP_\theta H_0{}^3\frac{\partial P_0}{\partial\lambda}) +$$
$$\frac{\partial}{\partial\varphi}(\frac{\partial Q}{\partial\Theta_0}P_0H_0{}^3\frac{\partial\widetilde{P}_0}{\partial\varphi}) + \frac{\partial}{\partial\lambda}(\frac{\partial Q}{\partial\Theta_0}P_0H_0{}^3\frac{\partial\widetilde{P}_0}{\partial\lambda}) + \frac{\partial}{\partial\varphi}(\frac{\partial Q}{\partial\Theta_0}\widetilde{P}_0H_0{}^3\frac{\partial P_0}{\partial\varphi}) +$$
$$\frac{\partial}{\partial\lambda}(\frac{\partial Q}{\partial\Theta_0}\widetilde{P}_0H_0{}^3\frac{\partial P_0}{\partial\lambda}) + 3QP_0H_0{}^3\frac{\partial P_0}{\partial\varphi}\frac{\partial}{\partial\varphi}(\frac{H_\theta}{H_0}) + 3QP_0H_0{}^3\frac{\partial P_0}{\partial\lambda}\frac{\partial}{\partial\lambda}(\frac{H_\theta}{H_0}) +$$
$$3\frac{\partial}{\partial\varphi}(\frac{\partial Q}{\partial\Theta_0}P_0H_0{}^2\widetilde{H}_0\frac{\partial P_0}{\partial\varphi}) + 3\frac{\partial}{\partial\lambda}(\frac{\partial Q}{\partial\Theta_0}P_0H_0{}^2\widetilde{H}_0\frac{\partial P_0}{\partial\lambda}) + 3\Lambda\frac{H_\theta}{H_0}\frac{\partial(P_0H_0)}{\partial\varphi} =$$
$$\Lambda\frac{\partial}{\partial\varphi}(P_0H_\theta + P_\theta H_0) + i2\Lambda\Omega(P_0H_\theta + P_\theta H_0) \tag{20}$$

$$H_\theta = \sin(\varphi - \theta_0) \tag{21}$$

According to the coordinate system at the bearing midplane as illustrated in Figure 3, by simultaneously solving the nonlinear Equations (18)–(21) using the partial derivative method with iteration procedure, the dynamic stiffness coefficients K_{ij} and dynamic damping coefficients D_{ij} of gas journal microbearing for fractal rough surface can be calculated by the following formula:

$$\begin{cases} -\frac{R}{B}\iint_A P_E \cos\varphi d\varphi d\lambda = K_{y\varepsilon} + i\Omega D_{y\varepsilon} \\ \frac{R}{B}\iint_A P_E \sin\varphi d\varphi d\lambda = K_{x\varepsilon} + i\Omega D_{x\varepsilon} \\ -\frac{R}{B}\iint_A P_\theta \cos\varphi d\varphi d\lambda = K_{y\theta} + i\Omega D_{y\theta} \\ \frac{R}{B}\iint_A P_\theta \sin\varphi d\varphi d\lambda = K_{x\theta} + i\Omega D_{x\theta} \end{cases} \tag{22}$$

The dynamic coefficients in the Cartesian coordinate system are given by the transformation matrix A.

$$\begin{cases} K_{ij} = \begin{pmatrix} K_{xx} & K_{xy} \\ K_{yx} & K_{yy} \end{pmatrix} = A\begin{pmatrix} K_{x\varepsilon} & K_{x\theta} \\ K_{y\varepsilon} & K_{y\theta} \end{pmatrix} \\ D_{ij} = \begin{pmatrix} D_{xx} & D_{xy} \\ D_{yx} & D_{yy} \end{pmatrix} = A\begin{pmatrix} D_{x\varepsilon} & D_{x\theta} \\ D_{y\varepsilon} & D_{y\theta} \end{pmatrix} \end{cases}, A = \begin{pmatrix} -\sin\theta_0 & -\cos\theta_0 \\ \cos\theta_0 & -\sin\theta_0 \end{pmatrix} \tag{23}$$

It was confirmed that the dynamic characteristics of the gas journal bearing can greatly affect the critical speed, unbalance response, and instability threshold of a hydrodynamic gas-lubricated bearing-rotor system after long-term research and practice. The comprehensive analysis of dynamic coefficients is important in the design of gas microbearings in MEMS applications, the key findings presented in the next section will reveal some insights into the ultra-thin gas film lubrication problems.

4. Results and Discussion

On the basis of fractal geometry theory and the Boltzmann slip correction factor, the combined effects of gas rarefaction and surface roughness on the static and dynamic characteristics of ultra-thin film gas lubrication in journal microbearings are investigated in detail. The pressure distribution, load carrying capacity, friction coefficient and attitude angles of bearing, dynamic stiffness and damping properties are analyzed concerning the fractal dimension and bearing geometric parameters in this section, the primary design parameters shown in Figure 3 are R = 1mm, B = 200 μm, c = 1 μm, $p_a = 1.033 \times 10^5$ N/m^2, and the aspect ratio is $B/D = 0.1$.

4.1. Steady-State Film Pressure

To verify the correctness of the developed theoretical model and program code employed in this paper, the non-dimensional pressure profile for the middle cross-section along the sliding direction obtained by the current solution is compared with Zhang et al. in Reference [22]. As indicated in Figure 4, the simulation results are in close agreement with the numerical predictions reported by Zhang et al. for ε = 0.7, B/D = 0.075, c = 12 μm, $\omega = 5 \times 10^5$ rpm. It can also be seen that the pressure randomly fluctuates when the effect of surface roughness is considered.

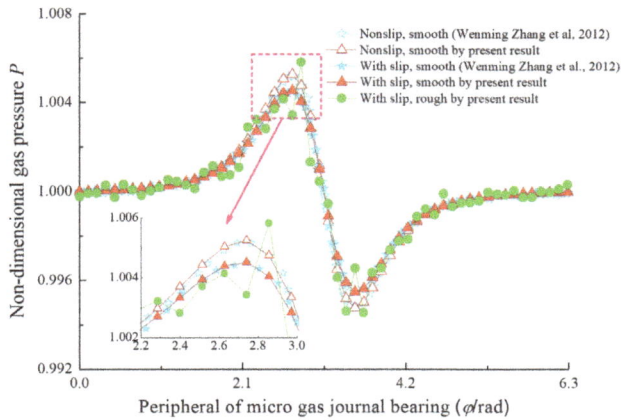

Figure 4. Comparison of dimensionless gas film pressure with Zhang et al. [20].

The influence of bearing number Λ, eccentricity ratio ε, and surface roughness on pressure distributions P is shown in Figure 5. An important observation exhibited by Figure 4a,b is that the maximum gas film pressure becomes larger as Λ increases for fixed fractal parameters G and D. This phenomenon can be explained by the enhanced aerodynamic effect, which indicates that higher angular velocity of the journal dilutes the pressure diffusion. In Figure 4a,c, increasing the eccentricity ratio ε leads to the higher pressure distribution and magnitude of pressure fluctuations decrease. In comparison with the smooth surface case, the roughness effect increases the pressure profile for the rough surface at the same bearing number and eccentricity ratio, and the random surface roughness makes the pressure distribution across the entire lubricating film unpredictable. The contour plots show more clearly the detail and variation of pressure over the rough bearing surfaces. At greater bearing number Λ the influence of surface roughness decreases and the contour lines approach that of the smooth case, the fluctuations of the pressure contour plots are significant with the increase of fractal roughness in gas bearing.

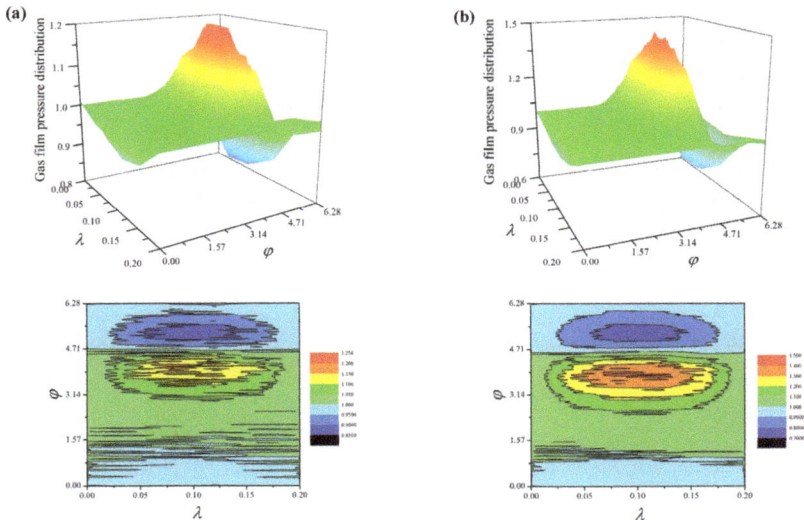

Figure 5. *Cont.*

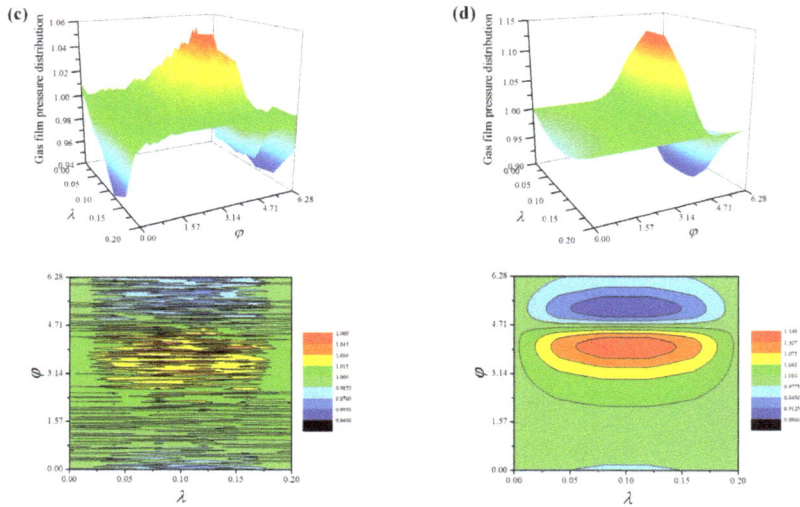

Figure 5. Pressure distributions and contour plots of the gas-lubricated microbearing for different bearing numbers and eccentricity ratios: (**a**) $\varepsilon = 0.6$, $\Lambda = 20$, $D_f = 2.3$, and $G = 1 \times 10^{-10}$ m, (**b**) $\varepsilon = 0.6$, $\Lambda = 60$, $D_f = 2.3$, and $G = 1 \times 10^{-10}$ m, (**c**) $\varepsilon = 0.3$, $\Lambda = 20$, $D_f = 2.3$, and $G = 1 \times 10^{-10}$ m, and (**d**) $\varepsilon = 0.6$, $\Lambda = 20$, smooth.

4.2. Load-Carry Capacity and Friction Coefficient

Figures 6 and 7 predict the variation of non-dimensional load carrying capacity and attitude angle as a function of eccentricity ratio for five different values of fractal dimension ($D_f = 2.2, 2.25, 2.3, 2.35, 2.4$). As the eccentricity ratio increases, which denotes the film thickness is thinner, the load capacity increases monotonically. The higher self-affine fractal dimensions yield the smaller roughness heights distribution on rough surfaces, and it is noted that the load carrying capacity is increased gradually when compared with the smooth bearing. The reason is that the increasing values of roughness heights reduces the sidewise leakage of airflow and the flow is restricted by the surface asperities. However, as the fractal dimension decreases further for $D_f = 2.2$, the load carrying capacity tends to decline under this condition. This is because the minimum air film clearance between rotor and bearing may become too small so that the surface roughness effect which increases the load-carrying capacity is weaker than the gaseous rarefaction effects which reduces the dimensionless load capacity, thus causing a decrease in the bearing load capacity. The attitude angle θ is found to decrease with the growing eccentricity ratio. The decrease in attitude angle is more accentuated for a rougher surface as compared to a nominally smooth surface.

The variation of friction coefficient with the eccentricity ratio for different values of fractal dimension with fixed values of $\Lambda = 20$ and $G = 1 \times 10^{-11}$ is depicted in Figure 8. It can be seen that friction coefficients monotonically increase as ε increases. Although the contact area between the rarefied gas flow and the surface asperities is larger when the homogeneous surface roughness becomes more and more obvious, the static friction coefficients show a slightly more gradual increase with fractal dimension. As the fractal dimension D_f is equal to 2.2, the friction coefficient is lower than that of smooth surface for the same reason that the gaseous rarefaction effect is more pronounced and thus friction coefficient drops.

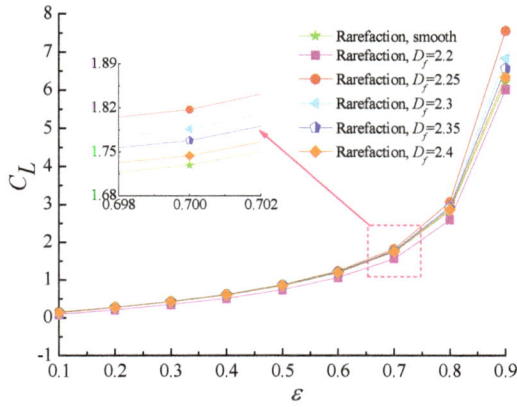

Figure 6. Non-dimensional load capacity versus eccentricity ratio for different fractal dimensions at $\Lambda = 20$, $G = 1 \times 10^{-11}$.

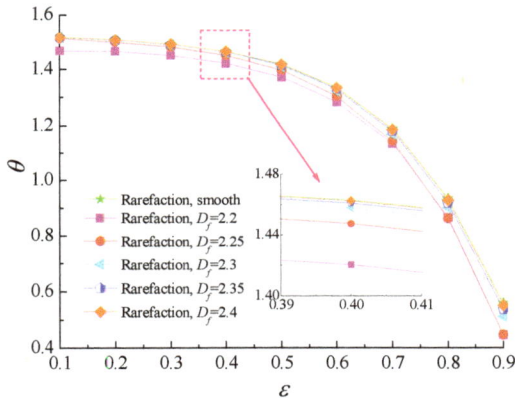

Figure 7. Attitude angle versus eccentricity ratio for different fractal dimensions at $\Lambda = 20$, $G = 1 \times 10^{-11}$.

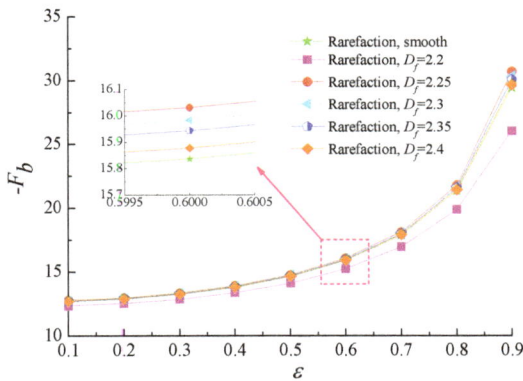

Figure 8. Static friction coefficient versus eccentricity ratio for different fractal dimensions at $\Lambda = 20$, $G = 1 \times 10^{-11}$.

Figures 9 and 10 describe the effect of increasing bearing number on the load carrying capacity and attitude angle with various fractal dimensions for $\varepsilon = 0.6$ and $G = 1 \times 10^{-10}$. It is found that increasing the values of the bearing number from $\Lambda = 3$ up to $\Lambda = 100$ increases the carrying capacity and reduces the corresponding attitude angle of gas journal microbearing. The surface roughness effect in aerodynamic lubrication enhances the load capacity as compared with the smooth-bearing case, especially for the bearing operating at high bearing number. Figure 11 shows the comparison of the static friction coefficients with bearing number for different fractal dimensions. The friction coefficients exhibit a near-linear increasing trend with increasing Λ, while the increase extent in static friction coefficient is even higher at smaller D_f values. Consequently, the strengthened gas-lubricated hydrodynamic effect and the bearing surface with roughness undulations have a significant influence on the skin friction at the interface.

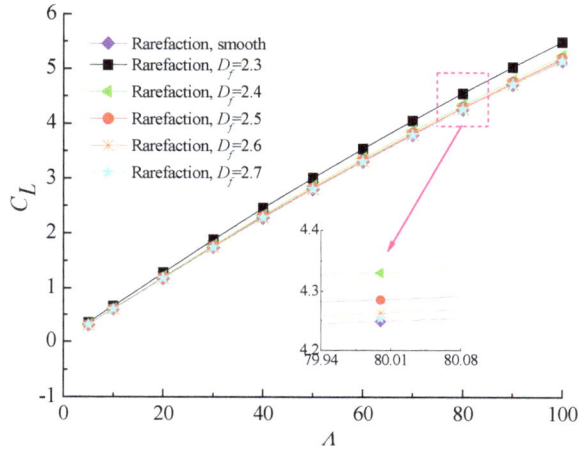

Figure 9. Non-dimensional load capacity versus bearing number for different fractal dimensions at $\varepsilon = 0.6$, $G = 1 \times 10^{-10}$.

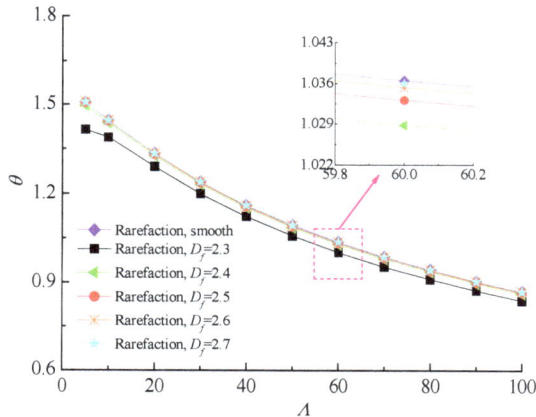

Figure 10. Attitude angles versus bearing number for different fractal dimensions at $\varepsilon = 0.6$, $G = 1 \times 10^{-10}$.

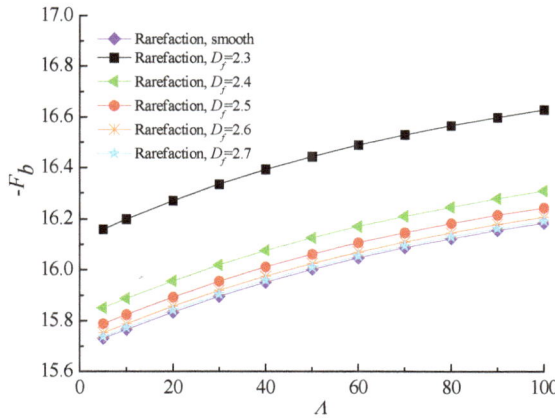

Figure 11. Static friction coefficient versus bearing number for different fractal dimensions at $\varepsilon = 0.6$, $G = 1 \times 10^{-10}$.

4.3. Dynamic Stiffness and Damping Coefficients

The variation of dynamic stiffness and damping coefficients with dimensionless perturbation frequency Ω for different values of fractal dimensions is plotted in Figures 12 and 13. The principal stiffness coefficients K_{xx} and K_{yy} increases as the perturbation frequency increases and the K_{yy} is much greater than K_{xx} because of the rarefied gas lubricating film supports the weight of journal in the vertical direction. The cross-couple stiffness K_{xy} increases at first, then decreases slightly with the growth of Ω, while K_{yx} decreases quickly at lower perturbation frequencies. Furthermore, the enhanced dynamic stiffness coefficients are seen for the rough bearing surface as compared to that of smooth bearing case. When $\Omega > 2$, all the dynamic stiffness coefficients of micro gas-lubricated journal bearing at the fractal dimension $D_f = 2.3$ are obviously greater than other rough surfaces. This is mainly due to the fact that the larger asperity heights lead to an increased Poiseuille flow component along the sliding direction and the side flow suffers the constriction resistance caused by homogeneous surface roughness. The influence of perturbation frequency on dynamic damping coefficients for different fractal roughness parameters can be observed from Figure 13. The principal damping coefficient D_{xx} first increases with increasing dimensionless perturbation frequency, reaches a maximum, then starts to decline slowly. The absolute values of the cross-coupling terms of damping coefficients D_{xy} and D_{yx} decrease quickly at low Ω, then approaches to zero. The result show that the damping coefficients D_{xx}, D_{yx} increase with the increase in the isotropic and homogeneous roughness heights of gas slider bearing surface, and the D_{xy} and D_{yy} are first decreases and then dramatically increases as the fractal dimension D_f decreases, whereas the difference in damping coefficients of rough and smooth cases appear to converge at higher values of Ω.

Figures 14 and 15 display the relationship between the dynamic coefficients of gas microbearing and eccentricity ratio with various fractal dimensions. It is found that the dynamic stiffness coefficients increase as the eccentricity ratio increases for fixed values of $\Lambda = 20$ and $G = 1 \times 10^{-11}$. The higher ε corresponds to the thinner gas film thickness, which results in the increase of Knudsen number K_n. With the growth of the fractal surface roughness, the stiffness coefficients first increase gradually, then decrease significantly at the same ε. As illustrated in Figure 15, the damping coefficients increase marginally as the eccentricity ratio increases and the effect of fractal dimension on damping coefficient D_{xy} is negligible at lower eccentricity ratios. The damping coefficients become more sensitive to surface roughness at higher eccentricity ratios about $\varepsilon > 0.7$. It can be seen that increasing the random roughness heights increase the effect of gas rarefaction in small spacing.

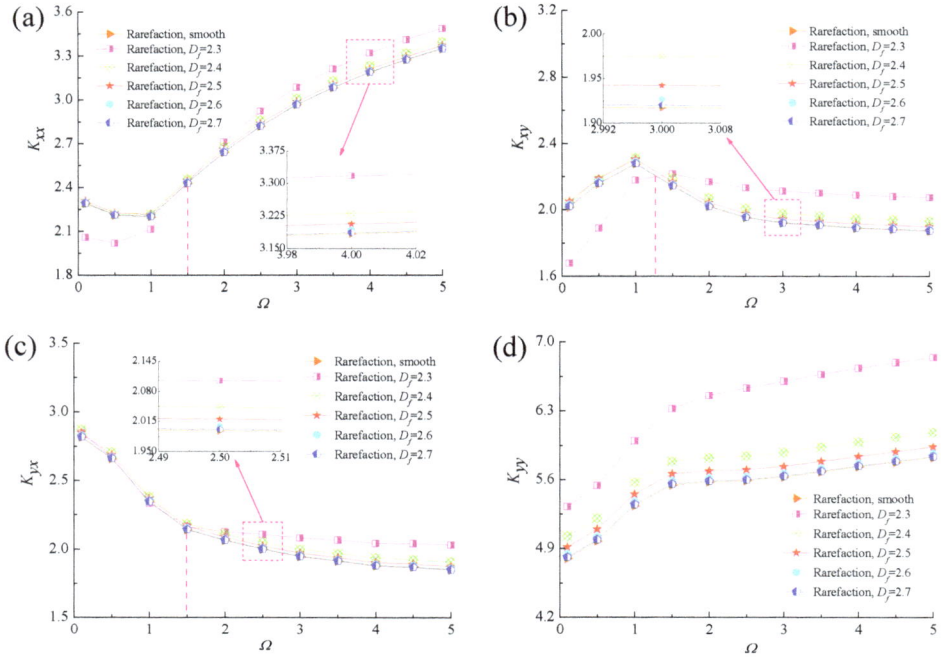

Figure 12. Effect of perturbation frequency on dynamic stiffness coefficients with different fractal dimensions. (**a**) K_{xx} vs. Ω; (**b**) K_{xy} vs. Ω; (**c**) K_{yx} vs. Ω; (**d**) K_{yy} vs. Ω.

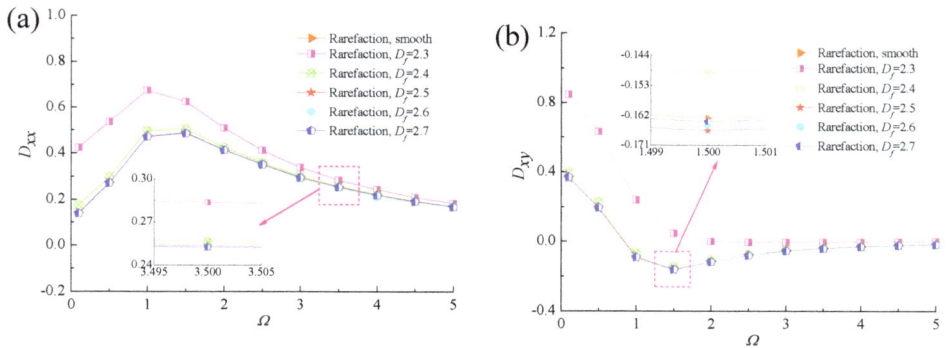

Figure 13. *Cont.*

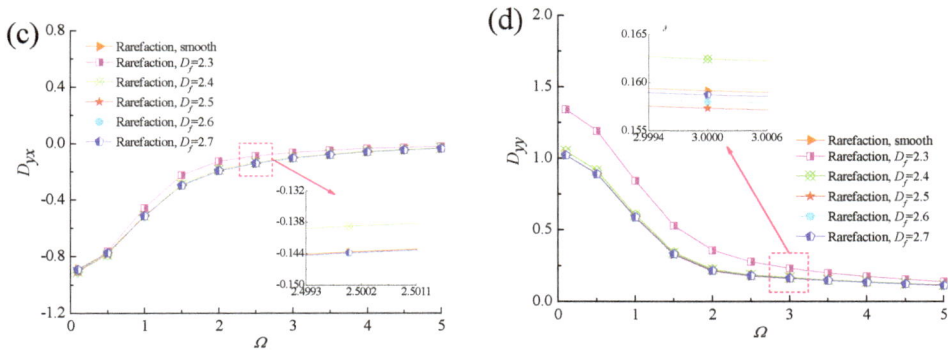

Figure 13. Effect of perturbation frequency on dynamic damping coefficients with different fractal dimensions. (**a**) D_{xx} vs. Ω; (**b**) D_{xy} vs. Ω; (**c**) D_{yx} vs. Ω; (**d**) D_{yy} vs. Ω.

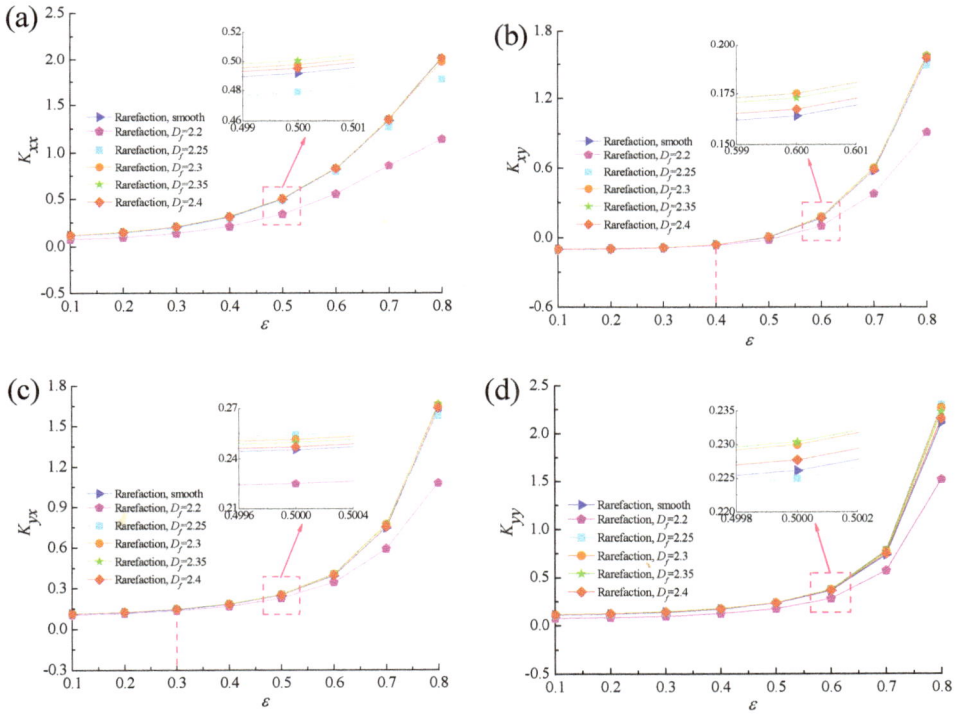

Figure 14. Effect of eccentricity ratio on dynamic stiffness coefficients with different fractal dimensions. (**a**) K_{xx} vs. ε; (**b**) K_{xy} vs. ε; (**c**) K_{yx} vs. ε; (**d**) K_{yy} vs. ε.

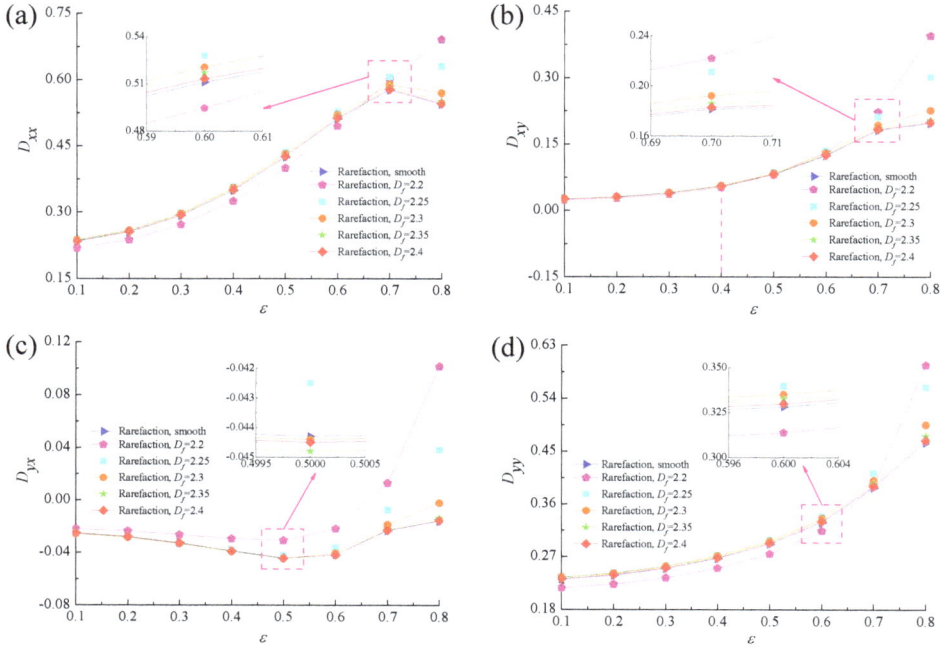

Figure 15. Effect of eccentricity ratio on dynamic damping coefficients with different fractal dimensions. (**a**) D_{xx} vs. ε; (**b**) D_{xy} vs. ε; (**c**) D_{yx} vs. ε; (**d**) D_{yy} vs. ε.

Figures 16 and 17 show the comparisons of the dynamic characteristics between rough and smooth bearing surfaces with different bearing numbers for the fixed values of $\varepsilon = 0.6$ and $G = 1 \times 10^{-10}$. Increment of the bearing number means the larger operating conditions. The principal stiffness K_{yy} is near proportionally dependent on Λ and dynamic stiffness coefficients K_{xx}, K_{xy} and K_{yx} increase gradually with increasing bearing number. It is also observed that the increase in the dynamic stiffness coefficients is more accentuated for the fractal roughness surface as compared to the smooth surface bearing with the enhanced aerodynamic effect in gas journal bearings. The damping coefficients exhibit similar trends to the fractal dimension, namely the principal damping D_{xx}, D_{yy} and cross-couple damping D_{xy} for rough bearing surface become larger than the ones in the smooth bearing, whereas the roughness effect is rather marginal in the case of the cross-couple damping D_{yx} in Figure 17c. The damping coefficients D_{xx}, D_{yy} and D_{xy} increase quickly at first and then decreased, while the damping coefficient D_{yx} decreases with increasing bearing number. Therefore, the dynamic stiffness and damping characteristics of gas microbearings with isotropic and homogeneous roughness show relatively high values since the aerodynamic effect of the bearing is enhanced.

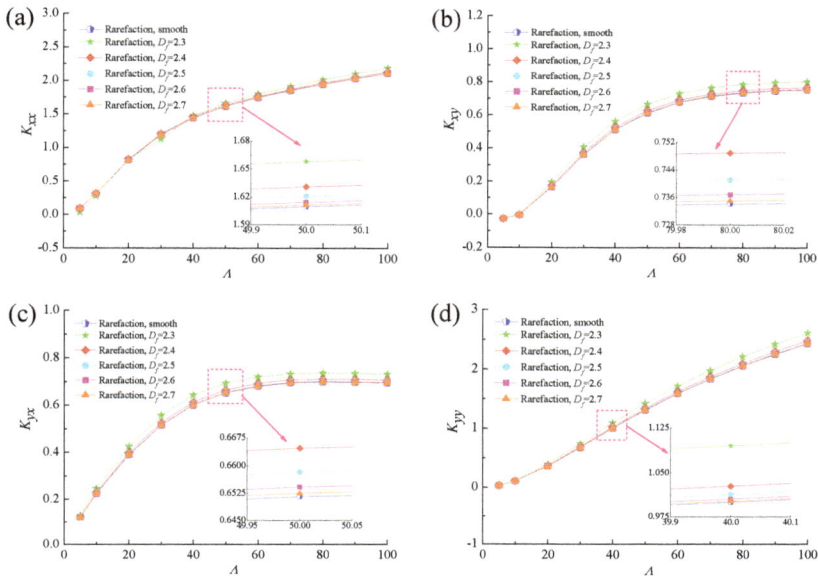

Figure 16. Effect of bearing number on dynamic stiffness coefficients with different fractal dimensions. (**a**) K_{xx} vs. Λ; (**b**) K_{xy} vs. Λ; (**c**) K_{yx} vs. Λ; (**d**) K_{yy} vs. Λ.

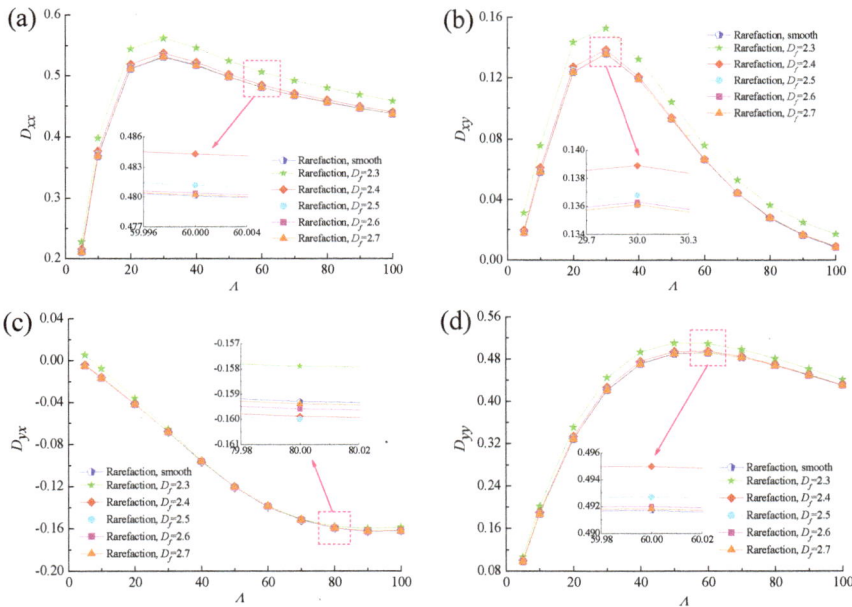

Figure 17. Effect of bearing number on dynamic damping coefficients with different fractal dimensions. (**a**) D_{xx} vs. Λ; (**b**) D_{xy} vs. Λ; (**c**) D_{yx} vs. Λ; (**d**) D_{yy} vs. Λ.

5. Conclusions

The coupled effects of gas rarefaction and surface roughness on the static and dynamic characteristics of gas microbearings are studied using fractal geometry theory and Boltzmann model for Poiseuille flow. Based on the results mentioned in previous section, the following conclusions have been drawn:

1. Surface roughness and gaseous rarefaction effects are of great importance to the lubrication performance of gas journal microbearings. At small asperity height distributions, the gas film pressure distributions increase for higher bearing numbers and eccentricity ratios, consequently leads to an increment of load-carrying capacity.

2. The attitude angle increases with the larger fractal dimensions compared with smooth surface, whereas the skin friction coefficient yields a reversed trend because of the long narrow ridges and furrows impose a series of constrictions on gas lubricant in the sliding direction.

3. The principal stiffness coefficients increase with the increase in dimensionless perturbation frequency and all the stiffness coefficients become larger and larger at higher eccentricity ratios and bearing numbers. The damping coefficients increase as the eccentricity ratio increases and the principal damping can be first magnified and then diminished with increasing bearing number.

4. The stiffness coefficients increase as the fractal dimension decreases at the same perturbation frequency and bearing number, while the stiffness coefficients first increase and then decrease for increasing eccentricity ratio. The damping coefficients increase with decreasing fractal dimension except for the D_{yx} at large bearing number values.

5. It seems that the fractal rough surface does not merely increases the steady-state and dynamic performance of gas microbearings. When the degree of rarefaction effect is more pronounced, the static load capacity and dynamic coefficients decrease quickly for larger values of distribution of asperity heights.

Author Contributions: Data curation, H.X.; Formal analysis, Y.W.; Investigation, Y.W.; Project administration, L.Y.; Software, T.X.; Validation, Y.W.; Writing—original draft, Y.W.; Writing—review & editing, Y.W.

Funding: This research was funded by the National Science Foundation of China (grant numbers 51575425, 11872288).

Conflicts of Interest: The authors declare no conflict of interest.

References

1. Prakash, J.; Sinha, P. Lubrication theory for micropolar fluids and its application to a journal bearing. *Int. J. Eng. Sci.* **1975**, *13*, 217–232. [CrossRef]
2. Lee, Y.B.; Kwak, H.D.; Kim, C.H.; Jang, G.H. Analysis of Gas-Lubricated Bearings with a Coupled Boundary Effect for Micro Gas Turbine. *ASLE Trans.* **2001**, *44*, 685–691. [CrossRef]
3. Piekos, E.S.; Breuer, K.S. Manufacturing Effects in Microfabricated Gas Bearings: Axially Varying Clearance. *J. Tribol.* **2002**, *124*, 815. [CrossRef]
4. Kleinstreuer, C.; Koo, J. Computational Analysis of Wall Roughness Effects for Liquid Flow in Micro-Conduits. *J. Fluids Eng.* **2004**, *126*, 1–9. [CrossRef]
5. Kim, D.; Lee, S.; Bryant, M.D.; Ling, F.F. Hydrodynamic Performance of Gas Microbearings. *J. Tribol.* **2004**, *126*, 711–718. [CrossRef]
6. Zhe, W.; Tong, Y.; Wang, Y. Promoting effect of silicon particles on gas-diffusion-reaction system: In-situ synthesis of AlN in Al-Si materials. *J. Alloys Compd.* **2017**, *735*, 13–25.
7. Zhe, W.; Wang, Y. Impact of convection-diffusion and flow-path interactions on the dynamic evolution of microstructure: Arc erosion behavior of Ag-SnO$_2$ contact materials. *J. Alloys Compd.* **2019**, *774*, 1046–1058.
8. Burgdorfer, A. The influence of the molecular mean free path on the performance of hydrodynamic gas lubricated bearings. *Trans. ASME* **1959**, *88*, 94.
9. Hsia, Y.T.; Domoto, G.A.; Hsia, Y.T.; Domoto, G.A. An Experimental Investigation of Molecular Rarefaction Effects in Gas Lubricated Bearings at Ultra-Low Clearances. *J. Tribol.* **1983**, *105*, 120–130. [CrossRef]
10. Mitsuya, Y. Modified Reynolds Equation for Ultra-Thin Film Gas Lubrication Using 1.5 Order Slip-Flow Model and Considering Surface Accommodation Coefficient. *Trans. Jpn. Soc. Mech. Eng.* **1992**, *58*, 3341–3346. [CrossRef]

11. Fukui, S.; Kaneko, R. A Database for Interpolation of Poiseuille Flow Rates for High Knudsen Number Lubrication Problems. *ASME J. Tribol.* **1990**, *112*, 78–83. [CrossRef]

12. Fukui, S.; Kaneko, R. Analysis of Ultra-Thin Gas Film Lubrication Based on Linearized Boltzmann Equation: First Report—Derivation of a Generalized Lubrication Equation Including Thermal Creep Flow. *J. Tribol.* **1988**, *110*, 253–261. [CrossRef]

13. Christensen, H.; Tonder, K. The Hydrodynamic Lubrication of Rough Bearing Surfaces of Finite Width. *J. Tribol.* **1971**, *93*, 324. [CrossRef]

14. Christensen, H.; Tonder, K. The Hydrodynamic Lubrication of Rough Journal Bearings. *Trans. ASME* **1973**, *95*, 166. [CrossRef]

15. Patir, N. A numerical procedure for random generation of rough surfaces. *Wear* **1978**, *47*, 263–277. [CrossRef]

16. Patir, N.; Cheng, H.S. Application of Average Flow Model to Lubrication Between Rough Sliding Surfaces. *Trans. ASME* **1979**, *101*, 220. [CrossRef]

17. Tripp, J.H. Surface Roughness Effects in Hydrodynamic Lubrication: The Flow Factor Method. *J. Lubr. Technol.* **1983**, *105*, 458–465. [CrossRef]

18. White, J.W.; Raad, P.E.; Tabrizi, A.H.; Ketkar, S.P.; Prabhu, P.P. A Numerical Study of Surface Roughness Effects on Ultra-Thin Gas Films. *J. Tribol.* **1986**, *108*, 171. [CrossRef]

19. Li, W.L.; Weng, C.; Hwang, C.C. Effects of roughness orientations on thin film lubrication of a magnetic recording system. *J. Phys. D Appl. Phys.* **1995**, *28*, 1011. [CrossRef]

20. Turaga, R.; Sekhar, A.S.; Majumdar, B.C. The effect of roughness parameter on the performance of hydrodynamic journal bearings with rough surfaces. *Tribol. Int.* **1999**, *32*, 231–236. [CrossRef]

21. Naduvinamani, N.B.; Siddangouda, A. Effect of surface roughness on the hydrodynamic lubrication of porous step-slider bearings with couple stress fluids. *Tribol. Int.* **2007**, *40*, 780–793. [CrossRef]

22. Zhang, W.M.; Meng, G.; Wei, K.X. Numerical Prediction of Surface Roughness Effect on Slip Flow in Gas-Lubricated Journal Microbearings. *Tribol. Trans.* **2012**, *55*, 71–76. [CrossRef]

23. Zhang, W.M.; Meng, G.; Peng, Z.K. Gaseous slip flow in micro-bearings with random rough surface. *Int. J. Mech. Sci.* **2013**, *68*, 105–113. [CrossRef]

24. Lin, J.R.; Hung, T.C.; Chou, T.L.; Liang, L.J. Effects of surface roughness and non-Newtonian micropolar fluids on dynamic characteristics of wide plane slider bearings. *Tribol. Int.* **2013**, *66*, 150–156. [CrossRef]

25. Jao, H.C.; Chang, K.M.; Chu, L.M.; Li, W.L. A Modified Average Reynolds Equation for Rough Bearings With Anisotropic Slip. *J. Tribol.* **2016**, *138*, 011702. [CrossRef]

26. Kalavathi, G.K.; Dinesh, P.A.; Gururajan, K. Influence of roughness on porous finite journal bearing with heterogeneous slip/no-slip surface. *Tribol. Int.* **2016**, *102*, 174–181. [CrossRef]

27. Quiñonez, A.F.; Morales-Espejel, G.E. Surface roughness effects in hydrodynamic bearings. *Tribol. Int.* **2016**, *98*, 212–219. [CrossRef]

28. Wang, X.; Qiao, X.; Ming, H.; Zhang, L.; Peng, Z. Effects of journal rotation and surface waviness on the dynamic performance of aerostatic journal bearings. *Tribol. Int.* **2017**, *112*, 1–9. [CrossRef]

29. Mandelbrot, B. How long is the coast of britain? Statistical self-similarity and fractional dimension. *Science* **1967**, *156*, 636–638. [CrossRef] [PubMed]

30. Hwang, C.C.; Fung, R.F.; Yang, R.F.; Weng, C.I.; Li, W.L. A new modified Reynolds equation for ultrathin film gas lubrication. *IEEE Trans. Magn.* **1996**, *32*, 344–347. [CrossRef]

31. Duwensee, M. Numerical and Experimental Investigations of the Head/Disk Interface. Ph.D. Thesis, The University of California, San Diego, CA, USA, 2007.

32. Yang, L.; Li, H.; Yu, L. Dynamic stiffness and damping coefficients of aerodynamic tilting-pad journal bearings. *Tribol. Int.* **2007**, *40*, 1399–1410.

33. Yang, L.; Qi, S.; Yu, L. Analysis on Dynamic Performance of Hydrodynamic Tilting-Pad Gas Bearings Using Partial Derivative Method. *J. Tribol.* **2009**, *131*, 011703.

34. Lie, Y.U.; Shemiao, Q.I.; Geng, H. A generalized solution of elasto-aerodynamic lubrication for aerodynamic compliant foil bearings. *Sci. China Ser. E Eng. Mater. Sci.* **2005**, *48*, 414–429.

micromachines

MDPI

Communication

Estimation of Air Damping in Out-of-Plane Comb-Drive Actuators

Ramin Mirzazadeh [†] and Stefano Mariani *

Department of Civil and Environmental Engineering, Politecnico di Milano, Piazza L. da Vinci 32, 20133 Milano, Italy; ramin.mirzazadeh@polimi.it
* Correspondence: stefano.mariani@polimi.it; Tel.: +39-02-2399-4279
† Currently with: Engineering Ingegneria Informatica S.p.A., 20090 Assago, Italy.

Received: 6 February 2019; Accepted: 16 April 2019; Published: 19 April 2019

Abstract: The development of new compliant resonant microsystems and the trend towards further miniaturization have recently raised the issue of the accuracy and reliability of computational tools for the estimation of fluid damping. Focusing on electrostatically actuated torsional micro-mirrors, a major dissipation contribution is linked to the constrained flow of air at comb fingers. In the case of large tilting angles of the mirror plate, within a period of oscillation the geometry of the air domain at comb-drives gets largely distorted, and the dissipation mechanism is thereby affected. In this communication, we provide an appraisal of simple analytical solutions to estimate the dissipation in the ideal case of air flow between infinite plates, at atmospheric pressure. The results of numerical simulations are also reported to assess the effect on damping of the finite size of actual geometries.

Keywords: resonant micro-electromechanical-systems (MEMS); micro-mirrors; out-of-plane comb actuation; fluid damping; analytical solution; FE analysis

1. Introduction

Starting from the late 1990s, the fields of application of micro-electromechanical-systems (MEMS) have been progressively expanding [1,2], leading to state-of-the-art devices like accelerometers and gyroscopes [3], gas sensors [4], bio-MEMS [5], and micro-opto-electromechanical-systems (MOEMS) [6], which are now available at low cost.

For a resonating system, energy dissipation plays an important role in modulating its response; such dissipation is customarily measured via the quality factor of the structure. High-resonance frequencies, achievable with miniaturization, and low dissipation can provide a better system response; nevertheless, moving from the macro-scale to the micro-scale, surface effects such as the viscous and electrostatic effects become dominant and need to be accurately quantified.

Comb-actuated resonant micro-mirrors (see, e.g., [7,8]) are a good example of systems designed to work at high frequencies to attain high scanning resolutions, despite the considerable dissipation arising from the interaction of the resonating structure with the surrounding air at atmospheric pressure. Besides the need to interact with the optical beams, these devices are preferred to operate at atmospheric pressure since fluid damping can avoid the risks of excessive dynamic amplifications and sharp frequency responses, and so ultimately of possible structural failures. The fluid behavior and, consequently, the dissipation induced on such devices might not be the same as those at the macroscale, due to rarefaction effects [2,9–11]. By considering the steady-state flow of a thin gas film between oscillating parallel plates, boundary conditions (BCs) should be modified to consider a finite slip at the fluid-structure interface. A so-called first-order finite slip model was reported in [12] to account rather accurately for the rarefaction effects at large values of the Knudsen number Kn, which is defined in the considered case as the ratio between the mean free path of air molecules and the width of the film. Even for geometries and working pressures leading to values of Knudsen number approaching $Kn = 1$, such slip BCs give rise to errors in terms of damping smaller than 10%.

For micro-mirrors actuated by comb finger arrays, the damping observed between two closely placed plates having a relative sliding motion is of major importance, due to the small gap between rotors and stators for actuation purposes. This type of fluid damping at the microscale was already investigated, e.g., in [13,14], by running experiments on parallel plates and modelling the relevant dissipation mechanisms; an empirical formula was also proposed in [15] for specific geometries. In [16,17], the Stokes equations were numerically solved for three-dimensional geometries via a fast Fourier transform (FFT) boundary element method, considering both the standard no-slip, or stick BCs and the rarefaction effects (see also [18]). These studies and those based on Navier–Stokes modeling approaches (see, e.g. [19,20]) focused on damping in the case of relatively small amplitudes of oscillations. For scanning reasons, micro-mirrors are instead designed to have large tilting angles (in the order of 15° or even more), which alter the fluid flow in comparison to the small oscillation case: the flow turns out to be far more complex and needs to be modelled numerically.

In order to provide an estimation for the real solution in terms of damping, namely, in terms of the ratio between the energy loss and the maximum energy stored within a period of oscillation of the mechanical system, a simplified analytical model is here discussed and validated against finite element simulations. Referring to the free vibrations of a (linear) system, in the ideal case of no damping the energy conservation law states that there must be a continuous switch between a purely internal or elastic contribution (when, e.g., the torsional velocity of the considered micro-mirror is zero) and a purely kinetic one (when, e.g., the torsional angle is zero). In the presence of damping, the sum of the elastic and kinetic contributions decays in time; in the viscous case, an effective damping parameter d quantifies such decay, as induced by the surrounding air on the motion of the movable structure. This parameter can be combined with the effective mass and stiffness of the structure, to provide the aforementioned quality factor.

Moving from already reported analyses, a solution is presented in Section 2 for the constrained flow of a fluid in a narrow gap between two parallel infinite plates. Two different solutions are detailed: a so-called disengaged one, when both plates are moving in-phase; and a so-called engaged one, when only one plate moves. Keeping aside the possible effects of finite slip, in the former case the velocity of the fluid is imposed equal at the boundaries with both the lateral surfaces; in the latter case, the velocity is instead non-zero at the boundary with one lateral surface only. These two solutions have been considered representative of the flow conditions occurring at the comb fingers of a compliant torsional micro-mirror, as schematically depicted in Figure 1: terms engaged and disengaged, respectively, refer to the solution corresponding to the torque angle being zero or maximum. In the engaged case (Figure 1a), rotors and stators are perfectly inter-digitated; in the disengaged case (Figure 1c), the local out-of-plane motion of the rotor does not allow the two sets of surfaces to directly interact. All the other conditions between these two extrema provide a partial engagement of the surfaces; accordingly, solutions for the studied cases provide a kind of bilateral bounding on any possible real configuration occurring during the torsional vibration of the moving structure. The estimated values of the damping coefficient in the two conditions are discussed in Section 3, at varying frequency of the oscillations; further to that, the effect of a finite size of the plates is evaluated numerically, so that charts can be made available for the estimation of dissipation in any geometry. Some concluding remarks are gathered in Section 4.

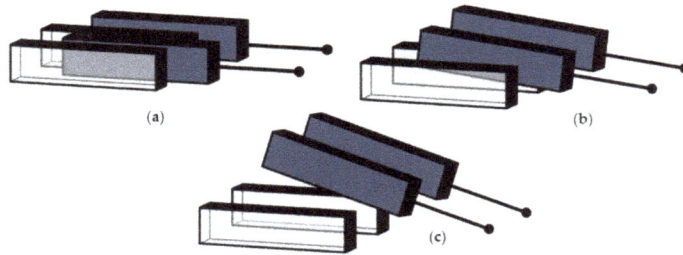

Figure 1. Schematics of stator/rotor configurations at comb fingers, during the torsional motion of a micro-mirror: (**a**) engaged, (**b**) partially engaged, and (**c**) disengaged phases.

2. Theoretical Analysis of the Constrained Shear Flow

Let two flat surfaces, bounding a gas film, be separated by a gap H and move in a direction parallel to each other. Referring to the configurations depicted in Figure 1, this geometry is relevant to a cross section of the comb-drives; the torsional motion of the micro-mirror induces a sliding between the lateral surfaces of rotor and stator, measured by the considered motion along the x-axis.

Starting from the Navier–Stokes equations, by neglecting the inertial forces and assuming that a quasi-static flow forms between the plates, a frequency-independent solution can be obtained. Here, we are instead interested in a vibrating structure, and so a time- and frequency-dependent velocity field must be computed for the fluid film. Some assumptions and restrictions are introduced to deal with the problem analytically: (i) the amplitude of oscillations is small in comparison with H; (ii) the thickness and length of the plates are large with respect to H; (iii) the velocity of the surfaces is low enough to prevent heating of the fluid; (iv) surfaces are ideally smooth at the molecular level, so that the angles of incidence and reflection of gas molecules colliding with them are identical, and the relevant tangential momentum does not change. The effects of assumption (ii) will be specifically assessed in Section 3, via numerical simulations dealing with finite plate geometries; assumption (iii) is instead introduced to prevent any additional thermal dissipation effect. Due to assumption (i), a switch between the engaged and disengaged configurations cannot be dealt with within a single solution; we therefore study the two different cases separately, in order to provide an estimation of the dissipation valid for small perturbations within the two regimes. The analytical model is thus incapable to provide an assessment of dissipation coping with the structural vibrations within a complete cycle.

In the reference frame depicted in Figure 2, with the x-axis aligned with the motion of the fluid and the y-axis perpendicular to the moving surface(s), the dynamics of the gas between the plates are modeled with the following one-dimensional diffusion equation [20,21]:

$$\frac{\partial u(y)}{\partial t} = v\frac{\partial^2 u(y)}{\partial y^2} \tag{1}$$

where u is the fluid velocity, function of the y coordinate only in the present ideal case; t is time; and $v = \mu/\rho$ is the kinematic viscosity of the fluid, μ its coefficient of dynamic viscosity, and ρ its density. In Equation (1), it has been assumed that the velocity of the oscillating plate(s) and the frequency of oscillations are such that air can be treated as incompressible, see [22,23]; recall also that a limitation on the value of the velocity u is enforced by assumption (iii) [12].

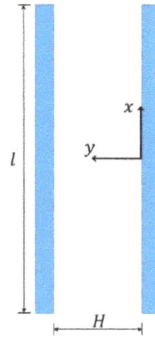

Figure 2. Lateral view of the parallel plates and of the considered fluid film in between: notation.

Equation (1) is solved in the case of a steady-state sinusoidal excitation at one or both borders of the film, featuring an amplitude $\bar{u}$ and a circular frequency ω. The solution, therefore, can be written as follows:

$$u(y) = C_1 \sinh(qy) + C_2 \cosh(qy) \tag{2}$$

where $q = \sqrt{j\omega/\nu}$ is the complex frequency variable and j is the imaginary unit. Constants C_1 and C_2 vary, depending on the BCs at the borders; solutions are therefore distinguished next, to deal with the disengaged and engaged cases separately.

The shear stress τ acting on the rotor (moving) wall is given by

$$\tau = -\mu \left. \frac{\partial u(y)}{\partial y} \right|_{y=H} \tag{3}$$

where the minus sign depends on the relative orientation of surface motion and velocity gradient, and states that the shear stress actually provides the source of dissipation. The quantity

$$\xi = \frac{\tau}{\bar{u}} \tag{4}$$

that is, the ratio between the shear stress acting on the rotor surface and the imposed surface velocity, is the damping admittance ξ. For a periodic steady-state solution, the real part of this complex quantity yields the actual damping coefficient per unit area.

2.1. Disengaged Solution: Two Moving Surfaces

We investigate first the flow between two plates oscillating in-phase with the same velocity $\bar{u}$. In case of flow at low frequencies, the velocity u turns out to be uniform all over the domain $0 \leq y \leq H$ and equal to the imposed one at the boundaries. As the frequency increases, the inertial forces assume importance and the local velocity becomes dependent on the position too, showing a Poiseuille-type distribution [24].

For stick BCs, namely, for $u(0) = u(H) = \bar{u}$, the velocity distribution in the gap turns out to be

$$u(y) = \bar{u} \left[\frac{1 - \cosh(qH)}{\sinh(qH)} \sinh(qy) + \cosh(qy) \right] \tag{5}$$

For slip BCs, namely, for $u(0) = \bar{u} + \lambda \left.\frac{\partial u(y)}{\partial y}\right|_{y=0}$ and $u(H) = \bar{u} - \lambda \left.\frac{\partial u(y)}{\partial y}\right|_{y=H}$, the velocity gradient $\frac{\partial u(y)}{\partial y}$ at the two surfaces and the mean free path of gas molecules λ affect the solution. Out of the considered ideal specular reflection condition, the solution also depends on the tangential momentum accommodation coefficient (see, e.g., [25]). The relevance of slip BCs is determined by the Knudsen number Kn, or by the ratio between the aforementioned mean free path λ and the characteristic physical length-scale H of the problem: when the two variables become comparable, stick BCs lead to an overestimation of damping, and a correction of continuum methods via slip BCs looks necessary. For this type of BCs, the velocity distribution thus becomes

$$u(y) = \bar{u}[K'\sinh(qy) + (1 - \lambda qK')\cos h(qy)] \tag{6}$$

where

$$K' = \frac{1 - \cosh(qH) - \lambda q\sin h(qH)}{\sinh(qH) + 2\lambda q\cos h(qH) + \lambda^2 q^2 \sinh(qH)} \tag{7}$$

Flows are studied for $H = 3$ and 12 µm; these values of the gap have been chosen as being representative of the micro-mirror geometry considered in [7,8]. As already discussed, micro-mirrors typically work at ambient pressure, so that $\lambda = 69$ nm, $\mu = 18.3 \times 10^{-6}$ kg/ms, $\rho = 1.185$ kg/m³.

Results are provided in Figure 3, in terms of the profile of the maximum amplitude of the velocity between the two oscillating surfaces. Having defined the corner frequency as $f_d = \nu/(2\pi H^2)$ [12], for which $|qH| = 1$, it can be seen that for frequencies smaller than f_d the inertial forces do not have any effect on the flow, and the fluid moves uniformly with the same velocity of the bounding surfaces. For frequencies larger than f_d, the inertial forces change the velocity profile to approach one mimicking a pressure-driven Poiseuille flow. The ratio f/f_d, also termed normalized frequency in the following, can therefore be considered as an implicit measure of the Reynolds number of the solution. At even higher frequencies, inertial forces get dominant and only a small part of the gas film, close to the moving surfaces, moves along: this portion of the gap gets narrower as the frequency increases. While the shape of the profiles is not affected by the value of H, it can be seen that the difference between the solutions relevant to stick and slip BCs becomes negligible for $H = 12$ µm. It must be anyway kept in mind that, by increasing the gap from $H = 3$ µm to $H = 12$ µm, the corner frequency decreases from $f_d = 323,615$ Hz to $f_d = 20,225$ Hz; this means that the inertial forces provide consistent effects on the solution for lower frequencies, in case of larger gaps.

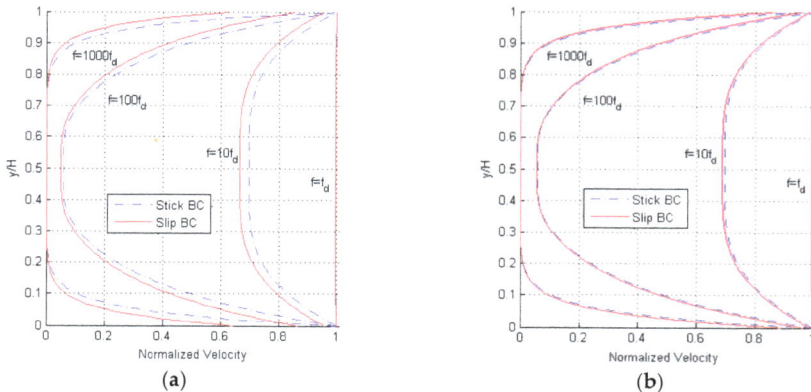

Figure 3. Disengaged solution, (a) $H = 3$ µm and (b) $H = 12$ µm: maximum amplitude of the velocity profile at varying frequency f of oscillation of the two plates, for both stick and slip boundary conditions (BCs).

In Figure 4, snapshots of the velocity profile during a cycle of the steady-state flow are depicted; these curves are shown for a frequency $f = 100f_d$, to testify the effects of inertia. Once more, it can be seen that the effects of slip BCs become negligible for $H = 12$ μm. The overall shape of the profile is again unaffected by H, as the solution is scale-independent in this dimensionless representation.

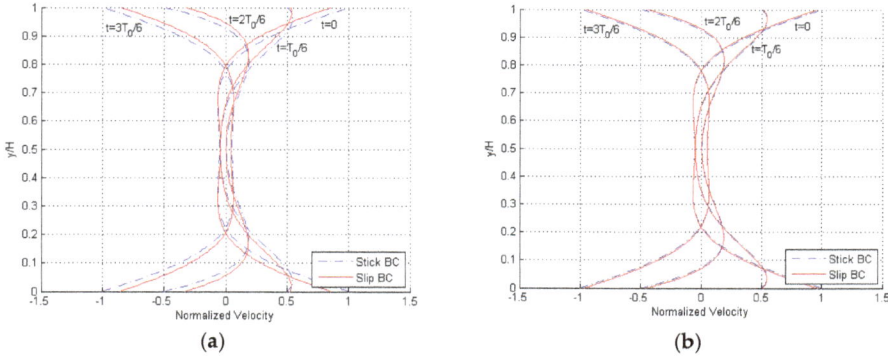

Figure 4. Disengaged solution, (**a**) $H = 3$ μm and (**b**) $H = 12$ μm: snapshots of the velocity profile during a single cycle of oscillation of the two plates, for $f = 100f_d$ and both stick and slip BCs.

2.2. Engaged Solution: One Moving Surface

We investigate now the flow between the same two plates, assuming the surface at $y = 0$ to be fixed and the one at $y = H$ to move as before with an assigned velocity $\bar{u}$. In the case of stick BCs, namely, for $u(0) = 0$, $u(H) = \bar{u}$, the velocity distribution reads as follows:

$$u(y) = \bar{u}\frac{\sinh(qy)}{\sinh(qH)} \tag{8}$$

For slip BCs, namely, for $u(0) = \lambda \left.\frac{\partial u(y)}{\partial y}\right|_{y=0}$ and $u(H) = \bar{u} - \lambda \left.\frac{\partial u(y)}{\partial y}\right|_{y=H}$, the velocity profile results instead are as follows [12]:

$$u(y) = \bar{u}\frac{\sinh(qy) + q\lambda\cos h(qy)}{(1 + q^2\lambda^2)\sinh(qH) + 2q\lambda\cos h(qH)} \tag{9}$$

Results are reported in Figure 5 in terms of the profile of the maximum amplitude of the velocity at varying frequency of oscillations of the plate at $y = H$, for $f \geq f_d$ and both stick and slip BCs. Some snapshots within a single cycle of the steady-state solution are depicted in Figure 6, for $f = 100f_d$. Profiles are again provided for $H = 3$ and 12 μm, and, in accordance with the disengaged solution, it can be seen that by increasing the gap, the effects of the velocity slip at the boundaries get smaller, becoming almost negligible for the larger value. The only feature to notice in the graphs, at variance with the former solution, can be observed at $y = 0$: for frequencies much larger than f_d, the gradient of the solution $u(y)$ close to the fixed surface becomes zero and so the velocity slip at the boundary becomes null as well. In Figure 6, the solution is thus displayed to be always zero at the bottom of the plots, independently of the time instant within the cycle.

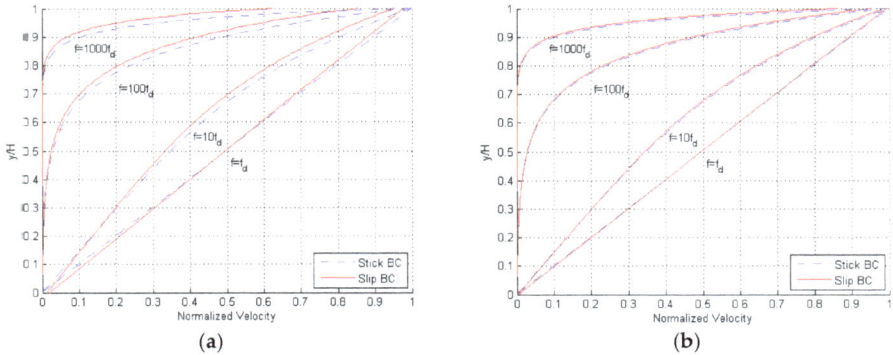

Figure 5. Engaged solution, (**a**) $H = 3$ µm and (**b**) $H = 12$ µm: maximum amplitude of the velocity profile at varying frequency f of oscillation of the moving plate at $y = H$, for both stick and slip BCs.

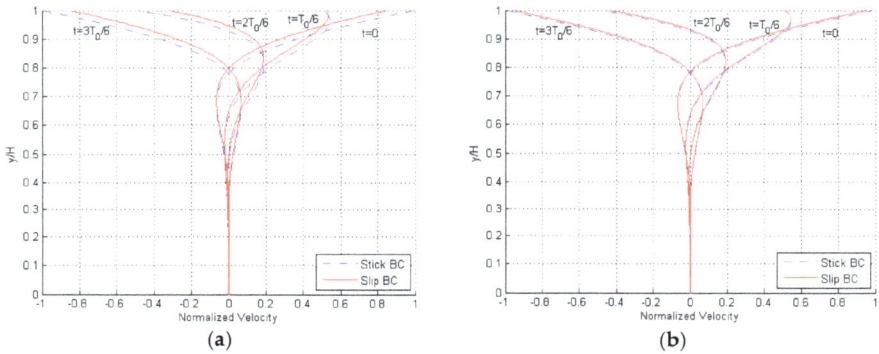

Figure 6. Engaged solution, (**a**) $H = 3$ µm and (**b**) $H = 12$ µm: snapshots of the velocity profile during a single cycle of oscillation of the plate at $y = H$, for $f = 100 f_d$ and both stick and slip BCs.

3. Estimation of the Damping Coefficient

The goal of the analytical model is, as said, an estimation of damping in the considered cases of shear flow. If the real part of the admittance ξ, which is already computed per unit area via Equation (4), is normalized by μ / H, the effects of the oscillation frequency on the damping are as shown in Figure 7 for the engaged case; in these plots, as before, we have considered the two values $H = 3$ and 12 µm for the gap, and both stick and slip BCs. By means of the handled nondimensionalization for the damping coefficient, a direct comparison is provided with the dissipation induced by the Couette flow with no-slip BCs, a linear profile of the fluid velocity along the y-axis, and a constant gradient equal to $\bar{u}/H$. For brevity, results are not reported for the disengaged case, but a comparison between the solutions obtained with the two types of BCs is shown next.

In Figure 7, it can be seen that damping at frequencies smaller than f_d is constant, due to the null effect of the fluid inertia on the shape of the velocity profile. The values of the normalized damping for slip BCs are slightly smaller than those relevant to stick BCs, for all the scanned frequencies; this outcome is obviously linked to the lag in the solution induced by the slip at the boundaries. Additionally, this effect gets almost negligible in the case $H = 12$ µm, except for very high-frequency values, above $1000 f_d$, for which the two curves tend to diverge.

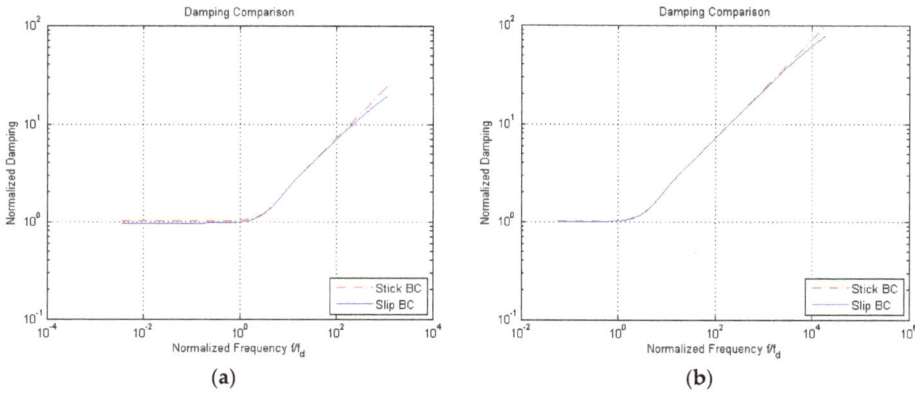

Figure 7. Engaged solution, (a) $H = 3\,\mu m$ and (b) $H = 12\,\mu m$: normalized damping at varying frequency f, for both stick and slip BCs.

Due to the small difference between the solutions relevant to the two types of BCs over a broad range of frequencies, we refer now to the stick BCs only. Figure 8 shows the normalized damping values for the engaged and disengaged solutions (respectively, denoted in the graph as one surface and two surfaces moving), as a function of the vibration frequency f. Damping for the disengaged case is negligible at low frequencies; by increasing the frequency, damping values for the two cases get closer and closer and then show the same trend. This outcome is due to the fact that at high frequencies only a portion of the air film moves with the wall(s), and the solution in terms of the damping coefficient thus becomes independent of the type of BCs.

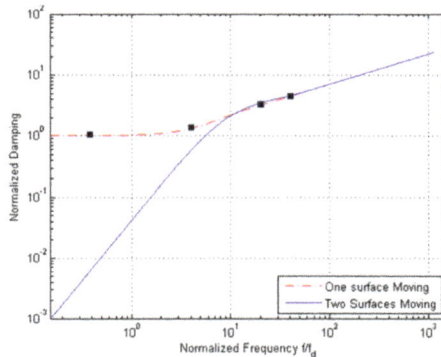

Figure 8. $H = 3\,\mu m$, stick BCs: comparison between the values of normalized damping for engaged and disengaged solutions.

In Figure 8, the black square symbols refer to the data reported in [12]: to validate the solution, results of numerical analyses were reported for the engaged case and a geometry characterized by $l/H = 1000$, allowing for finite slips at the boundaries. Having thereby minimized the fringe effects and set the corner frequency as $f_d = 2.5\,\text{MHz}$, the numerical results turn out to be in good agreement with the analytical solution, featuring a discrepancy of a few % only in the whole inspected frequency range. To also assess the accuracy of the solution in the disengaged case, and the effects of fringe flow on the damping behavior of real geometries with shorter plates, numerical analyses have been carried out. The solution reported in Figure 9 has been computed with ANSYS CFX (v13.0, Ansys Inc.,

Canonsburg, PA, USA) [26] for frequencies equal to $10f_d$, $100f_d$, and $500f_d$, at a varying length of the two plates.

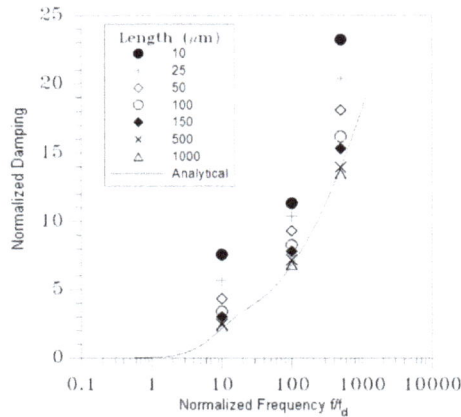

Figure 9. $H = 3\,\mu\text{m}$, disengaged solution, stick BCs: effect of surface length on the normalized damping value.

The finite-length numerical model has been setup with the fluid between the two plates and also surrounding their front and rear surfaces, to catch the aforementioned fringe effects during a cycle of the assigned sinusoidal motion. The width of the additional lateral fluid regions has been selected to attain a width-independent solution, and thus avoid any numerical artifact. Open BCs have been adopted on the two outer sides of the model, to guarantee uniform in and out flows. The space discretization of the fluid domain, for any frequency, has been checked to assure mesh-independent results. To cross-validate the analytical and numerical approaches, a relatively long surface featuring $l = 1000\,\mu\text{m}$ has been first adopted: results in Figure 9 show that the analytical solution is perfectly matched up to frequencies on the order of $100f_d$; the numerically computed damping is instead slightly smaller than the analytical one at $500f_d$, but still within the accuracy margin already reported for the engaged case. Next, a series of simulations has been carried out at varying plate length, or gap over length ratio, in the range $H/l = 0.003 - 0.3$. The computed values of the damping coefficient obviously turn out to be larger for the shorter plates, since air flow is affected more by the fringe flow. Results relevant to $H = 3\,\mu\text{m}$ show that, for $H/l = 0.03$, the damping is increased by 57%, 21%, and 16% at frequencies $10f_d$, $100f_d$, and $500f_d$, respectively. Such an increase with respect to the shear-dependent analytical baseline has to be linked to the pressure-dependent effects related to the front and rear surfaces of the plates, which are not included in the analytical solution; the interplay of these effects was already assessed numerically in, e.g., [22,23].

The reported variation of the damping coefficient turns out to be proportional to the ratio between the gap and the surface length, as for $H/l = 0.12$ it grows up to 163%, 53%, and 49% at the same frequencies. To cope with these variations of damping away from the analytical solution, in [12] an empirical law was proposed for the engaged case, to define an effective gap width in the form of $H^* = H^*(H, H/l)$. This solution may turn out to be case-dependent, since the actual three-dimensional geometry of the moving structure and its kinematics should be fully accounted for in H^*. Alternatively, compact modeling approaches resting on an equivalent electric circuit for the fluid film can be formulated as in [12]; again for the engaged case, this perspective was shown to provide very accurate solutions for frequencies below the corner one, while for higher values some fluctuations were reported for the frequency-dependent damping. In this communication, we do not aim to discuss methods to cope with case-dependent geometries; instead, analytical results related to the dependence of damping on the resonance frequency of the movable structure are intended to provide a simple estimation to

Micromachines **2019**, *10*, 263

help in the design of new devices. More accurate investigations, even supported by charts accounting for the finite geometry of the moving objects and built upon data like those reported in Figure 9, can then allow one to tune the damping of the whole structure and better fit the performances required for specific applications.

4. Conclusions

In this paper, we have provided an appraisal of analytical solutions to estimate energy dissipation in the case of shear flow within a thin air film bounded by two infinite plates. The values of the damping coefficient, measuring energy dissipation in the considered dynamic solution, have been compared with those obtained with finite element simulations, run to assess the effects of a finite length of the moving plate(s) and, thus, of fringe flow. Through these analyses, the amplification of the analytical damping coefficient has been discussed, as a function of the vibration frequency and of the ratio between the plate length and the gap between the plates.

For micro-mirrors characterized by a compliant movable structure due to scanning reasons, we have discussed how the fluid damping can be difficult to compute at comb-drives: as large values of the oscillation angle induce a finite change of the film configuration during each cycle of actuation, mesh updating procedures are necessary in cumbersome finite element simulations [7]. To provide a guideline for the design of new devices, at least for what concerns the dissipation at the said comb-drives, the analytical solutions here detailed can be considered as representative of states close to a null rotation of the micro-mirror (termed engaged case) and close to the maximum tilting angle (termed disengaged case). The analytical solutions relevant to infinitely long films can then be corrected through a set of numerical simulations giving information on how the front and rear sides of each finger affect the dissipation. Referring to the operational conditions and to a target layout, such corrective terms have to be computed only for specific values of the frequency of oscillations.

In the proposed analysis, the geometry of the fluid domain has been considered deterministically known. Stochastic effects induced by micro-fabrication, see, e.g., [27–31], have been therefore disregarded, though they may modify the solution by affecting the gap between rotor and stator. Within the current analytical frame, by partially allowing for the finite geometry of comb-drives, stochastic effects induced by microfabrication can be accounted for rather easily, to provide confidence intervals for the damping coefficient. The mentioned intervals would help in the design of new devices, especially with an eye towards further miniaturization and towards the development of high-frequency piezoelectric resonators, featuring performance indices to be finally compared with case-specific experimental data.

Author Contributions: Authors contributed equally to this work.

Funding: This research was funded by STMicroelectronics.

Conflicts of Interest: The authors declare no conflict of interest.

References

1. KO, W. Trends and frontiers of MEMS. *Sensors Actuators A Phys.* **2007**, *136*, 62–67. [CrossRef]
2. Gad-el-Hak, M. (Ed.) *The MEMS Handbook*; CRC Press: Boca Raton, FL, USA, 2002.
3. Corigliano, A.; Ardito, R.; Comi, C.; Frangi, A.; Ghisi, A.; Mariani, S. *Mechanics of Microsystems*; John Wiley and Sons: Hoboken, NJ, USA, 2018.
4. Kühne, S.; Graf, M.; Tricoli, A.; Mayer, F.; Pratsinis, S.E.; Heirlemann, A. Wafer-level flame-spray pyrolysis deposition of gas-sensitive layers on microsensors. *J. Micromech. Microeng.* **2008**, *18*, 035040. [CrossRef]
5. Ardito, R.; Bertarelli, E.; Corigliano, A.; Gafforelli, G. On the application of piezolaminated composites to diaphragm micropumps. *Compos Struct.* **2013**, *99*, 231–240. [CrossRef]
6. Lee, C.; Yeh, J.A. Development and evolution of MOEMS technology in variable optical attenuators. *J. Micro/Nanolithogr. MEMS MOEMS* **2008**, *7*, 021003.

7. Mirzazadeh, R.; Mariani, S.; Ghisi, A.; De Fazio, M. Fluid damping in compliant, comb-actuated torsional micromirrors. In Proceedings of the 2014 15th International Conference on Thermal, Mechanical and Mulit-Physics Simulation and Experiments in Microelectronics and Microsystems (EuroSimE), Ghent, Belgium, 7–9 April 2014; pp. 1–7.
8. Mirzazadeh, R.; Mariani, S.; De Fazio, M. Modeling of fluid damping in resonant micro-mirrors with out-of-plane comb-drive actuation. In Proceedings of the 1st International Electronic Conference on Sensors and Applications, 1–16 June 2014; pp. 1–16.
9. Jeong, N. Lattice Boltzmann approach for the simulation of rarefied gas flow in the slip flow regime. *J. Mech. Sci. Technol.* **2013**, *27*, 1753–1761. [CrossRef]
10. Bao, M.; Yang, H. Squeeze film air damping in MEMS. *Sensors Actuators A Phys.* **2007**, *136*, 3–27. [CrossRef]
11. Younis, M.I. *MEMS Linear and Nonlinear Statics and Dynamics*; Springer: New York, NY, USA, 2011.
12. Veijola, T.; Turowski, M. Compact damping models for laterally moving microstructures with gas-rarefaction effects. *J. Microelectromech. Syst.* **2001**, *10*, 263–273. [CrossRef]
13. Tang, W.C.; Nguyen, T.-C.H.; Howe, R.T. Laterally driven polysilicon resonant microstructures. *Sens. Actuators* **1989**, *20*, 25–32. [CrossRef]
14. Cho, Y.H.; Kwak, B.M.; Pisano, A.P.; Howe, R.T. Viscous energy dissipation in laterally oscillating planar microstructures: A theoretical and experimental study. In Proceedings of the IEEE Micro Electro Mechanical Systems, Fort Lauderdale, FL, USA, 10 February 1993; pp. 93–98.
15. Zhang, X.; Tang, W. Viscous air damping in laterally driven microresonators. In Proceedings of the IEEE Micro Electro Mechanical Systems An Investigation of Micro Structures, Sensors, Actuators, Machines and Robotic Systems, Oiso, Japan, 25–28 January 1994; pp. 199–204.
16. Ye, W.; Wang, X.; Hemmert, W.; Freeman, D.; White, J. Air damping in laterally oscillating microresonators: A numerical and experimental study. *J. Microelectromech. Syst.* **2003**, *12*, 557–566.
17. Ding, J.; Ye, W. A fast integral approach for drag force calculation due to oscillatory slip Stokes flows. *Int. J. Numer. Meth. Eng.* **2004**, *60*, 0235–1567. [CrossRef]
18. De, S.K.; Aluru, N.R. Coupling of hierarchical fluid models with electrostatic and mechanical models for the dynamic analysis of MEMS. *J. Micromech. Microeng.* **2006**, *16*, 1705–1719. [CrossRef]
19. Braghin, F.; Leo, E.; Resta, F. The damping in MEMS inertial sensors both at high and low pressure levels. *Nonlinear Dynam.* **2008**, *54*, 79–92. [CrossRef]
20. Sorger, A.; Freitag, M.; Shaporin, A.; Mehner, J. CFD analysis of viscous losses in complex microsystems. In Proceedings of the International Multi-Conference on Systems, Signals & Devices, Chemnitz, Germany, 20–23 March 2012; pp. 1–4.
21. Cho, Y.; Pisano, A.P.; Howe, R.T. Viscous damping model for laterally oscillating microstructures. *J. Microelectromech. Syst.* **1994**, *3*, 81–87. [CrossRef]
22. Kainz, A.; Hortschitz, W.; Schalko, J.; Jachimowicz, A.; Keplinger, F. Air damping as design feature in lateral oscillators. *Sens. Actuators A Phys.* **2015**, *236*, 357–363. [CrossRef]
23. Kainz, A.; Hortschitz, W.; Steiner, H.; Schalko, J.; Jachimowicz, A.; Keplinger, F. Accurate analytical model for air damping in lateral MEMS/MOEMS oscillators. *Sens. Actuators A Phys.* **2017**, *255*, 154–159. [CrossRef]
24. Thurston, G.R. Theory of oscillation of a viscoelastic medium between parallel planes. *J. Appl. Phys.* **1959**, *30*, 1855–1860. [CrossRef]
25. Agrawal, A.; Prabhu, S.V. Survey on measurement of tangential momentum accommodation coefficient. *J. Vac. Sci. Technol.* **2008**, *26*, 634–645. [CrossRef]
26. ANSYS. *ANSYS Academic Research, Release 15, ANSYS Mechanical APDL Coupled-Field Analysis Guide*; ANSYS, Inc.: Cannesburg, PA, USA.
27. Mirzazadeh, R.; Eftekhar Azam, S.; Mariani, S. Micromechanical characterization of polysilicon films through on-chip tests. *Sensors* **2016**, *16*, 1191. [CrossRef]
28. Mirzazadeh, R.; Mariani, S. Uncertainty quantification of microstructure-governed properties of polysilicon MEMS. *Micromachines* **2017**, *8*, 248. [CrossRef]
29. Mirzazadeh, R.; Eftekhar Azam, S.; Mariani, S. Mechanical characterization of polysilicon MEMS: A hybrid TMCMC/POD-kriging approach. *Sensors* **2018**, *18*, 1243. [CrossRef]

30. Mirzazadeh, R.; Ghisi, A.; Mariani, S. Statistical investigation of the mechanical and geometrical properties of polysilicon films through on-chip tests. *Micromachines* **2018**, *9*, 53. [CrossRef] [PubMed]
31. Mariani, S.; Ghisi, A.; Mirzazadeh, R.; Eftekhar Azam, S. On-Chip testing: a miniaturized lab to assess sub-micron uncertainties in polysilicon MEMS. *Micro Nanosyst.* **2018**, *10*, 84–93. [CrossRef]

micromachines

MDPI

Review

In-Situ Measurements in Microscale Gas Flows—Conventional Sensors or Something Else?

Juergen J. Brandner

Staff Position Microstructures and Process Sensors (MPS), Institute of Microstructure Technology (IMT), Karlsruhe Institute of Technology (KIT), Hermann-von-Helmholtz-Platz 1, 76344 Eggenstein-Leopoldshafen, Germany; juergen.brandner@kit.edu; Tel.: +49-721-6082-3963

Received: 11 February 2019; Accepted: 26 April 2019; Published: 29 April 2019

Abstract: Within the last few decades miniaturization has a driving force in almost all areas of technology, leading to a tremendous intensification of systems and processes. Information technology provides now data density several orders of magnitude higher than a few years ago, and the smartphone technology includes, as well the simple ability to communicate with others, features like internet, video and music streaming, but also implementation of the global positioning system, environment sensors or measurement systems for individual health. So-called wearables are everywhere, from the physio-parameter sensing wrist smart watch up to the measurement of heart rates by underwear. This trend holds also for gas flow applications, where complex flow arrangements and measurement systems formerly designed for a macro scale have been transferred into miniaturized versions. Thus, those systems took advantage of the increased surface to volume ratio as well as of the improved heat and mass transfer behavior of miniaturized equipment. In accordance, disadvantages like gas flow mal-distribution on parallelized mini- or micro tubes or channels as well as increased pressure losses due to the minimized hydraulic diameters and an increased roughness-to-dimension ratio have to be taken into account. Furthermore, major problems are arising for measurement and control to be implemented for *in-situ* and/or *in-operando* measurements. Currently, correlated measurements are widely discussed to obtain a more comprehensive view to a process by using a broad variety of measurement techniques complementing each other. Techniques for correlated measurements may include commonly used techniques like thermocouples or pressure sensors as well as more complex systems like gas chromatography, mass spectrometry, infrared or ultraviolet spectroscopy and many others. Some of these techniques can be miniaturized, some of them cannot yet. Those should, nevertheless, be able to conduct measurements at the same location and the same time, preferably *in-situ* and *in-operando*. Therefore, combinations of measurement instruments might be necessary, which will provide complementary techniques for accessing local process information. A recently more intensively discussed additional possibility is the application of nuclear magnetic resonance (NMR) systems, which might be useful in combination with other, more conventional measurement techniques. NMR is currently undergoing a tremendous change from large-scale to benchtop measurement systems, and it will most likely be further miniaturized. NMR allows a multitude of different measurements, which are normally covered by several instruments. Additionally, NMR can be combined very well with other measurement equipment to perform correlative *in-situ* and *in-operando* measurements. Such combinations of several instruments would allow us to retrieve an "information cloud" of a process. This paper will present a view of some common measurement techniques and the difficulties of applying them on one hand in a miniaturized scale, and on the other hand in a correlative mode. Basic suggestions to achieve the above-mentioned objective by a combination of different methods including NMR will be given.

Keywords: miniaturization; gas flows in micro scale; measurement and control; integrated micro sensors; advanced measurement technologies

Micromachines **2019**, *10*, 292

1. Introduction

The interest to gas flows in miniaturized systems has grown tremendously in the last couple of years. Driven by areas such as the automotive industry, semiconductors, the chemical and pharmaceutical industry, modeling, precise measurement and control of the flow of gaseous compounds, mixtures and reactive systems through mini- and micro-structured devices gained importance for various applications like heat transfer [1–16], gas-liquid or gas-solid contacting [17–24] and chemical reactions [25–30]. Amongst the correct design and manufacturing of the miniaturized devices, all the named topics need precise measurement and control of the processes taking place inside microstructures. The final objective of these efforts is to provide an almost comprehensive description of the process to be able to model and simulate as well as to predict by software. The following description will focus on measurement and close out control, to make the process is not too exhaustive.

While measurement of gaseous flows in macroscale is not trivial but manageable, it turns out to be much more problematic in the micro scale. There are several reasons for this. The fact that gas is a compressible medium per se makes it more difficult to measure certain behavior and parameters of a flow inside confined or miniaturized systems [31–35]. Additionally, the scale down of conventionally-sized macro scale tubes or channels into the micro scale makes it more complex to measure. This is, on one hand, due to sensors which are simply too large to be inserted into the micro devices. Figure 1a,b show examples for this. For temperature measurement in macro-scale tubes, a thermocouple is simply located into the flow, measuring the gas flow temperature. This is changed by scaling down the tube diameter. While Figure 1a shows a mini heat pipe cut open, Figure 1b provides a view to a so-called "micro-thermocouple". It is quite obvious that the thermocouple will block the inner diameter to a major extent and, therefore, completely change the fluidic behavior. Thus, no precise measurement would be possible. More problems occur due to the low density of gases, the change in viscosity, the small specific heat capacity and the need for increased leak-tightness of microstructures while handling gases (and here, depending to the gas, the acceptable leak rate Q_L can vary in orders of magnitude!). However, in many cases Micro Electromechanical Systems (MEMS) have been applied as measurement tools to be implemented into microstructure devices [33]. While this is often a practical solution, in other cases it is not, as will be shown in the later discussion.

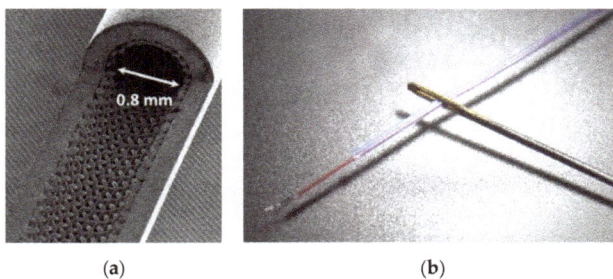

Figure 1. Microstructure fluidic device. (**a**) Micro heat pipe, opened to see the inside structure. Inner diameter is 0.8 mm. From: KIT-IMVT; (**b**) Miniaturized Type K thermocouple. Although this sensor has an outer diameter of 0.4 mm only, a major part of the structure in (**a**) would have been blocked. Picture from: www.ninomiya-ew.co.jp.

Nevertheless, from 1964 until today, roughly about 45,000 papers have been published on miniaturized gas sensors and sensor systems [36]. A large variety of relevant technologies is available which will not be described in this paper in detail, more can be found in comprehensive textbooks or in reference [37].

The following will present an overview on parameters of gas flows in microstructures to be measured as well as measurement methods for these. The presented overview is by no means complete. The envisaged field of measurement is very much in flux; thus, new technologies and improved methods are rapidly developing, especially by improving the sensitivity and selectivity of the measurements. One of these developments is the use of nanomaterials for sensing opportunities. Here, numerous different technologies based on nanostructured materials or nanostructures in specific materials have been described and are used now for sensing in gases. Examples are given in references [38–47]. The sensitivity and selectivity of nanostructure/nanomaterials sensors has improved significantly in the last few years [48–50].

Another trend is a combination of sensing elements in micro or nano scale for measurement of biological parameters or environment (biosensing, [51–56]). This field is relatively new, and lots of developments are to expect in the next future.

As mentioned before, an additional point is the correlation of a multitude of measurement methods to achieve a more dense "information cloud" and, therefore, reach a better understanding of the effects and behavior in gas flows. This leads to the necessity of *in-situ* and *in-operando* combination of several measurement methods with similar (ideally: identical) timely and spatial resolution. All measurements have to be taken simultaneously and at the same location to provide the highest possible information density of the process taking place at this point. As an example, for such a process a heterogeneously catalyzed gas phase reaction could be taken. Here, gas flow, gas composition, catalyst-gas-surface interaction, temperature, pressure and product concentration should be measured at the same time and the same location, to name just a few of the parameters of such a process.

The data obtained from each of such monitoring systems will then be correlated to those of all others, in time as well as in space, to get a description of an n-dimensional parameter space that is as comprehensive as possible. Within the overview presented here, several possible technologies to correlate measurements are very briefly and rudimentarily described. Possibilities, advantages and disadvantages of each method are presented, and a possible use in combination with other methods is evaluated. This overview is not intended to give a vast description of the measurement methods, their advantages and disadvantages, their working principles or underlying functionalities—this is left to much more specified papers or textbooks. This collection, without delving deeply into details of the methods themselves, should provide a quick overview for possible correlative measurement technologies only. For this objective, the following section deals with the parameters which most regularly are measured, as well as with some most common methods for those and with scaling, the problem of miniaturization itself [57–59].

2. Parameters, Measurement Methods and Scaling

Numerous parameters of gas flows in micro scale can be measured. However, many of them are only useful in specific applications. Amongst those are parameters like chemical reactivity, pH, solubility in liquids, speed of sound, etc. This publication will focus on more common parameters like temperature (or temperature difference), heat flux, pressure (or pressure difference), mass flow, flow velocity, mixture composition and concentration (concentration gradients, respectively) of species inside micro scale gas flows.

The measurement methods can in general be split in electrical measurements using conventional wire systems [60,61], electrical measurements using micro-electromechanical systems (MEMS) [62], spectroscopic measurements or optical measurement methods [63]. There are numerous papers on each of those different measurement methods, thus, it shall not be the objective of this publication to provide more detailed information on them.

Wire-system based methods use conventional sensors which are, in many cases, macroscopic. In general, these are thermocouples or thermistors (see below), pressure sensors, mass flow meters and similar, which are well known measurement devices in style as well as in behavior. These sensors are mounted on the process tubing, e.g., thermocouples before and after a possible reaction vessel as well as pressure sensors or mass flow meters. An *in-situ* or *in-operando* monitoring is hardly possible.

MEMS-based measurements show the possibility to be implemented for *in-situ* and *in-operando* measurements. Pressure sensors, temperature sensors, flow sensors and others more are available off the shelf. They are, in general, Si-based, small, provide a short response time, good reliability and long lifetime. Additionally, most of those systems are cheap to produce. However, they need to be coupled with a visualization module or a data logger to make the measurements available for the operator. This is regularly done by wiring. There are some cases of wireless MEMS sensors [64]. However, due to the continuous developments in semiconductor technology, wireless solutions are becoming more and more popular, and their use is greatly enhanced by, i.e., radio frequency identification (RFID) and near field communication (NFC) technology [65–69].

What was mentioned above for wire-based electrical sensor in principle holds also for optical sensors, where either an optical fiber is implemented into the measurement location, or an active optical component (light emitting diode LED, laser or other light source) is integrated there—in many cases an optical detector like a photo diode or similar could also be used. Examples are presented in references [70–76].

Most of the measurement methods of the sensors named above have been derived from macroscopic standards, adapted and improved for the mini scale and then applied and even further revised for the micro scale; and almost all of them (but the wireless MEMS devices) share the same problem of scaling. This holds for all type of sensors, whether electrical or optical.

Scaling a measurement method simply down from macro to micro scale will normally not generate the desired results. Either is the fluid influenced too strongly by the measurement method (as was shown in the example given in Figure 1), or the precision of measurement suffers from the miniaturization. It is, in many cases, not easy to decide where in the fluid flow to place the sensors to obtain correct measurements. A good example is the measurement of temperatures. If a MEMS-based measurement system is placed into the sidewall of a microfluidic system, the wall temperature is measured, or maybe the temperature of the gas flow near the wall. The same result will be obtained with an optical element placed inside the sidewall of a microfluidic system. The temperature in the core of the flow is not determined in any way. Moreover, the MEMS system may alter the wall-gas interaction (because the material might be different), and therefore not even provide a representative measurement signal for the sidewall temperature of the gas flow [77]. This might not be so much the case for the optical sensor. Figure 2 shows an example of a silicon MEMS sensor system used for temperature measurement inside a microchannel gas flow under ambient or slightly rarefied conditions [78].

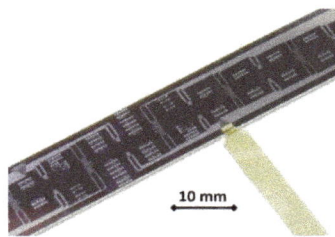

Figure 2. Silicon microstructure chip for temperature measurement in a gas flow. The chip is inserted into a sealed housing and forms the fourth side of a rectangular microchannel. Numerous thermopiles have been implemented into the chip, to measure the temperature of a gas flow in combination with precision resistors [78]. www.gasmems.eu.

If the process is well known, the core temperature of the flow might be calculated correctly. Not knowing the process precisely, it might be necessary to measure the core flow temperature—which is possible with a sensor implemented into the flow. Placing a very thin wired sensor or a thin optical fiber inside the flow core may lead to a more precise measurement of the temperature there, but also could lead to a deflection of the measurement sensor due to insufficient stiffness, or to generate local turbulences and eddies and, therefore, disturb the flow. The same holds for a wireless sensor, which has to be mounted and fixed somehow in the flow. In any case, scaling down fluidic systems for gas flows into the micro scale needs a very careful consideration of the location and the precision of the sensors applied. This holds for all parameters to be measured.

2.1. Temperature Measurement

One of the most common tasks in gas flows is the measurement of temperatures, as mentioned above. At the same time, it is a relatively complex measurement, which is, in many cases, underestimated [60,79–84]. In the following, some possible methods will be described and evaluated as a general example for other parameters to be measured at a micro scale.

2.1.1. Measurements Using Conventional Intrusive Sensors

Conventional sensors in this case means either thermistors or thermocouples. Optical sensors shall not be considered here, because they are more delicate to handle in this case and are generally more expensive [85–87]. Both thermocouples and thermistors are commercially available in various shapes, sizes and forms, which makes them flexible and handy tools. Moreover, efforts to obtain some results are limited. In both cases, wired standard solutions as well as wired MEMS are available.

Thermistors are temperature-dependent electrical resistors. They can be separated in devices with a positive temperature coefficient retrieving a higher resistance with increasing temperature, or a with negative temperature coefficient, which lowers the resistance by increasing temperature. In any case, the dependency between the change of resistance and the temperature is very linear (or can be linearized in an easy way) [88]. The most common example is the PT100, a platinum resistor with a nominal value of 100 Ω at standard conditions (variations with 10 Ω or 1000 Ω are also common). The PT-sensor is used as part of a Wheatstone bridge circuit, which makes a very precise measurement possible [89].

Temperature sensors based on the resistor principle are wide-spread in any application, because they can be manufactured as discrete devices as well as integrated circuits in semiconductor technology. Thus, extremely small sensors can be generated using standard semiconductor manufacturing processes. This also holds, in some cases, for the second common temperature measurement principle, which is thermocouples.

A thermocouple is an electrical device consisting of two dissimilar electrical conductors forming electrical junctions at differing temperatures. A thermocouple produces a temperature-dependent voltage as a result of the thermoelectric effect, and this voltage can be interpreted to measure a temperature [78,90]. Some types of thermocouples can also be produced in a micro- or even nano scale by use of semiconductor technology, while others cannot [37]. However, the two main disadvantages of thermocouples are their limited signal strength and the need for a reference temperature. While the signal strength can be enhanced by creating thermopiles (numerous thermocouples circuited in a row, see Figure 2), the reference temperature need cannot be avoided. This fact makes it difficult to apply thermocouples in miniaturized equipment, which regularly shows a more or less isothermal behavior, short distances in the mm or µm range and very short temperature spreading times. Thus, the difference between reference temperature and measured temperature might be diminished rapidly.

Both measurement principles mentioned above are regularly carried out with wired sensors. Wireless temperature sensors, especially in micro scale, are not so widely common yet, but will be more common in the future [91–93]. However, all the sensors presented in this subsection need a wire connection to the outer environment, thus, sealings and interconnections are necessary. This might be, depending on the supervised process, a source of leakage and uncertainty. Therefore, wireless measurement would be a better option, as was mentioned before. Additionally, it was mentioned that introducing a sensor into the gas flow would disturb the flow mode or even change the process parameters significantly. Thus, non-intrusive methods have to be used.

2.1.2. Measurements Using Non-Intrusive Methods

As the example of the use of conventionally sized resistors or thermocouples (presented in Figure 1b) shows, this is, in many cases, not an option, mainly for size reasons as described before.

Thus, it has to be asked which other techniques and methods could be used for such a simple measurement task such as retrieving the temperature?

One option is the use of non-intrusive optical systems (i.e., infrared (IR) radiation, ultra violet (UV) fluorescence, others) or ultrasound [94–96]. Here, surface temperatures are measured, providing a non-intrusive possibility to acquire data on the thermal state of a body. However, these methods have several disadvantages. They are not applicable for all cases, since certain materials are not transparent for IR radiation (i.e., common glasses, metals etc.) or reflect or damp-out ultrasound very strongly (i.e., dense liquids, metals etc.). In such cases, a wrong signal, a strongly reduced signal or no signal at all can be obtained, leading to a very low signal-to-noise ratio (SNR). When this occurs, a measurement might be useless. A further disadvantage is that the measured signal represents the surface temperature of a device obtained by conduction, convection or radiation from the inner side, but not the core temperature. This means the inner parts of a body could differ significantly in temperature from the outer surface. This outcome cannot be seen and measured by the chosen measurement technology. The same holds for other measurement technologies that use optical fluorescence to measure temperatures [97].

Another option is Raman spectrometry [98]. With this technique, temperature distributions inside of miniaturized structures are easy to measure if the surrounding area is transparent for the Raman light. All the limits named before for IR and ultrasound measurements hold here also, because this is still an optical method. Efforts are high, and the Raman systems are expensive and complex to handle. If all these drawbacks are given for measurement of an all-day parameter like temperature, does this also hold for other measurement needs like pressure, flow, density or more complex sets of parameters like local concentration? Furthrmore, what might be feasible in terms of *in-situ*, *in-operando* and correlated measurements?

2.2. Measurements of Other Parameters

As pointed out in the previous sections, measurements of several parameters lead to application of different techniques at a single process. The process pressure is measured with pressure transducers, in most cases located outside the process vessel. The process flow through tubes or vessels is obtained by anemometers or similar systems. Electrochemical sensors are used to measure pH or conductivity. Viscosity is obtained with a rheometer (in general, outside of the process run), and so on. Additional analytical methods like gas chromatography (GC) [99,100], infrared (IR) spectrometry [101], ultraviolet (UV) spectrometry [102], the above mentioned Raman spectrometry [103], mass spectrometry (MS) [104,105] or Nuclear Magnetic Resonance spectroscopy (NMR) [106–110] are used to determine the components of the mixture, depending on the applicability of the respective method to the gas. Visualization of flow processes can be obtained by, for example, high speed videography [111–113]. An example of a microstructure used for high speed videography is given in Figure 3.

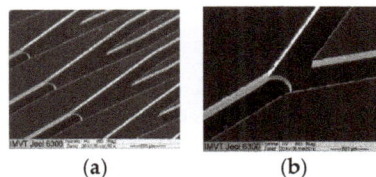

(a)   (b)

Figure 3. Bifurcative microchannel distribution system for liquids and gases. (**a**) Overview to a major part of the system; (**b**) Magnification of one of the splits from (**a**). A step between the single incoming channel and the two outgoing channels is clearly visible, which is due to manufacturing. Reproduced with permission from [111], published by J-STAGE, 2012.

All of these systems work, in most cases, independently, are time consuming and costly. *In-situ* measurements or *in-operando* measurements inside of micro scale systems are possible, but rarely done, in most cases the analysis is performed offline with a specific sample taken from the process flow. A correlation of results of these analytical methods is used to gain more complete process information, as was proposed in the introduction section, with measurement results of the process parameters obtained as described before being at least difficult [114–116], in many cases it is almost impossible.

Thus, it might be a good idea to have a combination of several complementary measurement and analytical systems which can provide all the information wanted within a single measurement campaign at once.

2.3. Combined and Correlated Measurement System

The objective mentioned above, namely to measure *in-situ*, *in-operando*, simultaneously at the same location, can be obtained by combining several measurement and analytical systems. However, aside of the tremendous efforts to be undertaken, this will result most likely in a spatial correlation only. Due to different measurement time constants of the different sensors and systems, a timely correlation is very unlikely to be obtained. All the methods named above show drawbacks for *in-situ*, *in-operando*, combined and correlated measurements. While Raman spectroscopy will need an optical access, IR spectroscopy as well as UV spectroscopy will need additionally specimen active in the named wavelength region. *In-situ* and *in-operando* GC or GC-MS is possible (see i.e., reference [117]), but in general is done in a macro scale. To scale those techniques down into micro scale might be feasible but is generally not available as an option yet. NMR measurement systems [110] are, in general, big machines with super-cooled magnets and a huge bunch of external equipment, which is one of the major drawbacks of this technology. Figure 4 shows an example of a 400 MHz (9.4 T) NMR device made by the Bruker company (Billerica, MA, USA).

Figure 4. 400 MHz (9.4 T) NMR system of the Bruker company.

However, with an NMR spectrometer like this, a sample can be characterized for various parameters in a single measurement campaign, correlating time and location with each of the separate measurements taken. The question arising now is whether such a system is able to characterize the desired parameters with a spatial and time resolution that is high enough to be useful in a micro scale.

For the spatial resolution, a value of about 10 μm is given in literature (see i.e., reference [118]). The time resolution is named to be around 20 ms (see i.e., reference [119]). This result is assumed to be valid for a multitude of measurement objectives performed by NMR, spatial as well as time resolution. The literature suggests that this result is in principle good enough for running NMR systems as well-chosen measurement and analytical tools for numerous applications, ranging from chemistry to materials research, pharmacy and food engineering to biomedical applications or physics. However, it has to be considered that many measurements will not be done after the 20 ms given above, but instead take much longer and average a multitude of single measurement shots.

Thus, beside the drawbacks of size, costs and running efforts for such a device in terms of liquid helium and nitrogen for cooling down the magnets etc., it is not clear yet whether all desired parameters can be measured in reasonable time, or if even all parameters important for a process to be characterized can be measured. Aside of this, not all materials can be used with NMR. The technology is very selective, but is limited to non-magnetic materials for devices. If process parameters can only be reached using a vessel made of magnetic material, then NMR measurement methods fail due to the magnetic properties of the device. Another drawback is that not all desired parameters can be measured directly, some can only be retrieved indirectly.

While the measurement of temperature with NMR is in some cases precise and simple [120], the process pressure can only be retrieved by pressure-sensitive materials [121]. Viscosity, density, mass flow and phase, phase changes or particle content can be measured directly (i.e., reference [122]), as well as the electrochemical potential of compounds (see reference [123]), concentrations of single mixture compounds [124] or the pH of a mixture.

Additionally, this technology allows us not only to measure different parameters non-intrusively in very small confinements like micro channels, but also in living tissues [125,126]. This is an add-on not provided by a multitude of the other techniques presented before. Another add-on is the possibility to visualize processes by magnetic resonance imaging (MRI, see reference [127]). An example of this is presented in Figure 5, showing a cut through a human head and imaging the brain.

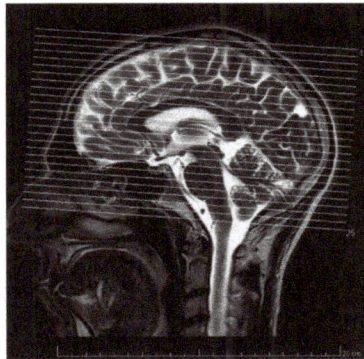

Figure 5. Magnetic resonance tomography (MRT) cut picture of a human head as an example for the MRI possibilities. From: Private.

Thus, NMR seems to provide the possibility to cover lots of measurement tasks of other methods in a single machine, being non-invasive, non-intrusive and non-detrimental to the measured object. With all the possibilities provided here, a discussion is needed on at least two topics: is measurement with an NMR correlative? Moreover, is NMR measurement the best possible solution for every measurement task?

3. Discussion

As was pointed out in Section 2, lots of different methods for measurement of process parameters are available and have been explored. Miniaturization of process systems results in major problems for scaling them down, as was mentioned and described before. However, with larger efforts, smart design and good planning of experiments, the scaling down of measurement systems and obtaining precise results of the process parameters is possible to a certain extent, using various methods for process parameter characterization. Thus, miniaturization and scaling is a problem for intrusive measurement methods, while precise measurement of the parameter distribution is a problem for non-intrusive methods. This has clearly been described in Section 2. In almost all cases, measurements are performed sequentially, having each of the measurement and analysis systems running after the one before. This type of performance, aside from the difficulties given by scaling, allows us to achieve good and reliable measurement results, but not to correlate them to each other and to obtain a more fundamental understanding of the interactions between process parameters and underlying principles. This is shown schematically in Figure 6, in which the current state-of-the-art measurement methodology is presented.

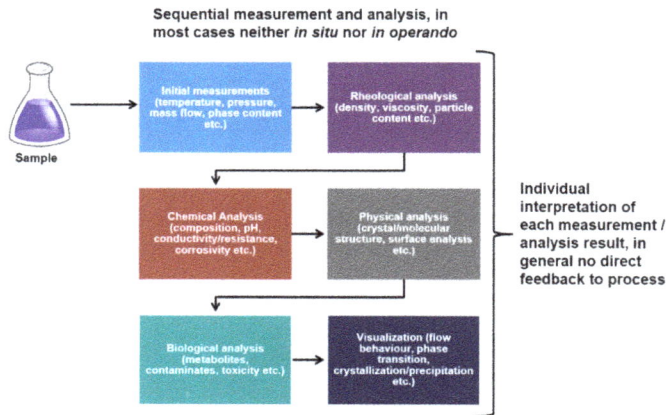

Figure 6. Current state-of-the-art sequential measurement and analysis method. The timely and locally separated measurement makes it difficult or even impossible to correlate measurement results to each other and retrieve by this deeper insight into the underlying mechanisms.

Thus, correlation of the measurement results is the major point to deal with. One of the future perspectives of measurement is to obtain correlated information *in situ* and *in operando* by as many different techniques as possible. Correlated measurement means timely and spatially, and with similar resolutions for all applied techniques. This is necessary to gain an exhaustive view of the examined process, to understand the underlying principles and actions, and therefore to make them accessible for modeling and simulation as well as in silico prediction and optimization. The latter will lead to major reductions in resource consumption. Figure 7 shows a scheme of a possible future measurement environment, in which a multitude of instruments are involved in parallel, obtaining means at the same time and location, measuring *in situ* and *in operando* and, therefore, delivering results which can directly be correlated. With this method, a much deeper understanding of links between actions and effects can be achieved in a more appropriate way. Thus, correlative measurement, generation of linked information and interpretation, management and feedback of connected data will be one of the main tasks in the future. The measurement structure schematically shown in Figure 7 can support this task.

Some of the measurement methods presented above are useful for stand-alone measurements only, while some can be performed simultaneously. However, the methods and techniques described in Sections 2.1 and 2.2 can hardly be done in a correlated way. It is possible to measure temperature, pressure and chemical composition of a flow at the same position using, i.e., thermocouples, pressure sensors and Raman spectrometry. To obtain a precise time correlation by using a trigger is difficult, mainly due to different time constants of the measurement methods. Additional visualization of the process adds some more problems, because for high speed videography as described in Section 2, very high light intensities are necessary, which might cope with other optical measurement methods; therefore, a combined high speed videography—Raman measurement is not an option.

All this holds for all of the conventional methods described above. Without precise triggering, correlative measurement is most likely impossible; and even with a time trigger, the location of measurement as well as the spatial resolution most likely differs largely.

NMR makes it partly easier in this particular point. As described before, in principle a multitude of parameters can be retrieved with NMR, all of them at the same location. Thus, spatial resolution is consistent. However, it is doubtful that a timely consistence can be reached because some measurements simply take much longer than others, as pointed out before. Thus, truly correlative measurements in terms of time and space cannot be generated for the most cases, meaning this problem remains. However, the data measured by NMR are to a certain extent per se correlated, because several parameters are measured with the same method and then averaged on a time period, which is a huge advantage compared to conventional measurement equipment. Process pressure has to be measured via pressure-sensitive materials (see description above and reference [121]), which makes this a special case to be dealt with. Additionally, with MRI a handy visualization tool is available, which provides an even more in-sight view to processes and flows. Thus, many of the problems attached to conventional measurement systems apply less for NMR. However, the time correlation problem is kept, and due to the limitation of applicability to non-magnetic device materials, NMR is not a panacea. It is a good toolbox for lots of possible measurement solutions correlating the obtained data directly internally. A future vision would be to retrieve all process parameters at the same time, with the same precision, the same spatial and time resolution, *in-situ* and *in-operando*, as was requested. Thus, NMR might be a nice additional tool for certain measurement applications, but it cannot be considered to be a stand-alone tool for generation of correlated data on processes yet. There's a need for combinations of methods, to obtain the desired "information cloud". Only with this, a reasonable combination of different measurement and analytical tools, might a more comprehensive descriptions of processes be feasible.

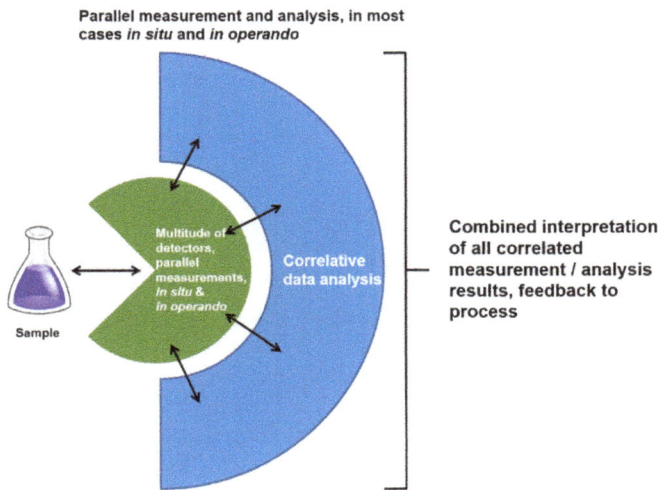

Figure 7. Future improved parallel measurement and analysis method. The detectors are triggered automatically in a way that they all measure at the same location, and deliver the measured data with the same spatial and time resolution. Data are then transferred to an analysis level, where they are correlated.

Recently ideas for further correlated measurements have emerged, like MR-optical combinations, or the integration of atomic force microscopy (AFM) and NMR into a single device. Moreover, proposals on combining NMR technology with additive manufacturing, i.e., 3D printing methods, came up. Here, the characterization as well as the measurement would be integrated in a single process, allowing us to generate miniaturized systems in a way that was not even considered ten years ago.

4. Conclusions

Macroscale measurement of gas flow parameters with conventional technology like thermistors, pressure sensors etc. is no longer a problem, but scaling those systems down into the mini or micro scale remains a challenge. Due to either insufficient size of classical sensors or insufficient possibilities to position micro sensors in the gas flow, the measurement largely influences the process itself. Non-intrusive methods are possible, but show other limitations, like optical accessibility, the need for high power light exposure or non-magnetic equipment materials.

One of the most interesting measurement problems is to achieve timely and spatially correlated results. By using conventional methods involving thermistors, GC, MS, Raman, IR or UV spectroscopy, pressure transducers etc., this is almost impossible. Either the measurement location largely differs, or the time resolution is simply not suitable. In addition, the combination of data from different measurement sources is not trivial. The use of nuclear magnetic resonance systems (NMR), which in principle allow measurement of a multitude of parameters at least at the same location, and in most cases with the same spatial resolution, eases this problem slightly. Here, results can be obtained from the same measurement location, show a consistent data format, and can therefore be easily correlated. Even visualization is relatively easily achieved by combining regular NMR with magnetic resonance imaging MRI, which adds some more valuable data to the parameter cloud obtained by the measurement. The scale range for NMR / MRI systems reaches from the macro to the nano scale. Spatial and time resolution do not depend on the scale but rather on the magnetic system, which is also an advantage compared to other measurement methods.

Recently, NMR and MRI systems themselves have been targets of miniaturization. Meanwhile, benchtop low field NMR systems up to 80MHz resonance frequency (about 1.88T) are available which can easily run in a conventional lab environment, used for stationary or flow-through data achievements. These systems are easy to handle and deliver a bunch of analytical possibilities. Nevertheless, miniaturization went on, and meanwhile developments are reaching into the direction of mini or micro NMR systems [125,127–129]. Even wireless systems are now being researched; thus, it can be expected that application of miniaturized and integrated NMR in correlative measurements will be greatly enhanced in the future.

A vision remains of having measurements of different parameters by NMR with the same time resolution, as well as the possibility of measuring all the desired parameters of a process. Thus, combinations of a multitude of analysis instruments will be necessary, ideally within a single system, controlled by a combined measurement electronic system, which internally triggers all methods.

Funding: This research received funding by the European Commission H2020 Marie Skłodowska-Curie Actions under the Grant Agreement (grant number 643095 "MIGRATE").

Acknowledgments: The author wish to acknowledge Jan Korvink and Neil MacKinnon of KIT-IMT for lots of fruitful discussions, which led to some of the thoughts briefly described in this paper.

Conflicts of Interest: The author declares no conflict of interest.

References

1. Kandlikar, S.G.; Colin, S.; Peles, Y.; Garimella, S.; Pease, R.F.; Brandner, J.J.; Tuckerman, D.B. Heat Transfer in Microchannels—2012 Status and Research Needs. *J. Heat Transf.* **2013**, *135*, 091001. [CrossRef]
2. Morini, G.L. Single-phase convective heat transfer in microchannels: A review of experimental results. *Int. J. Therm. Sci.* **2004**, *43*, 631–651. [CrossRef]
3. Mohammed, H.A.; Bhaskaran, G.; Shuaib, N.H.; Saidur, R. Heat transfer and fluid flow characteristics in microchannels heat exchanger using nanofluids: A review. *Renew. Sustain. Energy Rev.* **2011**, *15*, 1502–1512. [CrossRef]
4. Han, Y.; Liu, Y.; Li, M.; Huang, J. A review of development of micro-channel heat exchanger applied in air-conditioning system. *Energy Procedia* **2012**, *14*, 148–153. [CrossRef]
5. Dixit, T.; Ghosh, I. Review of micro- and mini-channel heat sinks and heat exchangers for single phase fluids. *Renew. Sustain. Energy Rev.* **2015**, *41*, 1298–1311. [CrossRef]
6. Rosa, P.; Karayiannis, T.G.; Collins, M.W. Single-phase heat transfer in microchannels: The importance of scaling effects. *Appl. Therm. Eng.* **2009**, *29*, 3447–3468. [CrossRef]

7. Mohammed Adham, A.; Mohd-Ghazali, N.; Ahmad, R. Thermal and hydrodynamic analysis of microchannel heat sinks: A review. *Renew. Sustain. Energy Rev.* **2013**, *21*, 614–622. [CrossRef]
8. Colin, S. Gas Microflows in the Slip Flow Regime: A Critical Review on Convective Heat Transfer. *J. Heat Transf.* **2011**, *134*, 020908. [CrossRef]
9. Yin, S.; Meyer, M.; Li, W.; Liao, H.; Lupoi, R. Gas Flow, Particle Acceleration, and Heat Transfer in Cold Spray: A review. *J. Therm. Spray Technol.* **2016**, *25*, 874–896. [CrossRef]
10. Asadi, M.; Xie, G.; Sunden, B. A review of heat transfer and pressure drop characteristics of single and two-phase microchannels. *Int. J. Heat Mass Transf.* **2014**, *79*, 34–53. [CrossRef]
11. Salman, B.H.; Mohammed, H.A.; Munisamy, K.M.; Kherbeet, A.S. Characteristics of heat transfer and fluid flow in microtube and microchannel using conventional fluids and nanofluids: A review. *Renew. Sustain. Energy Rev.* **2013**, *28*, 848–880. [CrossRef]
12. Zhai, Y.; Xia, G.; Li, Z.; Wang, H. Experimental investigation and empirical correlations of single and laminar convective heat transfer in microchannel heat sinks. *Exp. Therm. Fluid Sci.* **2017**, *83*, 207–214. [CrossRef]
13. Sahar, A.M.; Özdemir, M.R.; Fayyadh, E.M.; Wissink, J.; Mahmoud, M.M.; Karayiannis, T.G. Single phase flow pressure drop and heat transfer in rectangular metallic microchannels. *Appl. Therm. Eng.* **2016**, *93*, 1324–1336. [CrossRef]
14. Kim, B. An experimental study on fully developed laminar flow and heat transfer in rectangular microchannels. *Int. J. Heat Fluid Flow* **2016**, *62*, 224–232. [CrossRef]
15. Mokrani, O.; Bourouga, B.; Castelain, C.; Peerhossaini, H. Fluid flow and convective heat transfer in flat microchannels. *Int. J. Heat Mass Transf.* **2009**, *52*, 1337–1352. [CrossRef]
16. Sahar, A.M.; Wissink, J.; Mahmoud, M.M.; Karayiannis, T.G.; Ashrul Ishak, M.S. Effect of hydraulic diameter and aspect ratio on single phase flow and heat transfer in a rectangular microchannel. *Appl. Therm. Eng.* **2017**, *115*, 793–814. [CrossRef]
17. Hessel, V.; Angeli, P.; Gavriilidis, A.; Löwe, H. Gas–Liquid and Gas–Liquid–Solid Microstructured Reactors: Contacting Principles and Applications. *Ind. Eng. Chem. Res.* **2005**, *44*, 9750–9769. [CrossRef]
18. Doku, G.N.; Verboom, W.; Reinhoudt, D.N.; van den Berg, A. On-microchip multiphase chemistry—A review of microreactor design principles and reagent contacting modes. *Tetrahedron* **2005**, *61*, 2733–2742. [CrossRef]
19. Su, Y.; Zhao, Y.; Chen, G.; Yuan, Q. Liquid–liquid two-phase flow and mass transfer characteristics in packed microchannels. *Chem. Eng. Sci.* **2010**, *65*, 3947–3956. [CrossRef]
20. Woitalka, A.; Kuhn, S.; Jensen, K.F. Scalability of mass transfer in liquid–liquid flow. *Chem. Eng. Sci.* **2014**, *116*, 1–8. [CrossRef]
21. Kashid, M.N.; Renken, A.; Kiwi-Minsker, L. Gas–liquid and liquid–liquid mass transfer in microstructured reactors. *Chem. Eng. Sci.* **2011**, *66*, 3876–3897. [CrossRef]
22. Reichmann, F.; Tollkötter, A.; Körner, S.; Kockmann, N. Gas-liquid dispersion in micronozzles and microreactor design for high interfacial area. *Chem. Eng. Sci.* **2017**, *169*, 151–163. [CrossRef]
23. Yao, C.; Dong, Z.; Zhao, Y.; Chen, G. Gas-liquid flow and mass transfer in a microchannel under elevated pressures. *Chem. Eng. Sci.* **2015**, *123*, 137–145. [CrossRef]
24. Haase, S.; Murzin, D.Y.; Salmi, T. Review on hydrodynamics and mass transfer in minichannel wall reactors with gas–liquid Taylor flow. *Chem. Eng. Res. Des.* **2016**, *113*, 304–329. [CrossRef]
25. Yue, J. Multiphase flow processing in microreactors combined with heterogeneous catalysis for efficient and sustainable chemical synthesis. *Catal. Today* **2018**, *308*, 3–19. [CrossRef]
26. Tanimu, A.; Jaenicke, S.; Alhooshani, K. Heterogeneous catalysis in continuous flow microreactors: A review of methods and applications. *Chem. Eng. J.* **2017**, *327*, 792–821. [CrossRef]
27. Yao, X.; Zhang, Y.; Du, L.; Liu, J.; Yao, J. Review of the applications of microreactors. *Renew. Sustain. Energy Rev.* **2015**, *47*, 519–539. [CrossRef]
28. Kolb, G. Review: Microstructured reactors for distributed and renewable production of fuels and electrical energy. *Chem. Eng. Process. Process Intensif.* **2013**, *65*, 1–44. [CrossRef]
29. Pennemann, H.; Kolb, G. Review: Microstructured reactors as efficient tool for the operation of selective oxidation reactions. *Catal. Today* **2016**, *278*, 3–21. [CrossRef]
30. Rossetti, I.; Compagnoni, M. Chemical reaction engineering, process design and scale-up issues at the frontier of synthesis: Flow chemistry. *Chem. Eng. J.* **2016**, *296*, 56–70. [CrossRef]
31. Zhao, J.; Yao, J.; Zhang, M.; Zhang, L.; Yang, Y.; Sun, H.; An, S.; Li, A. Study of Gas Flow Characteristics in Tight Porous Media with a Microscale Lattice Boltzmann Model. *Sci. Rep.* **2016**, *6*, 32393. [CrossRef] [PubMed]

32. Harley, J.C.; Huang, Y.; Bau, H.H.; Zemel, J.N. Gas flow in micro-channels. *J. Fluid Mech.* **2006**, *284*, 257–274. [CrossRef]

33. Ho, C.-M.; Tai, Y.-C. Micro-electro-mechanical-systems (MEMS) and fluid flows. *Annu. Rev. Fluid Mech.* **1998**, *30*, 579–612. [CrossRef]

34. Hetsroni, G.; Mosyak, A.; Pogrebnyak, E.; Yarin, L.P. Fluid flow in micro-channels. *Int. J. Heat Mass Transf.* **2005**, *48*, 1982–1998. [CrossRef]

35. Goldstein, R. *Fluid Mechanics Measurements*, 2nd ed.; Taylor & Francis: Abingdon, UK, 2017.

36. Web of Science. *Gas Sensors*; Clarivate Analytics: Philadelphia, PA, USA, 2019.

37. Yunusa, Z.; Hamidon, M.N.; Kaiser, A.; Awang, Z. Gas Sensors: A Review. *Sens. Transducers* **2014**, *168*, 61–75.

38. Llobet, E. Gas sensors using carbon nanomaterials: A review. *Sens. Actuators B Chem.* **2013**, *179*, 32–45. [CrossRef]

39. Varghese, S.S.; Lonkar, S.; Singh, K.K.; Swaminathan, S.; Abdala, A. Recent advances in graphene based gas sensors. *Sens. Actuators B Chem.* **2015**, *218*, 160–183. [CrossRef]

40. Meng, F.-L.; Guo, Z.; Huang, X.-J. Graphene-based hybrids for chemiresistive gas sensors. *TRAC Trends Anal. Chem.* **2015**, *68*, 37–47. [CrossRef]

41. Pandey, S. Highly sensitive and selective chemiresistor gas/vapor sensors based on polyaniline nanocomposite: A comprehensive review. *J. Sci. Adv. Mater. Devices* **2016**, *1*, 431–453. [CrossRef]

42. Kuberský, P.; Syrový, T.; Hamáček, A.; Nešpůrek, S.; Stejskal, J. Printed Flexible Gas Sensors based on Organic Materials. *Procedia Eng.* **2015**, *120*, 614–617. [CrossRef]

43. Elhaes, H.; Fakhry, A.; Ibrahim, M. Carbon nano materials as gas sensors. *Mater. Today Proc.* **2016**, *3*, 2483–2492. [CrossRef]

44. Basu, S.; Bhattacharyya, P. Recent developments on graphene and graphene oxide based solid state gas sensors. *Sens. Actuators B Chem.* **2012**, *173*, 1–21. [CrossRef]

45. Toda, K.; Furue, R.; Hayami, S. Recent progress in applications of graphene oxide for gas sensing: A review. *Anal. Chim. Acta* **2015**, *878*, 43–53. [CrossRef] [PubMed]

46. Xu, K.; Fu, C.; Gao, Z.; Wei, F.; Ying, Y.; Xu, C.; Fu, G. Nanomaterial-based gas sensors: A review. *Instrum. Sci. Technol.* **2018**, *46*, 115–145. [CrossRef]

47. Bogue, R. Nanomaterials for gas sensing: A review of recent research. *Sens. Rev.* **2014**, *34*, 1–8. [CrossRef]

48. Zhang, J.; Liu, X.; Neri, G.; Pinna, N. Nanostructured Materials for Room-Temperature Gas Sensors. *Adv. Mater.* **2016**, *28*, 795–831. [CrossRef] [PubMed]

49. Joshi, N.; Hayasaka, T.; Liu, Y.; Liu, H.; Oliveira, O.; Lin, L. A review on chemiresistive room temperature gas sensors based on metal oxide nanostructures, graphene and 2D transition metal dichalcogenides. *Microchim. Acta* **2018**, *185*, 213. [CrossRef]

50. Liu, X.; Ma, T.; Pinna, N.; Zhang, J. Two-Dimensional Nanostructured Materials for Gas Sensing. *Adv. Funct. Mater.* **2017**, *27*, 1702168. [CrossRef]

51. Mabeck, J.T.; Malliaras, G.G. Chemical and biological sensors based on organic thin-film transistors. *Anal. Bioanal. Chem.* **2006**, *384*, 343–353. [CrossRef] [PubMed]

52. Liedberg, B.; Nylander, C.; Lunström, I. Surface plasmon resonance for gas detection and biosensing. *Sens. Actuators* **1983**, *4*, 299–304. [CrossRef]

53. Diamond, D.; Coyle, S.; Scarmagnani, S.; Hayes, J. Wireless Sensor Networks and Chemo-/Biosensing. *Chem. Rev.* **2008**, *108*, 652–679. [CrossRef]

54. Han, Z.J.; Mehdipour, H.; Li, X.; Shen, J.; Randeniya, L.; Yang, H.Y.; Ostrikov, K. SWCNT Networks on Nanoporous Silica Catalyst Support: Morphological and Connectivity Control for Nanoelectronic, Gas-Sensing, and Biosensing Devices. *ACS Nano* **2012**, *6*, 5809–5819. [CrossRef]

55. Emiliyanov, G.; Høiby, P.E.; Pedersen, L.H.; Bang, O. Selective Serial Multi-Antibody Biosensing with TOPAS Microstructured Polymer Optical Fibers. *Sensors* **2013**, *13*, 3242. [CrossRef] [PubMed]

56. Papkovsky, D.B. New oxygen sensors and their application to biosensing. *Sens. Actuators B Chem.* **1995**, *29*, 213–218. [CrossRef]

57. Temiz, Y.; Lovchik, R.D.; Kaigala, G.V.; Delamarche, E. Lab-on-a-chip devices: How to close and plug the lab? *Microelectron. Eng.* **2015**, *132*, 156–175. [CrossRef]

58. Walsh, D.I.; Kong, D.S.; Murthy, S.K.; Carr, P.A. Enabling Microfluidics: From Clean Rooms to Makerspaces. *Trends Biotechnol.* **2017**, *35*, 383–392. [CrossRef]

59. Chiu, D.T.; deMello, A.J.; Di Carlo, D.; Doyle, P.S.; Hansen, C.; Maceiczyk, R.M.; Wootton, R.C.R. Small but Perfectly Formed? Successes, Challenges, and Opportunities for Microfluidics in the Chemical and Biological Sciences. *Chem* **2017**, *2*, 201–223. [CrossRef]

60. Childs, P.R.N.; Greenwood, J.R.; Long, C.A. Review of temperature measurement. *Rev. Sci. Instrum.* **2000**, *71*, 2959–2978. [CrossRef]

61. Bacci da Silva, M.; Wallbank, J. Cutting temperature: Prediction and measurement methods—A review. *J. Mater. Process. Technol.* **1999**, *88*, 195–202. [CrossRef]

62. Azad, A.M.; Akbar, S.A.; Mhaisalkar, S.G.; Birkefeld, L.D.; Goto, K.S. Solid-State Gas Sensors: A Review. *J. Electrochem. Soc.* **1992**, *139*, 3690–3704. [CrossRef]

63. Hodgkinson, J.; Tatam, R.P. Optical gas sensing: A review. *Meas. Sci. Technol.* **2012**, *24*, 012004. [CrossRef]

64. Warneke, B.A.; Pister, K.S.J. MEMS for distributed wireless sensor networks. In Proceedings of the 9th International Conference on Electronics, Circuits and Systems, Dubrovnik, Croatia, 15–18 September 2002; Volume 291, pp. 291–294.

65. Lakafosis, V.; Rida, A.; Vyas, R.; Yang, L.; Nikolaou, S.; Tentzeris, M.M. Progress Towards the First Wireless Sensor Networks Consisting of Inkjet-Printed, Paper-Based RFID-Enabled Sensor Tags. *Proc. IEEE* **2010**, *98*, 1601–1609. [CrossRef]

66. Yao, W.; Chu, C.-H.; Li, Z. The Adoption and Implementation of RFID Technologies in Healthcare: A Literature Review. *J. Med Syst.* **2012**, *36*, 3507–3525. [CrossRef] [PubMed]

67. Abad, E.; Zampolli, S.; Marco, S.; Scorzoni, A.; Mazzolai, B.; Juarros, A.; Gómez, D.; Elmi, I.; Cardinali, G.C.; Gómez, J.M.; et al. Flexible tag microlab development: Gas sensors integration in RFID flexible tags for food logistic. *Sens. Actuators B Chem.* **2007**, *127*, 2–7. [CrossRef]

68. Kassal, P.; Steinberg, M.D.; Steinberg, I.M. Wireless chemical sensors and biosensors: A review. *Sens. Actuators B Chem.* **2018**, *266*, 228–245. [CrossRef]

69. Strömmer, E.; Hillukkala, M.; Ylisaukko-oja, A. Ultra-low Power Sensors with Near Field Communication for Mobile Applications. In Proceedings of the WSAN 2007: Wireless Sensor and Actor Networks, Albacete, Spain, 24–26 September 2007.

70. Kuswandi, B.; Nuriman; Huskens, J.; Verboom, W. Optical sensing systems for microfluidic devices: A review. *Anal. Chim. Acta* **2007**, *601*, 141–155. [CrossRef]

71. Choi, J.-R.; Song, H.; Sung, J.H.; Kim, D.; Kim, K. Microfluidic assay-based optical measurement techniques for cell analysis: A review of recent progress. *Biosens. Bioelectron.* **2016**, *77*, 227–236. [CrossRef]

72. Lin, S.-W.; Chang, C.-H.; Lin, C.-H. High-throughput Fluorescence Detections in Microfluidic Systems. *Genom. Med. Biomark. Health Sci.* **2011**, *3*, 27–38. [CrossRef]

73. Yang, F.; Li, X.-C.; Zhang, W.; Pan, J.-B.; Chen, Z.-G. A facile light-emitting-diode induced fluorescence detector coupled to an integrated microfluidic device for microchip electrophoresis. *Talanta* **2011**, *84*, 1099–1106. [CrossRef] [PubMed]

74. Geng, X.; Wu, D.; Wu, Q.; Guan, Y. Signal-to-noise ratio enhancement of the compact light-emitting diode-induced fluorescence detector. *Talanta* **2012**, *100*, 27–31. [CrossRef]

75. Xu, B.; Yang, M.; Wang, H.; Zhang, H.; Jin, Q.; Zhao, J.; Wang, H. Line laser beam based laser-induced fluorescence detection system for microfluidic chip electrophoresis analysis. *Sens. Actuators A Phys.* **2009**, *152*, 168–175. [CrossRef]

76. Wolfbeis, O.S. Fiber-Optic Chemical Sensors and Biosensors. *Anal. Chem.* **2008**, *80*, 4269–4283. [CrossRef]

77. Vittoriosi, A.; Brandner, J.J.; Dittmeyer, R. Integrated temperature microsensors for the characterization of gas heat transfer. *J. Phys. Conf. Ser.* **2012**, *362*, 012021. [CrossRef]

78. Vittoriosi, A.; Brandner, J.J.; Dittmeyer, R. A sensor-equipped microchannel system for the thermal characterization of rarefied gas flows. *Exp. Therm. Fluid Sci.* **2012**, *41*, 112–120. [CrossRef]

79. Kim, M.M.; Giry, A.; Mastiani, M.; Rodrigues, G.O.; Reis, A.; Mandin, P. Microscale thermometry: A review. *Microelectron. Eng.* **2015**, *148*, 129–142. [CrossRef]

80. Yang, Y.; Morini, G.L.; Brandner, J.J. Experimental analysis of the influence of wall axial conduction on gas-to-gas micro heat exchanger effectiveness. *Int. J. Heat Mass Transf.* **2014**, *69*, 17–25. [CrossRef]

81. Genix, M.; Vairac, P.; Cretin, B. Local temperature surface measurement with intrinsic thermocouple. *Int. J. Therm. Sci.* **2009**, *48*, 1679–1682. [CrossRef]

82. Makinwa, K.A.A. Smart temperature sensors in standard CMOS. *Procedia Eng.* **2010**, *5*, 930–939. [CrossRef]

83. Abram, C.; Fond, B.; Beyrau, F. Temperature measurement techniques for gas and liquid flows using thermographic phosphor tracer particles. *Prog. Energy Combust. Sci.* **2018**, *64*, 93–156. [CrossRef]

84. Haslam, R.T.; Chappell, E.L. The Measurement of the Temperature of a Flowing Gas. *Ind. Eng. Chem.* **1925**, *17*, 402–408. [CrossRef]

85. Hocker, G.B. Fiber-optic sensing of pressure and temperature. *Appl. Opt.* **1979**, *18*, 1445–1448. [CrossRef] [PubMed]

86. Bhatia, V.; Vengsarkar, A.M. Optical fiber long-period grating sensors. *Opt. Lett.* **1996**, *21*, 692–694. [CrossRef] [PubMed]

87. Morey, W.W.; Meltz, G.; Glenn, W.H. Fiber Optic Bragg Grating Sensors. In Proceedings of the Fiber Optic and Laser Sensors VII, Boston, MA, USA, 5–7 September 1989; Voume 1169.

88. Hayes, A. Available online: https://www.flowcontrolnetwork.com/rtds-vs-thermocouples/ (accessed on 1 April 2019).

89. Kibble, B.P. *Wheatstone Bridge*; McGraw-Hill Education: New York, NY, USA, 2014. [CrossRef]

90. Park, R.; Caroll, R.; Burns, G.B.; Desmaris, R.; Hall, F.; Herzkovitz, M.; MacKenzie, D.; McGuire, E.; Reed, R.; Sparks, L.; et al. *Manual on the Use of Thermocouples in Temperature Measurement*, 4th ed.; ASTM International: West Conshohocken, PA, USA, 1993. [CrossRef]

91. Carullo, A.; Corbellini, S.; Parvis, M.; Vallan, A. A Wireless Sensor Network for Cold-Chain Monitoring. *IEEE Trans. Instrum. Meas.* **2009**, *58*, 1405–1411. [CrossRef]

92. Yang, J. A Silicon Carbide Wireless Temperature Sensing System for High Temperature Applications. *Sensors* **2013**, *13*, 1884. [CrossRef] [PubMed]

93. Popovic, Z.; Momenroodaki, P.; Scheeler, R. Toward wearable wireless thermometers for internal body temperature measurements. *IEEE Commun. Mag.* **2014**, *52*, 118–125. [CrossRef]

94. Bagavathiappan, S.; Lahiri, B.B.; Saravanan, T.; Philip, J.; Jayakumar, T. Infrared thermography for condition monitoring—A review. *Infrared Phys. Technol.* **2013**, *60*, 35–55. [CrossRef]

95. Heyes, A.L.; Seefeldt, S.; Feist, J.P. Two-colour phosphor thermometry for surface temperature measurement. *Opt. Laser Technol.* **2006**, *38*, 257–265. [CrossRef]

96. Chen, T.-F.; Nguyen, K.T.; Wen, S.-S.L.; Jen, C.-K. Temperature measurement of polymer extrusion by ultrasonic techniques. *Meas. Sci. Technol.* **1999**, *10*, 139–145. [CrossRef]

97. Goss, L.P.; Smith, A.A.; Post, M.E. Surface Thermometry by Laser-Induced Fluorescence. *Rev. Sci. Instrum.* **1989**, *60*, 3702. [CrossRef]

98. Moya, F.; Druet, S.A.J.; Taran, J.P.E. Gas spectroscopy and temperature measurement by coherent Raman anti-stokes scattering. *Opt. Commun.* **1975**, *13*, 169–174. [CrossRef]

99. Bertsch, W. Two-Dimensional Gas Chromatography. Concepts, Instrumentation, and Applications—Part 1: Fundamentals, Conventional Two-Dimensional Gas Chromatography, Selected Applications. *J. High Resolut. Chromatogr.* **1999**, *22*, 647–665. [CrossRef]

100. Bertsch, W. Two-Dimensional Gas Chromatography. Concepts, Instrumentation, and Applications—Part 2: Comprehensive Two-Dimensional Gas Chromatography. *J. High Resolut. Chromatogr.* **2000**, *23*, 167–181. [CrossRef]

101. Suslick, K.S. *Kirk-Othmer Encyclopedia of Chemical Technology*; J. Wiley & Sons: New York, NY, USA. [CrossRef]

102. Hirt, R.C.; Vandenbelt, J.M. Ultraviolet Spectrometry. *Anal. Chem.* **1964**, *36*, 308–312. [CrossRef]

103. Schrötter, H.W. Update of reviews on Raman spectra of gases I. Linear Raman spectroscopy. *J. Mol. Struct.* **2003**, *661–662*, 465–468.

104. Armentrout, P.B. Mass Spectrometric Methods for the Determination of Thermodynamic Data. In *The Encyclopedia of Mass Spectrometry*; Gross, M.L., Caprioli, R.M., Eds.; Elsevier: Boston, MD, USA, 2016; pp. 231–239. [CrossRef]

105. Grayson, M.A. A History of Gas Chromatography Mass Spectrometry (GC/MS). In *The Encyclopedia of Mass Spectrometry*; Gross, M.L., Caprioli, R.M., Eds.; Elsevier: Boston, MD, USA, 2016; pp. 152–158. [CrossRef]

106. Danieli, E.; Perlo, J.; Duchateau, A.L.L.; Verzijl, G.K.M.; Litvinov, V.M.; Blümich, B.; Casanova, F. On-Line Monitoring of Chemical Reactions by using Bench-Top Nuclear Magnetic Resonance Spectroscopy. *ChemPhysChem* **2014**, *15*, 3060–3066. [CrossRef]

107. Guhl, S. Available online: https://opus4.kobv.de/opus4-bam/frontdoor/index/index/docId/39424 (accessed on 1 April 2019).

108. Killner, M.H.M.; Garro Linck, Y.; Danieli, E.; Rohwedder, J.J.R.; Blümich, B. Compact NMR spectroscopy for real-time monitoring of a biodiesel production. *Fuel* **2015**, *139*, 240–247. [CrossRef]

109. Kreyenschulte, D.; Paciok, E.; Regestein, L.; Blümich, B.; Büchs, J. Online monitoring of fermentation processes via non-invasive low-field NMR. *Biotechnol. Bioeng.* **2015**, *112*, 1810–1821. [CrossRef] [PubMed]

110. Sanders, J.K.M.; Hunter, B.K. *Modern NMR Spectroscopy: A Guide for Chemists*; Oxford University Press: Oxford, UK, 1988.

111. Brandner, J.J.; Maikowske, S.; Vittoriosi, A. A New Microstructure Device for Efficient Evaporation of Liquids. *J. Therm. Sci. Technol.* **2012**, *7*, 414–424. [CrossRef]

112. Henning, T.; Brandner, J.J.; Schubert, K. High-speed imaging of flow in microchannel array water evaporators. *Microfluid. Nanofluidics* **2005**, *1*, 128–136. [CrossRef]

113. Henning, T.; Brandner, J.J.; Schubert, K.; Lorenzini, M.; Morini, G.L. Low-Frequency Instabilities in the Operation of Metallic Multi-Microchannel Evaporators. *Heat Transf. Eng.* **2007**, *28*, 834–841. [CrossRef]

114. Brandner, J.J.; Emig, G.; Liauw, M.A.; Schubert, K. Fast temperature cycling in microstructure devices. *Chem. Eng. J.* **2004**, *101*, 217–224. [CrossRef]

115. Luther, M.; Brandner, J.J.; Kiwi-Minsker, L.; Renken, A.; Schubert, K. Forced periodic temperature cycling of chemical reactions in microstructure devices. *Chem. Eng. Sci.* **2008**, *63*, 4955–4961. [CrossRef]

116. Luther, M.; Brandner, J.J.; Schubert, K.; Renken, A.; Kiwi-Minsker, L. Novel design of a microstructured reactor allowing fast temperature oscillations. *Chem. Eng. J.* **2007**, *135*, S254–S258. [CrossRef]

117. Hellén, H.; Schallhart, S.; Praplan, A.P.; Petäjä, T.; Hakola, H. Using in situ GC-MS for analysis of C2–C7 volatile organic acids in ambient air of a boreal forest site. *Atmos. Meas. Tech.* **2017**, *10*, 281–289. [CrossRef]

118. Köckenberger, W.; Panfilis, C.D.; Santoro, D.; Dahiya, P.; Rawsthorne, S. High resolution NMR microscopy of plants and fungi. *J. Microsc.* **2004**, *214*, 182–189. [CrossRef] [PubMed]

119. Uecker, M.; Zhang, S.; Voit, D.; Karaus, A.; Merboldt, K.D.; Frahm, J. Real-time MRI at a resolution of 20 ms. *NMR Biomed.* **2010**, *23*, 986–994. [CrossRef] [PubMed]

120. Parker, D.L.; Smith, V.; Sheldon, P.; Crooks, L.E.; Fussell, L. Temperature distribution measurements in two-dimensional NMR imaging. *Med. Phys.* **1983**, *10*, 321–325. [CrossRef]

121. Kleinberg, R.L. Utility of NMR T2 distributions, connection with capillary pressure, clay effect, and determination of the surface relaxivity parameter ρ2. *Magn. Reson. Imaging* **1996**, *14*, 761–767. [CrossRef]

122. Thorn, R.; Johansen, G.A.; Hammer, E.A. Recent developments in three-phase flow measurement. *Meas. Sci. Technol.* **1997**, *8*, 691–701. [CrossRef]

123. Weber, H.W.; Kimmich, R. Anomalous segment diffusion in polymers and NMR relaxation spectroscopy. *Macromolecules* **1993**, *26*, 2597–2606. [CrossRef]

124. Akoka, S.; Barantin, L.; Trierweiler, M. Concentration Measurement by Proton NMR Using the ERETIC Method. *Anal. Chem.* **1999**, *71*, 2554–2557. [CrossRef]

125. Korvink, J.G.; Badilita, V.; Bordonali, L.; Jouda, M.; Mager, D.; MacKinnon, N. Nuclear magnetic resonance microscopy for in vivo metabolomics, digitally twinned by computational systems biology, needs a sensitivity boost. *Sens. Mater.* **2018**, *30*, 157–166. [CrossRef]

126. MacKinnon, N.; While, P.T.; Korvink, J.G. Novel selective TOCSY method enables NMR spectral elucidation of metabolomic mixtures. *J. Magn. Reson.* **2016**, *272*, 147–157. [CrossRef] [PubMed]

127. Badilita, V.; Kratt, K.; Baxan, N.; Anders, J.; Elverfeldt, D.; Boero, G.; Hennig, J.; Korvink, J.G.; Wallrabe, U. 3D solenoidal microcoil arrays with CMOS integrated amplifiers for parallel MR imaging and spectroscopy. In Proceedings of the 2011 IEEE 24th International Conference on Micro Electro Mechanical Systems, Cancun, Mexico, 23–27 January 2011; pp. 809–812.

128. Badilita, V.; Fassbender, B.; Kratt, K.; Wong, A.; Bonhomme, C.; Sakellariou, D.; Korvink, J.G.; Wallrabe, U. Microfabricated Inserts for Magic Angle Coil Spinning (MACS) Wireless NMR Spectroscopy. *PLoS ONE* **2012**, *7*, e42848. [CrossRef] [PubMed]

129. Spengler, N.; Höfflin, J.; Moazenzadeh, A.; Mager, D.; MacKinnon, N.; Badilita, V.; Wallrabe, U.; Korvink, J.G. Heteronuclear Micro-Helmholtz Coil Facilitates μm-Range Spatial and Sub-Hz Spectral Resolution NMR of nL-Volume Samples on Customisable Microfluidic Chips. *PLoS ONE* **2016**, *11*, e0146384. [CrossRef] [PubMed]

micromachines

MDPI

Article

Effect of the Pitot Tube on Measurements in Supersonic Axisymmetric Underexpanded Microjets

Sergey G. Mironov, Vladimir M. Aniskin *, Tatiana A. Korotaeva and Ivan S. Tsyryulnikov

Khristianovich Institute of Theoretical and Applied Mechanics, Siberian Branch of Russian Academy of Sciences, Novosibirsk 630090, Russia; mironov@itam.nsc.ru (S.G.M.); korta@itam.nsc.ru (T.A.K.); tsivan@itam.nsc.ru (I.S.T.)
* Correspondence: aniskin@itam.nsc.ru

Received: 15 February 2019; Accepted: 25 March 2019; Published: 6 April 2019

Abstract: This paper describes the results of methodical investigations of the effect of the Pitot tube on measurements of gas-dynamic parameters of supersonic axisymmetric underexpanded real and model microjets. Particular attention is paid to distortions of Pitot pressure variations on the jet axis associated with the wave structure of the jet and to distortions of the supersonic core length. In experiments with model jets escaping from nozzles with diameters ranging from 0.52 to 1.06 mm into the low-pressure chamber, the measurements are performed by the Pitot tubes 0.05 to 2 mm in diameter. The results are analyzed together with the earlier obtained data for real microjets escaping from nozzles with diameters ranging from 10 to 340 μm where the parameters of real microjets were determined by the Pitot microtube 12 μm in diameter. Interaction of the Pitot tube with an unsteady jet in the laminar-turbulent transition region is investigated; the influence of this interaction on Pitot pressure measurements is determined, and a physical interpretation of this phenomenon is provided.

Keywords: supersonic microjets; Pitot tube

1. Introduction

The classification of channels in terms of their size, which was proposed in [1] and which is based on Knudsen numbers, implies that microchannels are those having the characteristic size from 10 to 200 μm. Taking into account this classification and extending it to micronozzles, we can assume that microjets are those escaping from nozzles with the characteristic size smaller than 200 μm (diameter for axisymmetric nozzles and height for plane and rectangular nozzles).

The measurement of flow parameters in gas microjets is a difficult diagnostic problem. Because of the small sizes of microjets, the use of a number of known methods for diagnosing gas flows is limited, e.g., shadowgraphy and Schlieren diagnostics, laser-induced fluorescence, particle image velocimetry, and hot-wire anemometry. The most suitable tool for microjets is the Pitot tube. However, the main problem in using this method is fabrication of Pitot tubes much smaller than the characteristic scale of microjet flows to ensure local measurements of jet parameters and to avoid jet flow distortions.

There are only a few investigations of supersonic microjet flows [2–8]. The most comprehensive study of supersonic microjet flows was performed in [7,8], where supersonic jets escaping from micronozzles 10 to 340 μm in diameter in the range of Reynolds numbers from 300 to 27,000 were considered. The authors [7] managed to fabricate a Pitot tube of 12 μm in diameter, which is the smallest sensor of this kind at the moment. The authors [7] also obtained pioneering data on the size of the gas-dynamic structure cell (barrel) and supersonic core length. It was found that the barrel size in supersonic microjets do not differ much from the barrel size in supersonic macrojets. However, it was demonstrated that the supersonic core length in microjets and macrojets is significantly different. The authors [8] performed experiments with the use of hot-wire anemometry for studying integral fluctuations of the mass flow rate of supersonic microjets. It was shown that the increase in the supersonic core length in microjets is associated with the laminar flow in the mixing layer of the jets, whereas the drastic decrease in the supersonic core length is caused by turbulization of the mixing layer of the jets.

The main parameter in investigations of supersonic jets is the jet pressure ratio (JPR), which is defined as the ratio of the static pressure at the nozzle exit to the ambient pressure. The JPR for microjets escaping into the atmosphere with the ambient pressure of about 1 atm is rigorously related to the Reynolds number via the pressure (and, hence, density) in the settling chamber, i.e., each Reynolds number corresponds to a certain JPR value. In the case of escaping of supersonic microjets into the ambient atmosphere [7] from nozzles 10–340 μm in diameter, the Reynolds numbers ranged from 300 to 27,000, with JPR changing from 1 to 4. However, the same Reynolds numbers can be reached by organizing a supersonic flow from a macro-sized nozzle into a low-pressure ambient space. Having one macro-sized nozzle and choosing appropriate values of the ambient pressure and the pressure in the settling chamber, one can simulate a supersonic flow of microjets escaping from micronozzles of different diameters into the atmosphere. Such macrojets can be called low-Reynolds-number jets or model microjets. A certain pair of numbers (Reynolds number and JPR) reached in the case of a jet escaping from the macronozzle to the low-pressure region always corresponds to a pair of such numbers for the microjet escaping from the micronozzle of a certain diameter into the atmosphere. The diameter of this nozzle is called the model diameter, because it is used as a basis for flow modeling in terms of the Reynolds number (for the corresponding JPR value). The possibility of modeling the structure of microjets (barrel size and supersonic core length) by using macrojets in terms of the Reynolds number calculated on the basis of the macronozzle diameter and gas parameters at the nozzle exit was discussed in [9–11]. It was also demonstrated there that the Reynolds number at which the laminar jet flow transforms to the turbulent state is identical for both real microjets and model microjets.

Modeling of microjets on the basis of macrojet studies offers a possibility of extending investigations of real microjets. In particular, a problem important for practice can be solved—determining the maximum possible diameter of the Pitot tube with respect to the nozzle diameter, which would ensure correct measurements of the wave structure and correct determination of the supersonic core length in microjets. For solving this problem, we studied supersonic jets escaping from nozzles 0.52, 0.72, and 1.06 mm in diameter into a low-pressure chamber. The results were analyzed together with the data of [7,8], where the parameters of real microjets escaping from nozzles with diameters ranging from 10 to 340 μm were determined by the Pitot microtube 12 μm in diameter. Moreover, it was of interest to study the influence of an unsteady axisymmetric underexpanded microjet flow in the laminar-turbulent transition region on results measured by the Pitot tube.

2. Experimental Equipment

The experimental studies were performed for model air microjets escaping from sonic nozzles with diameters d = 0.52, 0.72, and 1.06 mm in a low-pressure jet setup. The nozzle scheme is shown in Figure 1. The nozzles had a different internal geometry.

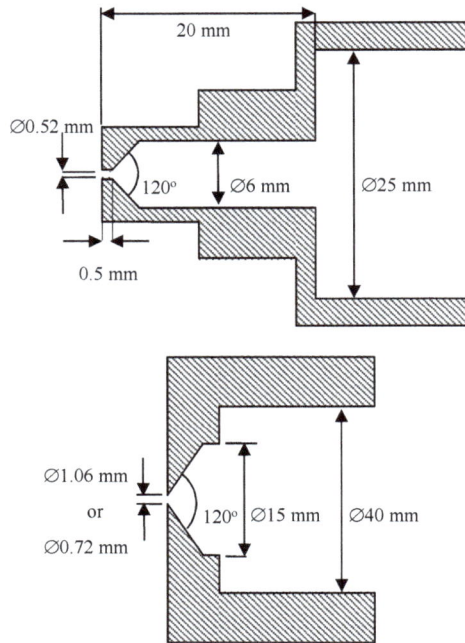

Figure 1. Nozzle scheme.

The low-pressure jet setup is described in detail in [9] and was a sealed chamber evacuated by a vacuum pump with a nozzle system containing a sensor for measuring the stagnation pressure P_0 of the jet being mounted in the chamber wall. Another pressure sensor was placed in the chamber for measuring the ambient pressure P_a of the medium to which the jet was exhausted. For example, for the jet escaping from the nozzle 1.06 mm in diameter to modeling the jet from the real size nozzle of 15.1 μm in diameter, the low-pressure ambient space P_a was set at 1500 Pa and P_0 ranged from 3000 to 17,000 Pa (that corresponded with JPR from 1.06 to 6 and Reynolds number from 460 to 2600).

The Pitot tube was mounted on a traversing gear, which ensured the possibility of probe motion in the interval from 0 to 200 mm. The Pitot tubes used in the present study are listed in Table 1.

Table 1. Pitot tube size and relaxation time.

D, mm	D_{int}, mm	τ, s
0.075	0.05	100
0.1	0.075	17
0.4	0.17	0.33
0.6	0.3	0.06
0.7	0.4	0.02
1.1	0.8	<0.01
2	1.5	<0.01
3	2	<0.01

The data in Table 1 are the outer diameter of the Pitot tube D, its internal diameter D_{int}, and characteristic relaxation time τ of pressure in the probe with this particular tube. The characteristic relaxation time τ was determined from the time dependence of the Pitot pressure P_0' in the case of rapid interruption/release of the jet. The Pitot tubes with the external diameters of 0.075 and 0.1 mm were made of glass, while the Pitot tubes of other diameters were made of stainless steel or copper.

The nozzle was mounted outside the low-pressure chamber and was equipped with microscrews for changing the nozzle position. With the use of these microscrews, the nozzle was mounted in such a way that the jet axis coincided with the line of Pitot tube motion. Their coincidence was monitored using transverse pressure distributions. The pressure was measured by Pitot tubes with external diameters of 0.075 and 0.1 mm and was performed at the points on the jet axis with allowance for the characteristic relaxation time. The pressure measurements by other Pitot tubes were performed with continuous motion of the tubes along the jet with a velocity of 0.25 mm/s. The velocity of Pitot tube motion was determined experimentally—it was chosen in such a way that the pressure distributions measured by the Pitot tube with the external diameter of 0.4 mm moving along the jet axis away from the nozzle and then back toward the nozzle were identical.

The pressures in the jet setup and in the Pitot tube were measured by TDM4-IV1 differential pressure probes (for pressures up to 0.4 bar) or TDM2-IV2 pressure probes (for pressures up to 1 bar). The electric signals from the pressure probes were digitized by a 12-bit analog to digital converter with frequency of 20 Hz.

The data determined in the experiments were the axial distributions of the Pitot pressure P_0' as functions of the nozzle diameter d, Pitot tube diameter D, and jet pressure ratio n. The JPR value is related to the pressure in the settling chamber by the formula $1/[1 + (\gamma - 1)/2]^{\gamma/(\gamma-1)}$, where γ is the ratio of specific heats of the gas in the jet. The flow at the supersonic jet axis transforms from supersonic to subsonic when P_0' reaches the value $P_a[1 + (\gamma - 1)/2]^{\gamma/(\gamma-1)}$, where P_a is the ambient pressure. The distance from the nozzle exit to the point where the supersonic flow transforms to the subsonic flow was determined as the supersonic core length L_C. The jet setup structure allowed made it possible to establish and fix the pressure of the surrounding space P_a when the jet flows into the chamber.

In addition, we also performed experiments recording the noise of the supersonic jet in the test section of the jet setup (both the free jet and the jet interacting with the Pitot tube). In particular, we measured the spectrum of acoustic oscillations in the ambient space around the jet. The spectral composition and the amplitude of the jet noise were correlated with the position of the frontal end of the Pitot tube and the value of the Pitot pressure P_0' as the Pitot tube was moved along the jet axis. The acoustic oscillations were detected by a PCB 132A31 piezoelectric sensor with a frequency range up to 500 kHz. The acoustic sensor of 3 mm in diameter was located at a distance of 10 mm from the jet axis and 27 mm from the nozzle. The axis of the direction indicatrix of the acoustic sensor was directed to the point on the jet axis located at a distance of 15 mm from the nozzle. The maximum level of acoustic oscillations of the supersonic jet was expected to be observed in the vicinity of this point.

3. Numerical Simulations

ANSYS Fluent commercial software (version 12) was used to simulate interaction of the Pitot tubes with outer diameters D = 0.4, 0.7, 1.1, and 2 mm with a supersonic underexpanded air jet, which escaped into the atmosphere from a nozzle with a diameter d = 0.5 and 1.06 mm at Reynolds numbers corresponding to modeling of a microjet escaping from a nozzle 16.1 μm in diameter and jet pressure ratio n = 2. Unsteady Reynolds-averaged Navier–Stokes equations were solved in an axisymmetric formulation with the use of the k-ω SST (shear stress transport) turbulence model. The solution procedure was performed by a "density-based" solver (term of the ANSYS Fluent software) and a second-order implicit scheme; convective fluxes were differentiated with the use of the Roe scheme. The computational domain included the settling chamber, the space around the jet, and the Pitot tube closed at the back side. The results of the computations were the unsteady and steady fields of Mach numbers and density around the jet and the Pitot tube, as well as the time-averaged pressure P_0' at the position of the frontal end of the tube at the jet axis at a distance equal to one half of the length of the first gas-dynamic cell of the jet structure. The computed data were compared with the measured results.

4. Results and Discussion

4.1. Effect of the Ratio of the Pitot Tube Diameter to the Nozzle Diameter

As an example, Figure 2 shows a typical distribution of pressure along the jet axis. The data was taken from a supersonic (real) microjet escaping from the nozzle 16.1 μm in diameter by the Pitot tube 12 μm in diameter. The pressure curve has a quasi-periodic form associated with the shock wave structure of the jet. The peaks of the axial distribution of pressure were located at the ends of the jet barrels. The barrel length is one of the basic characteristics of supersonic jets. In turbulent jets, the barrel size and their transverse sizes from the first to the last barrel decrease owing to the evolution of the mixing layer of the jet. The parameters usually considered by most researchers are the length of the first barrel, the mean size of the barrels (mean sizes of the second, third, and fourth barrels), position and size of the Mach disk, and supersonic core length.

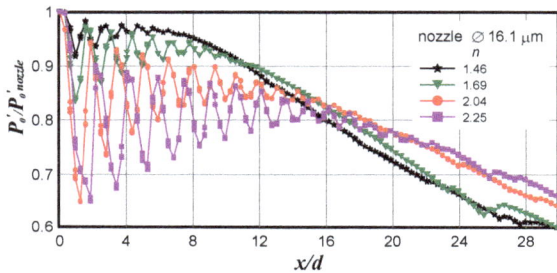

Figure 2. Axial pressure distribution measured by the Pitot tube in the jet escaping from the nozzle 16.1 μm in diameter.

The measurements of the first barrel length and the mean size of the barrels in real microjets performed by the Pitot tube 12 μm in diameter [7] showed that the mean barrel size was in excellent agreement with the first barrel size in macrojets [11] (except for those JPR values for which the mixing layer produces a significant effect). However, the first barrel size turned out to be greater than the first barrel size in macrojets almost for all jets, as shown in Figure 3, for all JPR values.

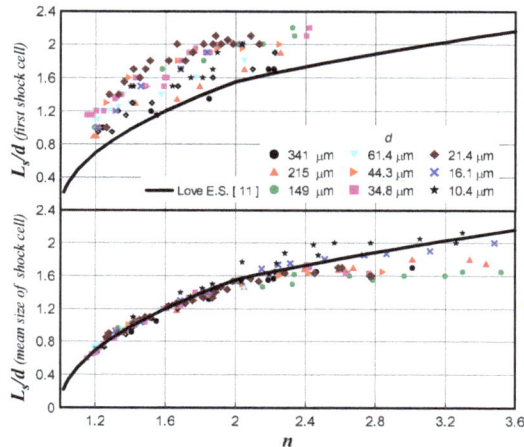

Figure 3. First barrel size and mean size of the barrels in real microjets.

To understand the reason for systematic overprediction of the first barrel size in microjets determined on the basis of the Pitot tube measurements, the following problem was solved. Exhaustion of a real supersonic microjet from the 16.1 μm nozzle into the ambient space (P_a = 1 atm, P_0 = 3.86 atm) at n = 2.04 was considered. The Pitot tube was located on the axis of the jet at different distances x/d. The flow around the Pitot tube was calculated and the pressure inside the Pitot tube was determined. Additionally, the P_0' in the free jet was calculated. The results of numerical simulation and experimental data are compared in Figure 4.

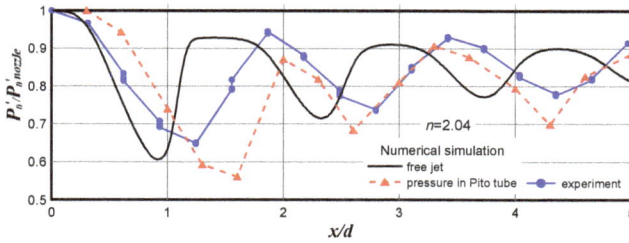

Figure 4. Comparison of the experimental and predicted data for a real microjet escaping from the nozzle 16.1 μm in diameter at n = 2.04.

It is seen that the calculated value of the pressure in the Pitot microtube located in the jet flow correlates fairly well with the experimental data. However, both these dependences were displaced with respect to the P_0' curve in the free jet approximately by 0.5 d. This shifting was induced by the detached shock wave formed on the Pitot microtube. As the Pitot tube measures the pressure behind the shock wave, we actually have a situation where x/d in Figure 4 corresponds to the Pitot tube position and P_0' corresponds to the shock wave position ahead of the Pitot tube. The distance between the Pitot tube and the shock wave ahead of the tube was not constant over the supersonic core of the jet; it varies periodically in proportion to the Mach number at the jet axis. The shift of the experimental data with respect to the P_0' curve for the free jet should be also affected by the ratio of the nozzle and Pitot tube diameters.

Numerical calculations were performed for free jets escaping from micronozzles 50, 75, 100, and 150 μm in diameter at n = 1.8 and also for the same jets impinging onto the Pitot tube 12 μm in diameter. The results are plotted in Figure 5.

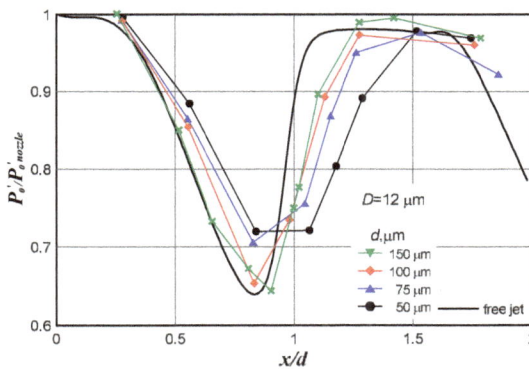

Figure 5. Calculated impingement of jets escaping from nozzles of different diameters onto the Pitot tube 12 μm in diameter.

The distributions of P_0' calculated for jets having an identical JPR value but escaping from nozzles of different diameters almost coincide in the dimensionless coordinates. This distribution is shown by the solid curve in Figure 5. As the ratio d/D increases, the effect of the Pitot tube diameter on the accuracy of determining the position of the first barrel of the jet becomes less pronounced.

Similar results were obtained in model microjet calculations. In particular, Figure 6 shows the results of calculating the jet escaping from the nozzle 1.06 mm in diameter, modeling jet exhaustion from the nozzle 16.1 µm in diameter at $n = 2$. Figure 6 shows the calculated results for the free jet (solid line) and the calculated pressure in the Pitot tube located in the jet. The two sizes of the Pitot tube diameters were used in calculations—1.1 mm/0.8 mm and 0.2 mm/0.085 mm (outer/internal diameter). It is seen that the results were significantly affected by the shock wave position.

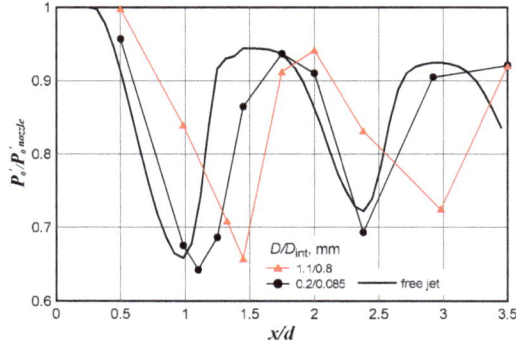

Figure 6. Calculation of the model microjet escaping from the nozzle 1.06 mm in diameter (modeling exhaustion from the nozzle 16.1 µm in diameter) and impinging onto the Pitot tubes of different diameters.

Figure 7 shows the P_0' distributions along the axis of the jet escaping from the nozzle with the diameter 1.06 mm and the jet pressure ratio $n = 2$. The pressure in the chamber was chosen in a way to model the flow from the nozzle 16.1 µm in diameter. The measurements were performed by the Pitot tubes with different diameters. The plots in Figure 7b are fragments from Figure 7a. The calculated data are shown by the solid black curve. The point of intersection of the curves with the dotted line corresponding to the pressure of the transition from the supersonic to subsonic flow shows the supersonic core length.

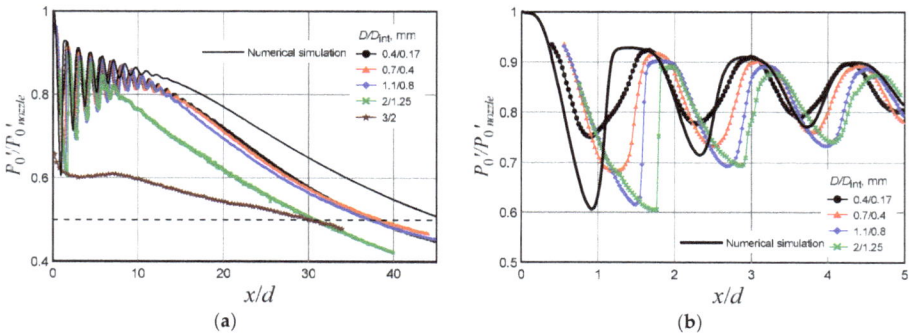

Figure 7. Pressure distribution in the jet escaping from the nozzle 1.06 mm in diameter (modeling exhaustion from the nozzle 16.1 µm in diameter). The plots in (**b**) are fragments from (**a**).

It is seen that the calculations predict a greater supersonic core length than that obtained in the experiments, as shown in Figure 7a. This difference is explained by the presence of a certain level of turbulence in the mixing layer of the real jet, as compared to the completely laminar model implied in the computations. The supersonic core lengths calculated from the experimental data obtained by the Pitot tubes with diameters up to 1.1/0.8 mm coincide with each other. For large Pitot tubes (2/1.25–3/2 mm), the supersonic core length was 20% smaller. The possibility of reliable determination of the supersonic core length by the Pitot tubes with comparatively large diameters was explained by the wide profile of the transverse pressure distribution in the jet cross section where the supersonic core length was determined. In this cross section, the maximum of the pressure distribution was fairly wide, and the tube diameter did not induce significant errors.

The pressure distribution near the nozzle exit was characterized by shifting of the experimental curves with respect to each other as the Pitot tube diameter increased, as shown in Figure 7b. As was demonstrated earlier, this shifting was determined by the detached shock wave. The greater the Pitot tube diameter, the greater the stand-off distance of the shock wave.

Though exhaustion of model microjets as a whole occurs in the continuum regime, there are local regions of reduced density in the jet, which correspond to the transitional flow regime from the viewpoint of the Knudsen number. In particular, the minimum density was observed in the middle of the first barrel, and it was in this region that the maximum influence of rarefaction on the results of Pitot tube measurements can be expected. It is worth mentioning that there was obvious disagreement in Figure 7b between the minimum of the pressure distribution in the first barrel of the jet obtained by the smallest Pitot tube (0.4/0.17 mm) and the calculated data.

At the point of the minimum pressure P_0' in the first barrel of the jet simulating exhaustion from the micronozzle with the diameter $d = 16.1$ μm, as shown in Figure 7b, the calculated gas density was $\rho = 0.013$ kg/m^3, which corresponds to the molecular concentration $n = 2.7 \times 10^{17}$ molecules/cm^3. Then the mean free path of molecules at this point was $\lambda = \frac{1}{\sqrt{2}n\sigma} = 2.5 \times 10^{-3}$ cm. Here σ is the molecule collision cross section ($\sigma \approx 10^{-15}$ cm^2). Thus, the local Knudsen number Kn directly calculated on the basis of the mean free path and Pitot tube size ($D = 0.4$ mm or $D_{int} = 0.17$ mm) is Kn $= \lambda/D$ 0.06 or 0.15, respectively. These values correspond to the transitional regime of the flow around the Pitot tube (from continuum to free-molecular). According to [12], this is responsible for reduction of the Pitot tube readings, which was actually observed in experiments with these Pitot tube sizes. Moreover, because of gas rarefaction, the thickness of the shock wave ahead of the Pitot tube increases; as a consequence, the dependence $P_0'(x/d)$ becomes less steep, which was again observed in experiments.

Despite the visible shift of the experimental data with respect to the true distribution of P_0' at the jet axis, the mean size of the jet barrels was determined fairly accurately, which was confirmed by the experimental data in Figure 3. From this viewpoint, the Pitot tube provides reliable results, even if the Pitot tube diameter was comparable with the nozzle diameter.

Figure 8a shows the normalized length of the supersonic core of the jet as a function of the ratio of the outer diameter of the Pitot tube to the nozzle diameter D/d for several values of n. The supersonic core length was normalized to its value measured by the Pitot tube with the diameter $D = 0.4$ mm. It was seen that the normalized length of the supersonic core decreases as the ratio D/d increases. If the measured data shown in Figure 8a are plotted in the coordinates $D/(dn)$, then all points fall on one decreasing curve, as is shown in Figure 8b.

In the region of the jet flow transition from the supersonic to subsonic state, the velocity profile in the jet becomes significantly expanded in the transverse direction, and the Pitot tube readings become less sensitive to the Pitot tube diameter. As a result, the condition for the Pitot tube diameter with respect to the nozzle diameter with 3% error is defined as $D/(dn) \leq 0.5$.

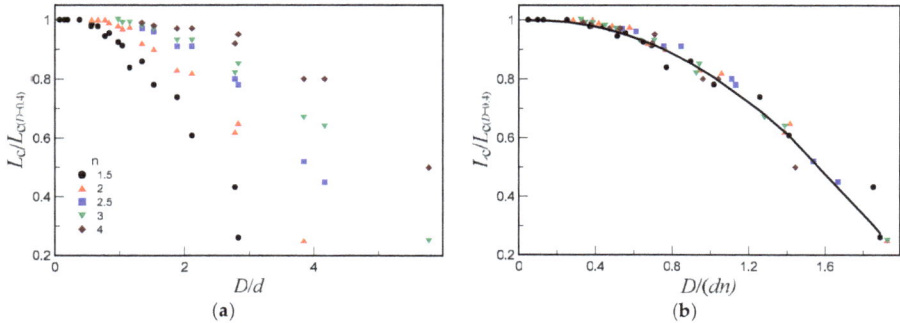

Figure 8. Normalized length of the supersonic core of the jet versus the ratio of the Pitot tube and nozzle diameters (**a**) and versus the ratio of the Pitot tube diameter D and nozzle diameter d multiplied by n (**b**).

4.2. Effect of Jet Flow Unsteadiness on Pitot Tube Measurements

The measurements of the Pitot pressure distributions $P_0'(x/d)$ on the jet axis at Reynolds numbers close to the conditions of the laminar-turbulent transition in the jet revealed significant steady fluctuations of the Pitot pressure. Such variations can be observed in both real, as shown in Figure 9, and model, as shown in Figure 10, microjets. In both real and model microjets, this effect was manifested for nozzle diameters 16–26 μm in a moderate range of JPR values.

As an example, Figure 9 shows the pressure distributions in real microjets escaping from the nozzles 21.4, as shown in Figure 9a, and 16.1 μm, as shown in Figure 9b, in diameter. All data in Figure 9 refer to the laminar regime of jet exhaustion, and minor pressure fluctuations were observed only in a small range of JPR values.

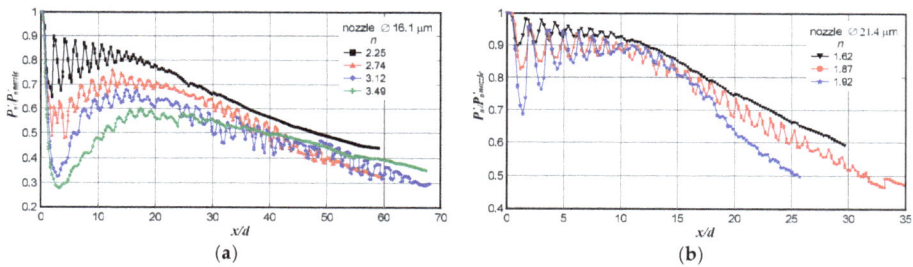

Figure 9. Pressure distributions in real microjets escaping from nozzles 21.4 (**a**) and 16.1 μm in diameter (**b**).

Figure 10 illustrates the emergence of pressure fluctuations with increasing JPR for the model microjet escaping from the nozzle with $d = 1.06$ mm (domain 1) and modeling jet exhaustion from the nozzle 16.1 μm in diameter. The measurements were performed by the Pitot tube with the diameter $D = 0.4$ mm. Exhaustion of the model microjet corresponds to the laminar flow regime at $n = 2$ and 2.2, to the transitional flow regime at $n = 2.35$, and to the turbulent flow regime at $n = 3$ (when the sharp drop-off of the pressure has occurred). Domain 2 in Figure 10b shows the measurements in the quasi-turbulent region of the jet, where random overshoots of the pressure P_0' were observed. It can be noted that the spatial period of pressure fluctuations at the jet axis in domain 1 coincides with the spatial period of pressure fluctuations on the cells of the wave structure of the jet near the nozzle.

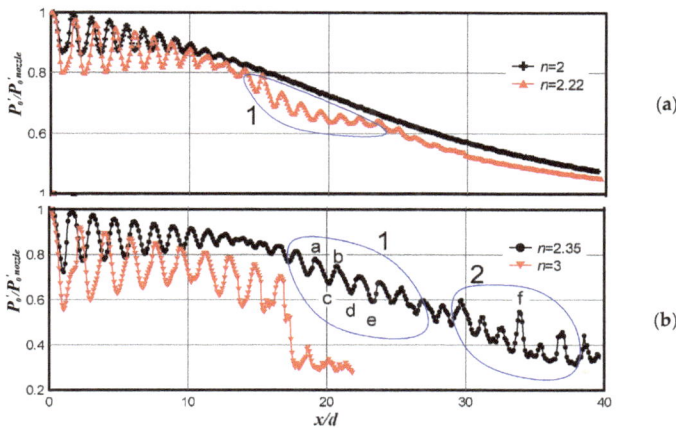

Figure 10. Pressure distribution in the model microjet escaping from the nozzle 1.06 mm in diameter (modeling exhaustion from the nozzle 16.1 μm in diameter). (**a**) $n = 2$ and 2.22, (**b**) $n = 2.35$ and 3.

The letters in Figure 10b indicate the points on the jet axis where the pressure evolution with time was measured. The measured results are shown in Figure 11. It is seen that the pressure at the measurement points in domain 1 were almost steady, whereas the pressure in domain 2 behaved randomly with time, which can be attributed to the transition of turbulent spots in the jet. Possibly, the frequency of passage of the turbulent spots (pressure jumps) was appreciably higher than that shown in the plot, but the pressure fluctuations in the experiments were averaged because of the finite characteristic time of pressure relaxation in the Pitot probe, and the sensor provided only an averaged pattern.

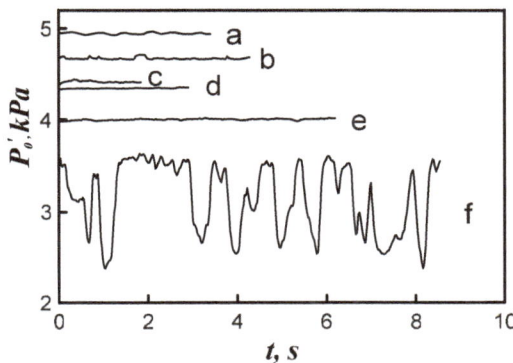

Figure 11. Time evolution of the pressure P_0' at the points (**a–f**) on the jet axis shown in Figure 10b.

This behavior of pressure in domain 1 was typical for the emergence of acoustic feedback in a steady flow of supersonic underexpanded jets where screw instability modes B and C develop near the laminar-turbulent transition region [13]. The frequency of these modes was close to the frequency of acoustic waves in the ambient space, which, in turn, was equal to the velocity of sound in the ambient space divided by the doubled length of the gas-dynamic cells of the wave structure of the jet. The evolution and enhancement of these disturbances is the reason for global instability of underexpanded jets.

In the plane of the normalized distance from the nozzle versus frequency, Figure 12 shows the spectra of acoustic oscillations detected by a piezoelectric sensor mounted near the jet in the case of motion of the Pitot tube with the diameter 0.4 mm for three values of n—1.75, 2.5, and 3. It is seen that there was one frequency of oscillations for $n = 1.75$, as shown in Figure 12a, and two frequencies for $n = 2.5$ and $n = 3$, as shown in Figure 12b,c. The frequency and amplitude of acoustic oscillations depends on the Pitot tube position on the jet axis.

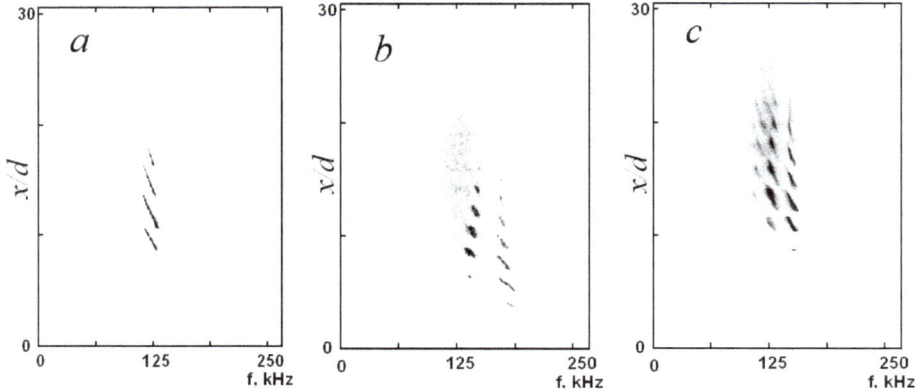

Figure 12. Spectra of acoustic oscillations generated by the underexpanded air jet in the plane of the normalized distance from the nozzle versus frequency for $n = 1.75$ (**a**), 2.5 (**b**), and 3 (**c**) for $d = 0.72$ mm and $D = 0.4$ mm.

The lower and upper frequencies of oscillations coincide with the frequencies of modes B and C of screw perturbations of the underexpanded jet, respectively. The periodicity of the emergence and vanishing of acoustic oscillations during the motion of the Pitot tube along the jet axis coincides with the periodicity of passing of the Pitot tube through the gas-dynamic cells of the wave structure of the jet. As the Pitot tube moves along the jet axis, Figure 12 in addition to variations of the amplitude of acoustic oscillations also shows periodic fluctuations of the frequency of these oscillations, which were noted in [14]. In the range $D = 0.4$–0.7 mm, a weak dependence of the amplitude of acoustic oscillations on the outer diameter of the Pitot tube was observed.

As an example, Figure 13 shows the pressure P_0' and acoustic root-mean-square fluctuations of pressure p' at the frequency of mode B in the ambient space as functions of the normalized distance from the nozzle. The pressure P_0' was measured by the Pitot tube with the diameter 0.4 mm. It is seen that the minimum points of P_0' coincide with the peaks of the acoustic oscillations of pressure in the ambient space of the jet.

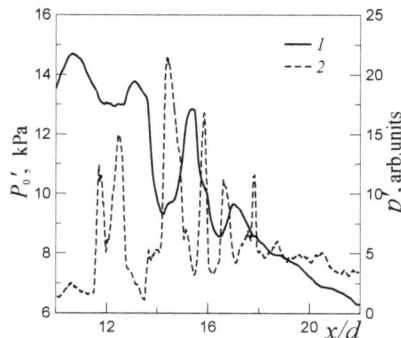

Figure 13. Pitot pressure P_0' (solid curve 1) and amplitude of acoustic oscillations p' of mode B (dotted curve 2) versus the normalized distance from the nozzle; $d = 0.72$ mm, $D = 0.4$ mm, and $n = 3$.

The results obtained offer a simple physical explanation for the emergence of intense fluctuations of the pressure P_0' in the flow region close to the laminar-turbulent transition in the underexpanded jet. Intense fluctuations of screw modes of instability occur and develop in this region of the jet, leading to high-frequency motions of the gas-dynamic cells of the wave structure of the jet in the radial and azimuthal directions with respect to the jet axis with a spatial period equal to the length of two cells of the wave structure of the jet. Such motions were described in many publications dealing with instability of supersonic underexpanded macrojets, e.g., [13,15–17]. Interaction of the Pitot tube tap with an unsteady jet generates acoustic oscillations in the ambient medium, which propagate toward the nozzle exit, thus, creating feedback between the jet and the ambient medium, and can induce either enhancement or suppression of global instability of the jet. Enhancement or suppression depends on the phase of acoustic oscillations at the nozzle exit. In turn, the phase of acoustic oscillations depends on the Pitot tube position in the gas-dynamic cells of the wave structure of the jet. The feedback produces the minimum effects when the Pitot tube was located in a position where the phase at the nozzle exit corresponds to the node of acoustic oscillations. If the phase at the nozzle exit corresponds to the peak of acoustic oscillations, the maximum feedback effect is observed, resulting in jet flow failure, acceleration of mixing with the ambient medium, and decrease in P_0'. These minimums and maximums were reached every time when the Pitot tube moved in one of the cells of the wave structure of the jet.

Thus, the motion of the Pitot tube along the jet axis generates periodic overshoots of acoustic oscillations, as shown in Figure 12. These acoustic oscillations provoke global instability of the underexpanded jet, which leads to acceleration of jet flow mixing with the ambient gas and to a decrease in P_0' with a spatial period equal to the length of the gas-dynamic cells of the wave structure of the jet. From this viewpoint, the amplitude of fluctuations of the pressure P_0' was actually overestimated because of the presence of the Pitot tube in the unsteady jet. Most probably, the true amplitude of Pitot pressure fluctuations would be appreciably lower if there were no Pitot tube in the jet.

5. Conclusions

The influence of the Pitot tube diameter on the pressure distribution $P_0'(x/d)$ and supersonic core length in an underexpanded jet was considered.

It was demonstrated that the experimental distributions of pressure along the axis of both real and model microjets were shifted with respect to the curves calculated for free microjets. It was found that this shifting was caused by the shock wave formed on the tube. As a result, there appears a significant error in determining the first barrel size on the basis of the pressure distribution. However, the shift of the experimental dependence does not affect the mean barrel sizes—they are determined exactly.

Though exhaustion of model jets as a whole occurs in the continuum regime, there were local regions of reduced density in the jet, which correspond to the transitional flow regime from the viewpoint of the Knudsen number, leading to significant distortions of the pressure distributions in the jet in these regions.

The Pitot tube diameter has a minor effect on determining the supersonic core length, which was explained by the wide profile of the transverse pressure distribution in the jet cross section where the supersonic core length was determined. In this cross section, the maximum of the pressure distribution was fairly wide, and the tube diameter does not induce significant errors. It was shown that acceptable accuracy of measurements of the supersonic core length can be provided if the inequality $D/(dn) \leq 0.5$ is satisfied.

The presence of the Pitot tube in the jet in the region of the laminar-turbulent transition was found to produce a significant effect on the results of $P_0'(x/d)$ measurements in the jet flow. A physical mechanism of the emergence of intense fluctuations of the Pitot pressure P_0' on the underexpanded jet axis near the laminar-turbulent transition was proposed. The relationship between these fluctuations and interaction of the unsteady jet with the Pitot tube tap was revealed.

Author Contributions: Conceptualization, V.M.A. and S.G.M.; methodology, S.G.M.; software, T.A.K. and I.S.T.; validation, V.M.A., S.G.M. and I.S.T.; formal analysis, S.G.M.; investigation, S.G.M.; resources, V.M.A.; data curation, S.G.M.; writing—original draft preparation, S.G.M.; writing—review and editing, V.M.A.; supervision, V.M.A.; project administration, V.M.A.; funding acquisition, V.M.A.

Funding: This research was funded by Russian Scientific Foundation, grant number 17-19-01157.

Conflicts of Interest: The authors declare no conflict of interest.

References

1. Kandlikar, S.G. Microchannels and minchannels—History, terminology, classification and current research needs. In Proceedings of the First International Conference on Microchannels and Minichannels, Rochester, NY, USA, 24–25 April 2003; ICMM2003-1000.

2. Scroggs, S.D.; Settles, G.S. An experimental study of supersonic microjets. *Exp. Fluids* **1996**, *21*, 401–409. [CrossRef]

3. Phalnicar, K.A.; Kumar, R.; Alvi, F.S. Experiments on free and impinging microjets. *Exp. Fluids* **2008**, *44*, 819–830. [CrossRef]

4. Hong, C.; Yoshida, Y.; Matsushita, S.; Ueno, I.; Asako, Y. Supersonic micro-jet of straight micro-tube exit. *J. Therm. Sci. Technol.* **2015**, *10*. [CrossRef]

5. Lempert, W.R.; Boehem, M.; Jiang, N.; Gimelshein, S.; Levin, D. Comparison of molecular tagging velocimetry data and direct simulation of Monte Carlo simulations in supersonic micro jet flows. *Exp. Fluids* **2003**, *34*, 403–411. [CrossRef]

6. Handa, T.; Mii, K.; Sakurai, T.; Imamura, K.; Mizuta, S.; Ando, Y. Study on supersonic rectangular microjets using molecular tagging velocimetry. *Exp. Fluids* **2014**, *55*, 1725. [CrossRef]

7. Aniskin, V.M.; Maslov, A.A.; Mironov, S.G. Investigation of the structure of supersonic nitrogen microjets. *Microfluid. Nanofluid.* **2013**, *14*, 605–614. [CrossRef]

8. Aniskin, V.M.; Maslov, A.A.; Mironov, S.G.; Tsyryulnikov, I.S. Supersonic axisymmetric micrijets: Structure and laminar-turbulent transition. *Microfluid. Nanofluid.* **2015**, *19*, 621–634. [CrossRef]

9. Aniskin, V.M.; Maslov, A.A.; Mironov, S.G. Flows of Supersonic Underexpanded Jets on the Range of Moderate Reynolds Numbers. *Fluid Dyn.* **2018**, *53*, 1–8. [CrossRef]

10. Rudyak, V.Y.; Aniskin, V.M.; Maslov, A.A.; Minakov, A.V.; Mironov, S.G. *Micro- and Nanoflows. Modeling and Experiment*; Thess, A., Ed.; Fluid Mechanics and Its Applications; Springer International Publishing: Cham, Switzerland, 2018; Volume 118, p. 258.

11. Love, E.S.; Grigsby, C.E.; Lee, L.P.; Woodling, M.J. *Experimental and Theoretical Studies of Axisymmetric Free Jets*; NASA Technical Report; NASA: Washington, DC, USA, 1959; pp. 1–292.

12. Beckwith, E.; Harvey, W.D.; Clark, F.L. *Comparisons of Turbulent Boundary Layer Measurements at Mach Number 19.5 with Theory and in Assessment of Probe Errors*; NASA TN D-6192; NASA: Washington, DC, USA, 1971.

13. Tam, C.K.W. Supersonic jet noise. *Annu. Rev. Fluid Mech.* **1995**, *27*, 17–43. [CrossRef]

14. Gorshkov, G.F.; Uskov, V.N. Self-excited oscillations in rarefied, impact supersonic jets. *J. Appl. Mech. Tech. Phys.* **1999**, *40*, 455–460. [CrossRef]

15. Merl, M. Emissions acoustiques associees aux juts d'air supersonique. *J. Mec.* **1965**, *4*, 305–307.

16. Westley, R.; Woolley, J.H. *The Near Field Sound Pressure of a Choked Jet When Oscillating in the Spinning Mode*; AIAA Paper 75-479; AIAA: Reston, VA, USA, 1975.

17. Raman, G. Cessation of screech in underexpanded jets. *J. Fluid Mech.* **1997**, *336*, 69–90. [CrossRef]

micromachines

MDPI

Article

Design Guidelines for Thermally Driven Micropumps of Different Architectures Based on Target Applications via Kinetic Modeling and Simulations

Guillermo López Quesada [1,2,*], **Giorgos Tatsios** [2], **Dimitris Valougeorgis** [2],
Marcos Rojas-Cárdenas [1], **Lucien Baldas** [1], **Christine Barrot** [1] **and Stéphane Colin** [1]

1   Institut Clément Ader (ICA), CNRS, INSA, ISAE-SUPAERO, Mines Albi, UPS, Université de Toulouse, 31400 Toulouse, France; marcos.rojas@insa-toulouse.fr (M.R.-C.); lucien.baldas@insa-toulouse.fr (L.B.); christine.barrot@insa-toulouse.fr (C.B.); stephane.colin@insa-toulouse.fr (S.C.)
2   Department of Mechanical Engineering, University of Thessaly, 38334 Volos, Greece; tatsios@mie.uth.gr (G.T.); diva@mie.uth.gr (D.V.)
*   Correspondence: lopezque@insa-toulouse.fr; Tel.: +34-618512360

Received: 15 March 2019; Accepted: 11 April 2019; Published: 14 April 2019

Abstract: The manufacturing process and architecture of three Knudsen type micropumps are discussed and the associated flow performance characteristics are investigated. The proposed fabrication process, based on the deposition of successive dry film photoresist layers with low thermal conductivity, is easy to implement, adaptive to specific applications, cost-effective, and significantly improves thermal management. Three target application designs, requiring high mass flow rates (pump A), high pressure differences (pump B), and relatively high mass flow rates and pressure differences (pump C), are proposed. Computations are performed based on kinetic modeling via the infinite capillary theory, taking into account all foreseen manufacturing and operation constraints. The performance characteristics of the three pump designs in terms of geometry (number of parallel microchannels per stage and number of stages) and inlet pressure are obtained. It is found that pumps A and B operate more efficiently at pressures higher than 5 kPa and lower than 20 kPa, respectively, while the optimum operation range of pump C is at inlet pressures between 1 kPa and 20 kPa. In all cases, it is advisable to have the maximum number of stages as well as of parallel microchannels per stage that can be technologically realized.

Keywords: Knudsen pump; thermal transpiration; vacuum micropump; rarefied gas flow; kinetic theory; microfabrication; photolithography; microfluidics

1. Introduction

The rapid development of the semiconductor industry has been followed, in the last decades, by huge progress in microfabrication processes, providing a large number of micro-electro-mechanical systems (MEMS). Some of these systems, such as lab-on-a-chip (LOC) or micro total analysis systems (µTAS) for gas sensing, analyzing and separation, as well as for drug delivery, require an external pumping system to move the gas samples through the various stages of the device [1]. Additionally, radio frequency switches, vacuum tubes, and other components that depend on electron or ion optics require a stable vacuum environment for proper operation. Since merely sealing these devices is not sufficient to guarantee long-term operation free of leakages and outgassing, miniaturized vacuum pumping components are needed to maintain proper functionality [2].

Thermal transpiration (or thermal creep) has been known for more than 100 years, since the first studies of Maxwell [3], Reynolds [4], and Knudsen [5,6]; however, functional pumping prototypes based on this phenomenon have been developed only during the last 20 years with the massive development and adoption of microfabrication techniques. The Knudsen pump, which is one of the devices exploiting the thermal transpiration phenomenon, is able to generate a macroscopic gas flow by applying exclusively a tangential temperature gradient along a surface without any moving parts or external pressure gradient [7–12]. The induced thermal transpiration flow is useful for technological purposes, provided that the flow is through microscale channels, since the phenomenon is intensified as the characteristic length of the system is decreased. Since the Knudsen pump only requires a temperature gradient for its operation, its architecture is quite simple and it does not require any moving parts, which provides high reliability and avoids any maintenance. Its advanced compactness allows low power consumption. Furthermore, since the direction of the flow can be reversed by inverting the thermal gradient in the microchannels, the Knudsen pump could provide significant benefits for sampling and separation devices [13,14]. The classic architecture of the Knudsen pump consists of a series of wide channels (or reservoirs) connected by microchannels [15] or a porous medium [16]. In order to avoid large temperature gradients, and since the pumping effect generated by a single stage with a moderate temperature gradient is not adequately strong, a multistage system is commonly applied, with a periodic temperature variation in each stage. The thermal transpiration flow produced in the microchannels or the porous medium is much larger than the counter one in the wide channels and therefore, a net pumping effect, which is expected to be increased with the number of stages, is obtained. This architecture presents various difficulties related to microfabrication and local thermal gradients control and therefore, the progress in the field has been limited.

Using advanced microfabrication techniques, a few functional prototypes of Knudsen pump have been recently developed. The first microfabricated single-stage Knudsen pumps were reported in [17], where a nanoporous aerogel was used as a transpiration membrane, and in [15], where the pump was formed of constant cross section microchannels. Some works have integrated Knudsen pumps in fully integrated MEMS and several multistage Knudsen pumps with various operation characteristics have been developed [16,18–20], demonstrating low power consumption (less than 1 W) [19] and long operating times (more than 11750 h of continuous operation) [16]. In order to further decrease the power consumption of these devices, the Knudsen pump can also be powered by passive heat recovery from other processes. Finally, a fully electronic micro gas chromatography system integrated with all its fluidic components has been recently achieved, including a bi-directional Knudsen pump with a mesoporous mixed cellulose-ester membrane sandwiched between two glass dies to provide the parallel flow channels, significantly reducing the fabrication complexity and cost [21,22].

In the present work, specific design guidelines for manufacturing Knudsen pumps by means of a low thermal conductivity bulk material and an innovative microfabrication process are described in order to further improve manufacturing and operational issues concerning flexibility, adaptability, thermal management and overall performance. Three Knudsen pump designs, depending upon the target application, are proposed. The associated numerical investigation is performed to demonstrate the versatility of the proposed designs.

2. Proposed Pump Designs, Manufacturing Materials, and Fabrication Process

Concerning the performance characteristics of Knudsen pumps and their corresponding targeted applications, there are, in general, three alternative designs, named below as designs A, B, and C. The first one (design A) targets applications such as micro-gas chromatography, where the needed mass flow rate is high while the corresponding pressure difference is small. This goal can be reached by designing a large number of channels in parallel to increase the overall mass flow rate. As there is no need to generate a strong pressure difference, this first configuration requires only one stage, which simplifies the design. The second design (B) targets applications such as vacuum maintenance in low-power devices where the needed pressure difference is high, while the corresponding mass flow rate is not so relevant. It requires a large number of stages to increase the pressure difference, which is limited for each stage by the available temperature difference. In order to increase compactness and decrease power consumption, each stage is made of a single channel, as a small mass flow rate is acceptable for this kind of application. Finally, the third design (C) is less specific and targets applications where both relatively high mass flow rates and pressure differences are needed. The third design may be employed in MEMS, as well as in other applications potentially substituting conventional pumps.

In this work, all three pump designs (Figure 1) are considered to address the corresponding performance characteristics:

- Pump A consists of an array of multiple parallel narrow microchannels in one single pumping stage to achieve high mass flow rate ($\dot{m}$) performance. The layout area is $a \times a$ with n parallel narrow microchannels of diameter d and length L.
- Pump B consists of a multistage system where each stage is formed by one single narrow pumping microchannel followed by one wide channel (where the reduced counter thermal transpiration flow will appear) to achieve high pressure difference (ΔP) performance. The layout area is $(W \times W) + (b \times b)$. The diameter of the narrow and wide channels are d and D, respectively, and the length of both channels is L.
- Pump C combines the two previous designs. More specifically, it consists of a multistage system and each stage is formed by an array of n parallel narrow pumping microchannels, followed by one wide channel where the reduced counter thermal transpiration flow will appear. This design provides high ΔP and $\dot{m}$ performances, due to the multi-stage cascade system and to the multiple narrow microchannels per stage, respectively. The layout area is $(W \times W) + (c \times c)$. The diameter of the narrow and wide channels are d and D, respectively, while the length of all channels is L.

Simulations of pumps A, B, and C have been performed via kinetic modeling, described in Section 3 and the performance characteristics are analyzed, in terms of the parameters affecting the flow, in Section 4.

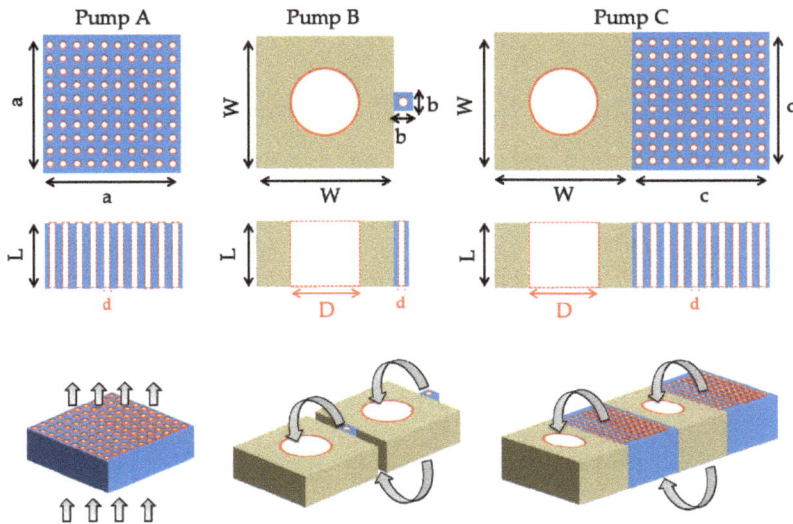

Figure 1. Representative view of single-stage pump A, and of two consecutives stages of multistage pumps B and C. Gray arrows denote the pumping flow direction, the hot and cold regions being at the top and the bottom of each stage, respectively.

Most high-precision microfabrication processes have been initially based on silicon micromachining as the semiconductor industry has been leading the market. However, silicon displays a high thermal conductivity that increases the heat transfer through the solid between the hot and cold reservoirs and results in a high power consumption, while increasing the difficulty of adequately controlling the temperature gradient along the microchannels. On the contrary, by using bulk materials with lower thermal conductivity, such as glass or polymers, the thermal management of the device is simplified and the performance can be improved by maintaining higher temperature gradients in simpler structures. Based on all the above information, we propose using polymer instead of silicon as the bulk material. On the other hand, the substrate used for connecting the heating and cooling elements, respectively, to the hot and cold regions at the top and the bottom of each stage, can be made of silicon to improve the temperature uniformity in these hot and cold regions.

The proposed Knudsen pump designs can be realized by the following innovative fabrication process. Instead of etching, the technique of the deposition of dry film (DF) photoresist layers [23] can be implemented in a cost- and time-effective way. The proposed DF photoresist approach is based on a negative epoxy, which is a low-cost material that can be combined with standard photolithography procedures and multilevel laminating by rolling the DF layers with a specific pressure and temperature. Each new layer can then be stacked on the previous one without damaging the patterns of other layers, which leads to the creation of 3D structures, as shown in Figure 2, and allows for fabricating multistage Knudsen pumps. The fabrication process is summarized as follows: (i) the DF photoresist layer (uncrosslinked DF) is laminated onto a planar substrate (glass or silicon wafer); (ii) following standard photolithography procedures, the DF layer is exposed to UV light and baked; (iii) the DF layer is developed (reticulated DF) in a bath of cyclohexanone that removes the material of the non-exposed areas during the photolithography process. This process is repeated for each of the successive layers, enabling the production of different patterns [23]. The alignment of each layer has been reported with a deviation of less than 1 μm. Also, since the typical thicknesses of the commercialized DF layers go from 5 to 100 μm, various layers can be stacked for a specific pattern, to increase and adjust the thickness of that particular pattern (i.e., the length of the channels).

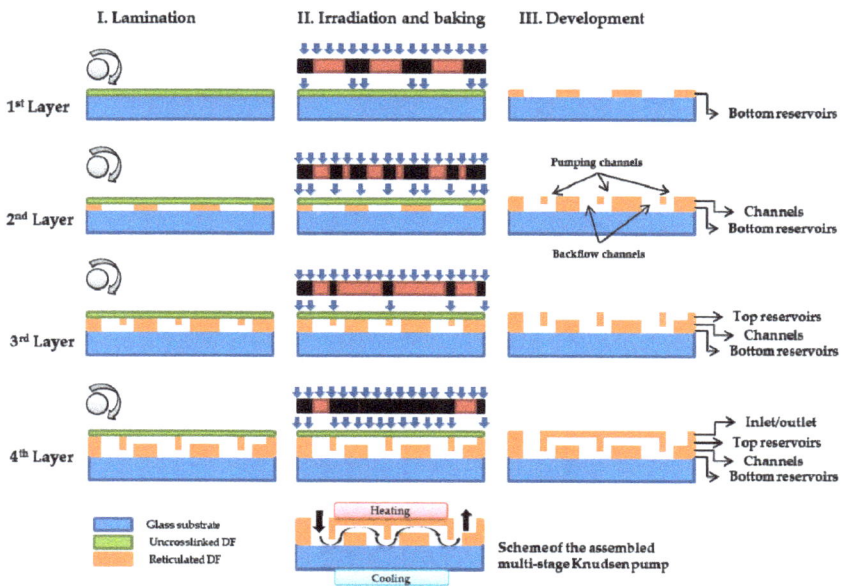

Figure 2. Schematic of fabrication process with the superposition of dry film (DF) photoresist layers and the use of lamination (grey cylinders) and lithography techniques. Typical thicknesses of the DF layers are 5, 25, 50, and 100 μm, while the thickness of the glass wafer is 500 μm. In the cooled reservoirs the glass substrate can be replaced with a silicon substrate to improve temperature uniformity.

Furthermore, the proposed pump designs are to be fabricated in such a way that successive stages are in the same block and the only connections of the pump are reduced to one inlet and one outlet. The possible sources of leakage, which are common in gas microsystems due to numerous connections, are thus drastically reduced. In addition, manufacturing the microchannels across the thickness of each layer keeps the hot and cold regions of the Knudsen pump spatially separate. Consequently, it simplifies the thermal management and allows easy bi-directional pumping. The proposed manufacturing process is quite flexible in terms of geometry, and Knudsen pump prototypes with various shapes, sizes and lengths can easily be produced. Actually, this approach with the channels through the thickness of the substrate and the hot and cold regions spatially separated, has already been employed for Knudsen pumps with high flow (with a design similar to that of pump A) in [20–22] and is generalized here in all three designs.

3. Kinetic Modeling

The flow configurations in the three proposed pump designs are modeled in order to obtain the expected performance for each case. Since the flow has a wide range of Knudsen number, kinetic modeling is introduced. Furthermore, since the length of the narrow microchannels is always much longer than the radius, the flow may be considered fully developed, i.e., the pressure varies only in the axial direction and remains constant in each cross section of the capillary. Thus, flow modeling is based on the infinite capillary theory, which is well known and established for pressure- and temperature-driven rarefied gas flows [24–30]. Additionally, in cases where the fully developed assumptions are not met (i.e., in the wide channels of diameter D subject to thermally driven back flow), the end effect correction is accordingly introduced [31,32]. The correction is introduced only in the pressure and not in the corresponding temperature-driven flow, since the mass flow rate in the former one is about one order of magnitude larger than in the latter.

Consider the fully developed thermal transpiration flow of a monatomic variable hard sphere molecule through a circular channel with length L and radius R (with $R << L$) that connects two reservoirs maintained at different temperatures T_C and T_H, with $T_C < T_H$. In addition to the temperature-driven flow, a pressure-driven flow is generated. Similar to the analysis performed in [30] for thermally driven flow through tapered channels, the net mass flow rate $\dot{m}$ may be computed based on the differential equation

$$\frac{dP}{dz} = -\frac{\dot{m}v_0(z)}{\pi R^3 G_P(\delta)} + \frac{G_T(\delta)}{G_P(\delta)}\frac{P(z)}{T(z)}\frac{dT}{dz}, \tag{1}$$

subject to the given pressures $P(0)$ and $P(L)$ at the channel inlet and outlet, respectively. Here, $z \in [0, L]$ is the coordinate along which the flow is directed, $v_0(z) = \sqrt{2R_g T(z)}$ is the probable molecular speed, R_g is the specific gas constant, $T(z)$ is the imposed linear temperature distribution along the channel wall, and $P(z)$ is the unknown pressure distribution and it is part of the solution. Additionally, $G_P(\delta)$ and $G_T(\delta)$ are the dimensionless flow rates, also known as kinetic coefficients [24,25,27,33] for the pressure- and temperature-driven flows, respectively. They depend on the local gas rarefaction parameter, defined as

$$\delta(z) = \frac{P(z)R}{\mu(z)v_0(z)}, \tag{2}$$

where $\mu(z)$ is the local dynamic viscosity. In the present work, the kinetic coefficients $G_P(\delta)$ and $G_T(\delta)$ are retrieved from a kinetic database, which has been developed based on the linearized Shakhov model subject to pure diffuse boundary conditions in the range $\delta \in [0, 50]$. Additionally, when $\delta > 50$, the analytical slip solution is used. For completeness, tabulated results and the analytical slip solution of the kinetic coefficients versus gas rarefaction parameter are presented in Appendix A.

Equation (1) is solved for $\dot{m}$ and $P(z)$ as follows: an initial value for the mass flow rate $\dot{m}$ is assumed and Equation (1) is integrated with initial condition $P_{in} = P(0)$ along $z \in [0, L]$. At each integration step along z, the values of G_P and G_T are updated based on the local $\delta(z)$. The computed outlet pressure is compared to the specified $P_{out} = P(L)$ and then the mass flow rate is corrected depending on the difference between the computed and specified outlet pressures. This procedure is repeated until $\dot{m}$ and the corresponding $P(z)$ converge to yield the specified $P(L)$.

The above description refers to a single channel and may be applied in a straightforward manner to pumps A, B, and C, provided that the carrier gas, the pump geometry, the temperature variation along the channels, and the inlet and outlet pressures are all specified. In each case the computed mass flow rate $\dot{m}$ with the associated pressure difference $\Delta P = P_{out} - P_{in}$ between the inlet and the outlet fully characterize the pump performance. It is noted that a very close kinetic type analysis of pump B, based on the ellipsoidal-statistical (ES) model, and the one presented here, based on the Shakhov model, has been performed in [10].

4. Results and Discussion

The performance characteristics of the three proposed pump designs are computationally investigated, taking into account manufacturing and operational constraints. Proper thermal management is of major importance. Previous experimental works on thermal transpiration flows through capillaries [34,35], supplemented by recent typical heat transfer simulations in film layers, clearly indicate that it is possible to provide temperature differences $\Delta T = T_H - T_C$ on the order of 100 K within a channel of length $L = 300$ μm by integrating active heating and cooling. In addition, via the proposed microfabrication process, microchannels with diameters ranging from 100 down to 5 μm are attainable with $L = 300$ μm by deposition of successive layers of dry film photoresist. This length corresponds to the thickness of each layer and is kept small in order to minimize the pump volume. In a later stage, smaller diameters may be fabricated by further reducing the channel length, i.e., the film layer thickness.

Based on the above, simulations for pump designs A, B, and C consisting of narrow and wide microchannels with diameters $d = 50, 20, 10, 5$ μm and $D = 100$ μm, respectively, always keeping the length $L = 300$ μm, have been performed. As pointed out before, the pumping effect is produced by the thermal transpiration flow in the narrow microchannels, while the counter thermal transpiration flow in the wide microchannels is much smaller due to the larger diameter of the channel and, therefore, there is a net flow in the pumping direction. The temperature difference is set to $\Delta T = 100$ K, maintaining the cold and hot temperatures at $T_C = 300$ K and $T_H = 400$ K, respectively. The gas considered here is argon with a specific constant $R_g = 208.1$ J kg^{-1}K^{-1} and a dynamic viscosity varying with temperature according to an inverse power-law model, which is consistent with the variable hard sphere molecule hypothesis [11]: $\mu(z) = \mu_0(T(z)/T_0)^{0.81}$ with a viscosity $\mu_0 = 2.211 \times 10^{-5}$Pa s at the reference temperature $T_0 = 273$ K.

Furthermore, taking into account the space needed between the channels from the fabrication point of view, it can be estimated that the minimum total square area allowing an array of at most $n = 400$ microchannels is 200×200 μm^2. A schematic view of the corresponding layouts to be examined is shown in Figure 3. As the microchannel diameter is reduced, the number n of parallel microchannels in the layout is increased, keeping the same area ratio between the flow and the layout cross section areas. In this way, the comparison of the performance characteristics of the different layouts always involves the same cross section flow area. Additional details of the layout geometry are provided in Table 1.

The computational investigation includes pumps A, B, and C and is presented in Section 4.1, Section 4.2, and Section 4.3, in terms of the inlet pressure $P_{in} \in [0.1 - 10^5]$ kPa, the diameter d of the narrow microchannels, and the number N of stages.

Table 1. Layout data in terms of diameter, number of microchannels, and elementary square area.

Total Layout Area $a \times a$ (μm × μm)	Microchannel Diameter d (μm)	Number n of Parallel Microchannels
200 × 200	50	4
200 × 200	20	25
200 × 200	10	100
200 × 200	5	400

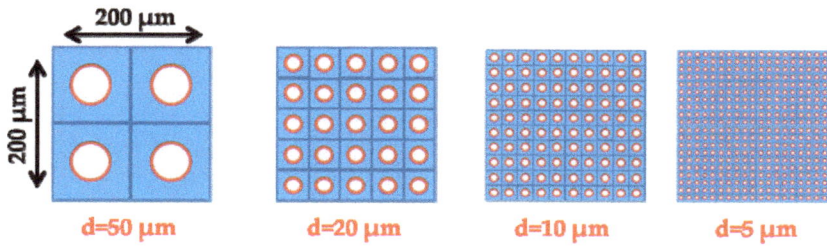

Figure 3. View of layouts with increasing the number n of microchannels and decreasing the channel diameter d, keeping the same ratio between the channels and the overall cross sections.

4.1. Pump A: One Pumping Stage with Multiple Parallel Microchannels

Pump A consists of a single pumping stage with n parallel microchannels, mainly targeting high mass flow rates $\dot{m}$ and small pressure differences. It is obvious that, as the channel diameter is reduced, the flow rate in a single channel, $\dot{m}$, is also reduced. This mass flow rate loss is partly compensated for, since as the diameter is decreased more channels are packed in parallel, keeping the same cross section area. The total mass flow rate will be equal to that of a single channel multiplied by the number n of parallel channels. On the contrary, the total pressure difference will be equal to that of each channel, which increases as the channel diameter decreases. Therefore, first, a comparison is performed between the performances of the one-stage pumping for the different layout geometries by varying the narrow diameter and the number of channels, while preserving the same total surface, as shown in Figure 3 and Table 1.

The investigation is performed by computing the total maximum mass flow rate $\dot{m}_n$ through the n parallel channels corresponding to zero pressure difference, i.e., to a pump with the same pressure at both ends, as well as the maximum pressure difference ΔP_n corresponding to zero net mass flow rate (closed system). For $n = 4, 25, 100$, and 400, $\dot{m}_n$ and ΔP_n are compared with the reference values $\dot{m}_1$ and ΔP_1 respectively, corresponding to the case of a single channel, $n = 1$. The channels diameters associated with $n = 1, 4, 25, 100$, and 400 are $d = 100, 50, 20, 10$, and 5 μm, respectively. The ratios $\dot{m}_1/\dot{m}_n$ and $\Delta P_n/\Delta P_1$ are computed to estimate the mass flow rate decrease and the pressure difference increase when the number of channels is increased and their diameter is decreased. The results are presented in Figure 4 as a function of the inlet pressure P_{in}.

As expected, as the diameter is decreased and the number of microchannels is increased, both $\dot{m}_1/\dot{m}_n$ and $\Delta P_n/\Delta P_1$ are monotonically increased, meaning that the total flow rate through the parallel channels decreases and the pressure difference increases. In terms of the inlet pressure, it is seen that at a low inlet pressure the ratios of the total mass flow rates $\dot{m}_1/\dot{m}_n$ are larger than the corresponding ratios of pressure differences $\Delta P_n/\Delta P_1$, while at moderate and high inlet pressures, the situation is reversed and the ratios $\dot{m}_1/\dot{m}_n$ become smaller than the corresponding $\Delta P_n/\Delta P_1$. This behavior clearly implies that, in the microchannels, the relative pressure difference increase compared to the corresponding total mass flow rate decrease is less significant in highly rarefied conditions, while in less rarefied conditions closer to the slip and hydrodynamic limit, it becomes more significant. Furthermore, the solid red circles indicate the inlet pressure values where the two ratios are equal. Then, at the left side of these points, the ratios of the mass flow rates $\dot{m}_1/\dot{m}_n$ are larger than the corresponding ratios of pressure differences $\Delta P_n/\Delta P_1$, while at the right side $\Delta P_n/\Delta P_1$ are larger than $\dot{m}_1/\dot{m}_n$. For $P_{in} \geq 5$ kPa, whatever the number n of parallel microchannels, the relative variation in pressure difference is much more significant than the relative variation in the total mass flow rate through the combined parallel channels.

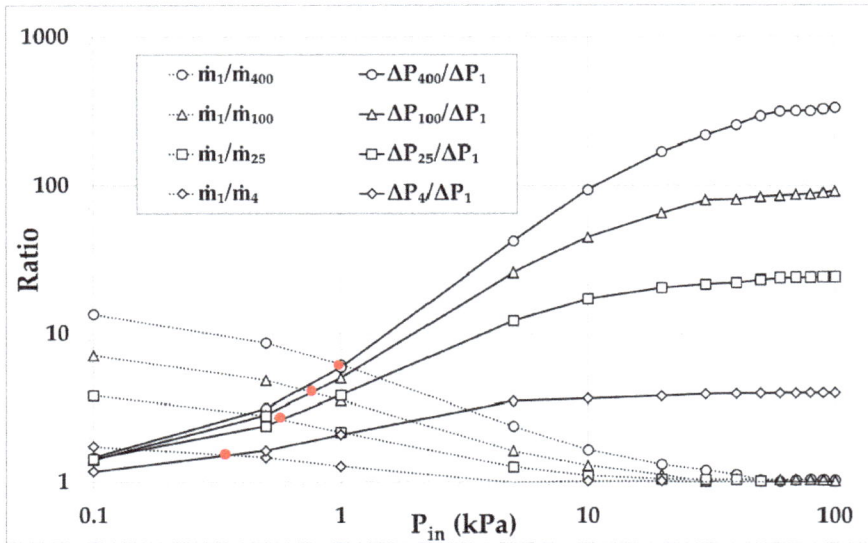

Figure 4. Ratios of total maximum mass flow rates $\dot{m}_1 / \dot{m}_n$ and associated maximum pressure differences $\Delta P_n / \Delta P_1$ for various one-stage layouts (numbers of microchannels $n = 4, 25, 100, 400$ and corresponding diameters $d = 50, 20, 10, 5$ μm) compared to the reference layout ($n = 1, d = 100$ μm).

To have a clear view of the performance characteristics of the one-stage Knudsen pump A in absolute quantities (not in relative ones, as previously), the pressure difference and the mass flow rate through a single channel are plotted in Figure 5, for microchannels with $d = 5, 10, 20$ μm, as a function of the inlet pressure P_{in}. In all cases, as the inlet pressure is increased, the pressure difference is initially rapidly increased, reaching its peak value, and then is slowly decreased. The peak values ΔP_{peak} occur, depending upon the diameter d, at about $P_{in} = 4 - 10$ kPa, corresponding to values of the gas rarefaction parameter $\delta = 3 - 4$, i.e., in the transition regime. This behavior has also been observed and reported in experimental studies and is independent of the working gas, the channel characteristic length, and the temperature difference [34]. Depending on the inlet pressure, the generated pressure difference ΔP varies from 10 to 420 Pa and, as expected, the obtained pressure difference is increased as the diameter is reduced with the maximum $\Delta P = 420, 200, 100$ Pa for $d = 5, 10, 20$ μm, respectively. Regarding the mass flow rate, it is rapidly increased as the inlet pressure is initially increased and then it keeps increasing at a much slower pace until it becomes constant in the hydrodynamic regime. For $P_{in} \geq 20$ kPa, the mass flow rate is almost stabilized and reaches values of about 0.5×10^{-12}, 2.1×10^{-12} and 8.5×10^{-12} kg/s for $d = 5, 10, 20$ μm, respectively. Multiplying these values with the corresponding number of microchannels, $n = 400, 100$ and 25, results in mass flow rates higher than 10^{-10} kg/s. These results, properly scaled, are in good agreement with the corresponding ones in [20], where a design similar to pump A was realized.

Overall, it may be stated that, for pump A, it is preferable to have a wider channel to increase the total mass flow rate rather than a large number of parallel microchannels with smaller diameters. However, since reducing the channels diameter results in huge gains of ΔP with only small reductions of $\dot{m}_n$, as shown in Figure 4, depending on the application, it might be advisable to reduce the diameter to boost ΔP while only slightly reducing the overall mass flow rate $\dot{m}_n$. Finally, it is preferable to operate the pump with moderate and high inlet pressures ($P_{in} > 5$ kPa), where the mass flow rate starts to stabilize at large values.

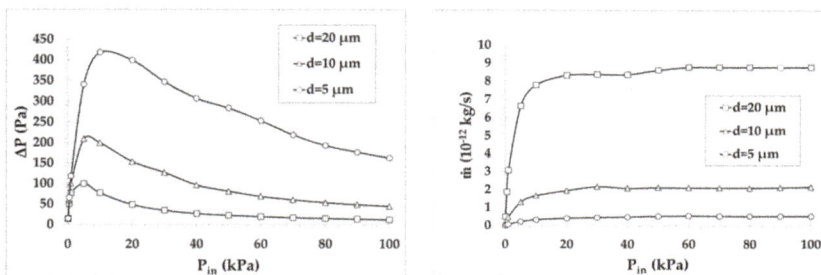

Figure 5. (**a**) Pressure difference and (**b**) mass flow rate versus inlet pressure for the one-stage pump A with $d = 5, 10, 20$ μm (mass flow rates are given for a single channel of the pump). The rarefaction parameter is ranging from 0.03 to 130, and it increases with the inlet pressure and the diameter, providing the maximum pressure difference at $\delta = 3 - 4$, in the transition regime.

4.2. Pump B: Multistage Pumping with One Narrow and One Large Channel Per Stage

Pump B is a multistage system, where each stage consists of one single pumping narrow microchannel with $d = 50, 20, 10, 5$ μm, followed by one wide channel with $D = 100$ μm, targeting high pressure differences and small mass flow rates (as there is only one microchannel per stage).

The maximum pressure difference corresponding to zero net mass flow rate (closed system) at various inlet pressures $P_{in} = 1, 5, 10, 20, 50, 100$ kPa are provided in Figure 6a,b versus the number of stages with $N \leq 1000$ and $N \leq 100$, respectively. The considered narrow and wide microchannels at each stage have diameters $d = 10$ μm and $D = 100$ μm, respectively. As seen in Figure 6a, the pressure difference, ΔP, increases with the number of stages in a qualitatively similar manner for all inlet pressures, except for the lowest inlet pressure $P_{in} = 1$ kPa and $N \leq 100$. It is also seen that for a small number of stages the pressure difference is rapidly increased and then it keeps increasing, but at a slower pace that gradually decreases as the number of stages increases. This is due to the fact that, for all the inlet pressures shown, except $P_{in} = 1$ kPa, the pumps, independently of N, always operate in the decreasing region of ΔP in terms of P_{in} (see Figure 5a). In all these cases, each time a stage is added its contribution to the overall ΔP is slightly reduced compared to the previous one, because the inlet pressure of the stage is increased. Therefore, the rate at which ΔP is increased is slowly reduced with increasing number of stages. In the specific case of $P_{in} = 1$ kPa the pump starts to operate in the increasing region of ΔP in terms of P_{in} (see Figure 5a). Consequently, starting at $P_{in} = 1$ kPa, every added stage, compared to the previous one, contributes with a larger ΔP until the pressure of $P_{in} = 5$ kPa, corresponding approximately to the peak value of the pressure difference ΔP_{peak}, is reached. Thus, a more detailed view for $N \leq 100$ is shown in Figure 6b. It is clearly seen that the pressure difference with respect to the number of stages for $P_{in} = 1$ kPa is qualitatively different compared to all other P_{in} values and is increased more rapidly. Of course, once a sufficient number of stages has been added and the inlet pressure for the next stage becomes higher than $P_{in} = 5$ kPa, then the pump operates in the decreasing region of ΔP in terms of P_{in} and it behaves qualitatively similar to all others. Overall, it is observed that for a multistage pump with $N \geq 100$ the pressure difference generated is very significant and may be of the same order as the inlet pressure or even several times higher, when $P_{in} \leq 20$ kPa. It is noted that the present results are in excellent qualitative agreement with the corresponding ones reported in [10]. Furthermore, running the present code for some of the geometrical and operational parameters in [10], it has been found that the deviation in the numerical results is small (about 10%) and is due to the corresponding deviation between the kinetic coefficients obtained by the ES and Shakhov models.

Although pump B targets high ΔP and the mass flow rate is not of major importance, it is interesting to observe its variation versus the number N of stages. In Figure 7, the maximum mass flow rate corresponding to zero pressure difference (i.e., a pump with the same pressure at both ends) is shown for the same parameters as in Figure 6a. The mass flow rate is of the order of 10^{-12} or 10^{-13} kg/s and it is low since only one narrow channel is considered per stage. As expected, $\dot{m}$ is decreased as P_{in} is decreased, but more importantly, $\dot{m}$ is kept constant as the number of stages is increased due to the fact that all stages have the same geometry and the same inlet and outlet pressures. It may be stated that for Knudsen pumps based on the architecture of pump B, it is desirable to add as many stages as possible and to operate the pump in low and moderate P_{in}, since ΔP grows with N, while the maximum $\dot{m}$ remains constant. The corresponding results for other narrow microchannel diameters have a similar qualitative behavior.

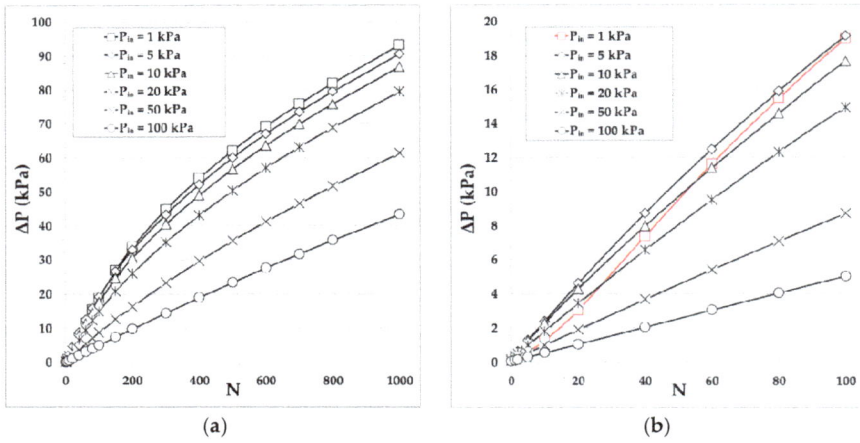

Figure 6. Maximum pressure difference (corresponding to zero net mass flow rate) of pump B with single narrow and wide microchannels of diameters $d = 10$ μm and $D = 100$ μm, respectively, in each stage, at various inlet pressures, versus the number of stages with (**a**) $N \leq 1000$ and (**b**) $N \leq 100$. The range of the rarefaction parameter in the narrow channel is (**a**) $\delta \approx 0.6 - 56$ and (**b**) $\delta \approx 0.6 - 12$ for $P_{in} = 1$ kPa, and (**a**) $\delta \approx 60 - 86$ and (**b**) $\delta \approx 60 - 62$ for $P_{in} = 100$ kPa. Always, δ is increasing with the inlet pressure and the number of stages.

The evolution of the pressure inside a system vacuumed with pump B connected at its inlet, while the outlet is at the atmospheric pressure ($P_{out} = 100$ kPa), is analyzed in Figure 8, showing the influence of the number of stages and of the narrow channels diameter $d = 5, 10, 20$ μm, with the diameter of the wide channel being always $D = 100$ μm. Three pressure drop regions can clearly be identified: at high pressures (green region), the pressure is reduced very slowly; at intermediate pressures (red region), the pressure is rapidly reduced; and finally, at low pressures (blue region), the pressure keeps reducing at a slower pace. This behavior can be understood by observing the corresponding results in Figure 5, where the pressure difference is very small at high inlet pressures, then becomes quite large at intermediate inlet pressures, where the maximum pressure difference is reported (red symbols in Figure 8) and, finally, becomes small again at low inlet pressures. These regions of Figure 5 correspond to the green, red, and blue regions in Figure 8. As expected, the required number of stages to reach the final low inlet pressure is increased by increasing the diameter of the narrow microchannels. These plots confirm that the optimal operating pressure range of the pumps B presented is, as pointed above, in the red zone (Figure 8), i.e., at moderate and low inlet pressures corresponding to values of the gas rarefaction parameter close to $\delta = 3 - 4$ where ΔP_{peak} is found (Figure 5).

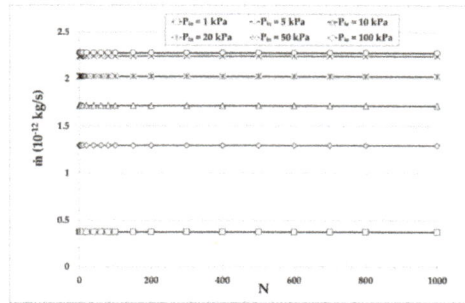

Figure 7. Maximum mass flow rate (corresponding to zero pressure difference) of pump B with single narrow and wide microchannels of diameters $d = 10$ μm and $D = 100$ μm, respectively, in each stage, at various inlet pressures, versus the number of stages.

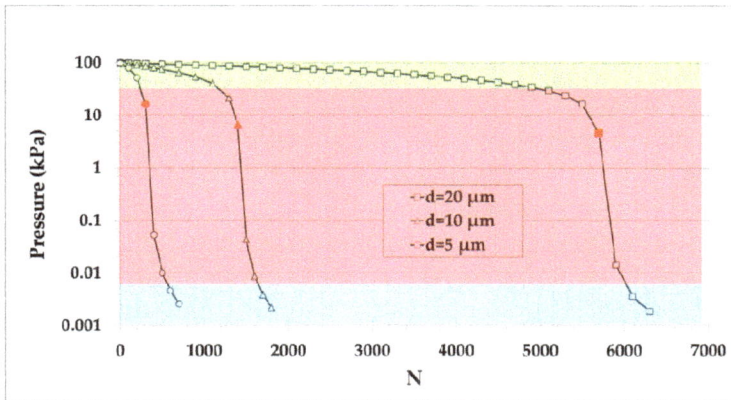

Figure 8. Inlet pressure evolution of a system connected to pump B as a function of its number of stages, considering narrow diameter channels $d = 5, 10, 20$ μm and a constant outlet pressure $P_{out} = 100$ kPa (red solid symbols represent the pressure and stage number, where the maximum slope, corresponding to the peak values of Figure 5a, is observed).

4.3. Pump C: Multistage Pumping with Multiple Parallel Microchannels Per Stage

Pump C is a multistage system, where each stage consists of n parallel pumping narrow microchannels, followed by one wide channel, targeting both high pressure differences and mass flow rates. The total pressure difference is obtained by adding the pressure difference of each stage (shown in Figure 6), while the total mass flow rate $\dot{m}_n$ is calculated by multiplying the mass flow rate of a single microchannel multistage system (shown in Figure 7) times the number n of parallel microchannels.

The performance curves of pump C, showing the variation of pressure difference ΔP versus the total mass flow rate $\dot{m}_n$, are plotted in terms of the number of stages N in Figure 9 for $N = 1, 5, 10, 20$ and in Figure 10 for $N = 40, 100, 200, 500, 1000$, with inlet pressures $P_{in} = 1, 5, 20, 100$ kPa. They are presented in different figures for clarity purposes, since going from $N = 1$ up to $N = 1000$ the pressure difference is increased about two orders of magnitude. The narrow microchannels and wide channel at each stage have diameters $d = 10$ μm and $D = 100$ μm, respectively, while the corresponding characteristic curves for other narrow microchannels diameters have a similar qualitative behavior. The values of ΔP for $\dot{m}_n = 0$ and of $\dot{m}_n$ for $\Delta P = 0$, denoted by ΔP_{max} and $\dot{m}_{n,max}$, respectively, are considered as the limiting cases corresponding to a closed system with no net mass flow rate and to an open system with the same pressure at both ends. The characteristic performance curves of the pump are defined by these two limiting values and all flow scenarios in between with $0 < \dot{m} < \dot{m}_{n,max}$ and $0 < \Delta P < \Delta P_{max}$. As expected, in all the cases the characteristic performance curves exhibit a pressure difference decrease as the mass flow rate is increased. Since the maximum pressure difference is increased with the number of stages, while the maximum mass flow rate is constant independently of the number of stages (see the discussion in Figures 6 and 7), the mean slope of the characteristic curves is increased with the number of stages. For a specified mass flow rate, the developed pressure difference is increased as the number of stages is increased; similarly, for a specified pressure difference, the produced mass flow rate is increased with the number of stages. The largest pressure differences are observed at $P_{in} = 1$ kPa and $P_{in} = 5$ kPa for $N \geq 100$ and $N < 100$, respectively, as shown in Figure 6. On the contrary, as expected, the mass flow rates are monotonically increased with P_{in}, with a gain of a factor of 5 when the inlet pressure is increased by a factor of 100. Furthermore, it is seen that the performance curves in Figure 9 are close to linear. This is justified, since in pumps with a small number of stages, the output pressure is relatively close to the inlet pressure and the corresponding variation of δ and $G_P(\delta)$ along the stages of the pump is small, resulting in almost constant terms in Equation (1). However, as N is increased, the outlet pressure is also increased, resulting in a more significant variation of $G_P(\delta)$ along the pump stages and therefore, the performance curves exhibit a nonlinear behavior, as the terms in Equation (1) develop a larger variation. This behavior starts to appear from $N = 20$ in Figure 9 and becomes more evident as N is further increased in Figure 10. Another interesting issue is that, as P_{in} is decreased and N is increased, the pressure difference ΔP decreases very slowly as the mass flow rate increases and then, when the mass flow rate approaches its maximal value $\dot{m}_{n,max}$, the pressure difference rapidly decreases. This behavior, clearly shown in Figure 10, with $P_{in} = 1$ kPa, is beneficial for pumps operating at low inlet pressures.

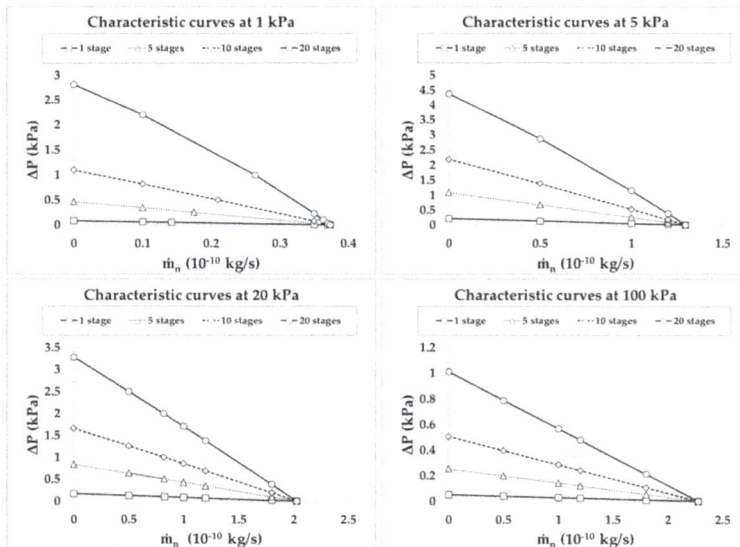

Figure 9. Performance characteristic curves of a Knudsen pump, based on pump C, for a number of stages $N = 1, 5, 10, 20$ with inlet pressures $P_{in} = 1, 5, 20, 100$ kPa when narrow microchannels diameter $d = 10$ μm and wide channel diameter $D = 100$ μm.

In order to maximize the performance of pump C, it is preferable to operate at low and moderate inlet pressures in the wide range of $P_{in} \in [1-20]$ kPa, with as many parallel microchannels per stage and number of stages as possible, to take advantage of the flattening of the performance curves but not too close to the maximum mass flow to avoid the abrupt decrease of the pressure difference.

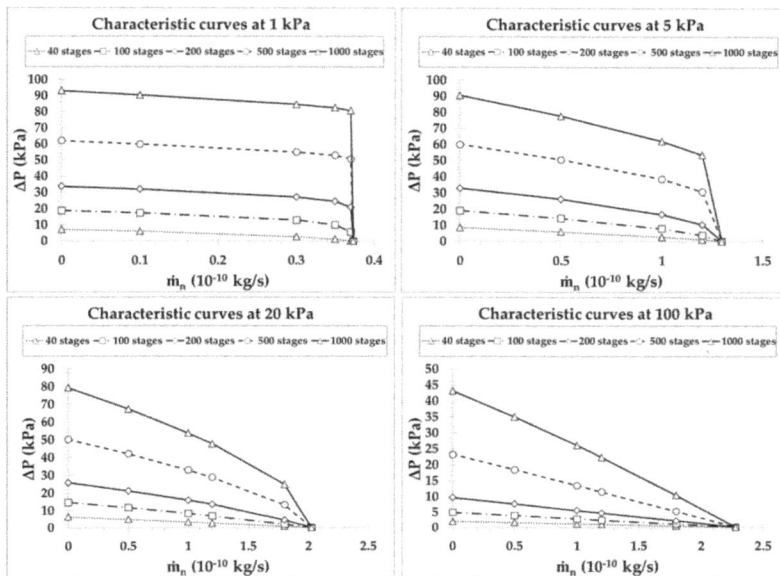

Figure 10. Performance characteristic curves of a Knudsen pump, based on pump C, for a number of stages $N = 40, 100, 200, 500, 1000$ with inlet pressures $P_{in} = 1, 5, 20, 100$ kPa when narrow microchannels diameter $d = 10$ μm and wide channel diameter $D = 100$ μm.

5. Concluding Remarks

The manufacturing process and the structure architecture of three thermally-driven micropump designs targeting specific applications have been proposed and the associated performances have been examined. The microfabrication process includes the development of low thermal conductivity dry film layers via lamination on glass, exposition to UV lighting and baking via photolithography procedures and finally, removal of the non-exposed material in cyclohexanone bath. The three pump designs include single-stage pumping through many parallel microchannels targeting high mass flow rates (pump A), multistage pumping with each stage composed of a single narrow microchannel and a wide channel targeting high pressure differences (pump B) and a combination of the former two configurations targeting both high mass flow rates and pressure differences (pump C). Modeling is based on kinetic theory via the infinite capillary approach.

The proposed microfabrication process is fast, easily realized, adaptive to specific applications and cost effective. The thermal management of the pumps is significantly improved because of the involved low thermal conductivity materials and the separation of the hot and cold areas in two different surfaces. The computational investigation has taken into account all foreseen manufacturing and operational constraints, and the optimum performance conditions as a function of the inlet pressure and pump geometry (number of parallel microchannels per stage and number of stages) have been identified. Micropumps based on the architecture of pumps A and B operate more efficiently at inlet pressures higher than 5 kPa and lower than 20 kPa, respectively. In addition, it is advisable to manufacture pump A with as many parallel microchannels as possible and pump B with as many stages as possible. Pump C should be built with both a high number of parallel microchannels per stage and a high number of stages. Furthermore, their optimum operation range is found to be at inlet pressures between 1 kPa and 20 kPa and a working regime not too close to the limiting maximum mass flow rate.

Using the proposed materials and fabrication process, Knudsen pump prototypes based on pump C architecture, with 2000 stages, 100 microchannels of diameter $d = 10$ μm, and one wide channel of diameter

$D = 100$ µm per stage, employing a total layout surface of 160 mm^2, are presently under construction. Since the theoretical pressure differences cover several orders of magnitude, the device could decrease the pressure of a system from atmospheric pressure $P_{in} = 100$ kPa down to 1 or 2 Pa, with associated mass flow rates higher than 10^{-10} kg/s. These theoretical performances shall be corrected to deal with specific operational constraints such as leakages, issues in thermal management of the device, etc.

Author Contributions: Conceptualization, G.L.Q., S.C., and D.V.; software, G.L.Q.; validation, G.T., M.R.-C., C.B., L.B., S.C., and D.V.; writing—original draft preparation, G.L.Q. and D.V.; writing—review and editing, G.T., M.R.-C., C.B., L.B., and S.C.; visualization, G.L.Q.; supervision, S.C. and D.V.; project administration, S.C. and D.V.

Funding: This project was funded by the European Union's Framework Programme for Research and Innovation Horizon 2020 (2014-2020) under the Marie Skłodowska-Curie (grant agreement number. 643095).

Acknowledgments: This work has received technical support from Laboratoire d'Analyse et d'Architecture des Systemes regarding the manufacturing process proposed.

Conflicts of Interest: The authors declare no conflict of interest.

Appendix A Tabulated Results of the Kinetic Coefficients

The linearized Shakhov model equation for fully developed pressure- and temperature-driven flows in a channel subject to purely diffuse boundary conditions has been computationally solved to deduce kinetic coefficients G_P and G_T for several values of the gas rarefaction parameter $\delta \in \left[5 \times 10^{-4}, 50\right]$. They have been obtained based on the discrete velocity method. These results are also available in the literature [24,25,27,33] and they are presented here, in Table A1, mainly for completeness, since they are implemented (linearly interpolated) in the solution of Equation (1) in order to compute the mass flow rate and the associated pressure distribution along the channel. Furthermore, for values of $\delta > 50$, which are associated with high inlet pressures, the kinetic coefficients are analytically obtained by the corresponding slip solutions as $G_P = \delta/4 + \sigma_P$ and $G_T = \sigma_T/\delta$, where the viscous and thermal slip coefficients are $\sigma_P = 1.018$ and $\sigma_T = 1.175$, respectively [36].

Table A1. Kinetic coefficients G_P and G_T for the pressure- and temperature-driven flows, respectively, in terms of the gas rarefaction parameter δ.

δ	0.0005	0.001	0.005	0.01	0.02	0.03	0.04	0.05
G_P	1.5023	1.5008	1.4904	1.4800	1.4636	1.4514	1.4418	1.4339
G_T	0.7502	0.7486	0.7366	0.7243	0.7042	0.6884	0.6752	0.6637
δ	0.06	0.07	0.08	0.09	0.1	0.2	0.3	0.4
G_P	1.4273	1.4217	1.4168	1.4127	1.4101	1.3911	1.3876	1.3920
G_T	0.6536	0.6444	0.6359	0.6281	0.6210	0.5675	0.5303	0.5015
δ	0.5	0.6	0.7	0.8	0.9	1.0	1.2	1.4
G_P	1.4011	1.4130	1.4270	1.4425	1.4592	1.4758	1.5158	1.5550
G_T	0.4779	0.4576	0.4367	0.4237	0.4092	0.3959	0.3721	0.3514
δ	1.6	1.8	2.0	3.0	4.0	5.0	6.0	7.0
G_P	1.5956	1.6373	1.6799	1.9014	2.1315	2.3666	2.6049	2.8455
G_T	0.3330	0.3165	0.3016	0.2439	0.2042	0.1752	0.1531	0.1359
δ	8.0	9.0	10.0	20.0	30.0	40.0	50.0	
G_P	3.0878	3.3314	3.5749	6.0492	8.5392	11.0360	13.4950	
G_T	0.1220	0.1106	0.1014	0.05426	0.03685	0.02785	0.02212	

References

1. Schomburg, W.K.; Vollmer, J.; Bustgens, B.; Fahrenberg, J.; Hein, H.; Menz, W. Microfluidic components in LIGA technique. *J. Micromech. Microeng.* **1994**, *4*, 186–191. [CrossRef]
2. Laser, D.J.; Santiago, J.G. A review of micropumps. *J. Micromech. Microeng.* **2004**, *14*, 35–64. [CrossRef]
3. Maxwell, J.C. On stresses in rarefied gases arising from inequalities of temperature. *Proc. R. Soc. Lond.* **1878**, *27*, 304–308.

4.	Reynolds, O. XVIII. On certain dimensional properties of matter in the gaseous state. *Philos. Trans. R. Soc. Lond. Ser. A* **1879**, *170*, 727–845.

5.	Knudsen, M. Eine Revision der gleichgewichtsbedingung der gase. Thermische Molekularströmung. *Ann. Phys.* **1909**, *336*, 205–229. [CrossRef]

6.	Knudsen, M. Thermischer molekulardruck der gase in Röhren. *Ann. Phys.* **1910**, *338*, 1435–1448. [CrossRef]

7.	Sone, Y.; Waniguchi, Y.; Aoki, K. One-way flow of a rarefied gas induced in a channel with a periodic temperature distribution. *Phys. Fluids* **1996**, *8*, 2227–2235. [CrossRef]

8.	Sone, Y. *Molecular Gas Dynamics: Theory, Techniques, and Applications*; Birkhäuser: Basel, Switzerland, 2007; ISBN 9780817643454.

9.	Aoki, K.; Mieussens, L.; Takata, S.; Degond, P.; Nishioka, M.; Takata, S. *Numerical Simulation of a Knudsen Pump Using the Effect of Curvature of the Channel*; Global Science Press: Hong Kong, China, 2007.

10.	Aoki, K.; Takata, S.; Kugimoto, K. Diffusion approximation for the knudsen compressor composed of circular tubes. *AIP Conf. Proc.* **2008**, 953–958. [CrossRef]

11.	Colin, S. Single-Phase Gas Flow in Microchannels. In *Heat Transfer and Fluid Flow in Minichannels and Microchannels*; Elsevier: Amsterdam, The Netherlands, 2014; pp. 11–102.

12.	Aoki, K.; Degond, P.; Takata, S.; Yoshida, H. Diffusion models for Knudsen compressors. *Phys. Fluids* **2007**, *19*, 117103. [CrossRef]

13.	Li, X.; Oehrlein, G.S.; Schaepkens, M.; Ellefson, R.E.; Frees, L.C. Spatially resolved mass spectrometric sampling of inductively coupled plasmas using a movable sampling orifice. *J. Vac. Sci. Technol. A* **2003**, *21*, 1971–1977. [CrossRef]

14.	Hamad, F.; Khulbe, K.C.; Matsuura, T. Comparison of gas separation performance and morphology of homogeneous and composite PPO membranes. *J. Membr. Sci.* **2005**, *256*, 29–37. [CrossRef]

15.	McNamara, S.; Gianchandani, Y.B. On-chip vacuum generated by a micromachined Knudsen pump. *J. Microelectromech. Syst.* **2005**, *14*, 741–746. [CrossRef]

16.	Gupta, N.K.; Gianchandani, Y.B. Porous ceramics for multistage Knudsen micropumps-modeling approach and experimental evaluation. *J. Micromech. Microeng.* **2011**, *21*, 095029. [CrossRef]

17.	Vargo, S.E. Initial results from the first MEMS fabricated thermal transpiration-driven vacuum pump. *AIP Conf. Proc.* **2001**, *585*, 502–509.

18.	Gupta, N.K.; An, S.; Gianchandani, Y.B. A Si-micromachined 48-stage Knudsen pump for on-chip vacuum. *J. Micromech. Microeng.* **2012**, *22*, 105026. [CrossRef]

19.	An, S.; Gupta, N.K.; Gianchandani, Y.B. A Si-Micromachined 162-stage two-part knudsen pump for on-chip vacuum. *J. Microelectromech. Syst.* **2014**, *23*, 406–416. [CrossRef]

20.	Qin, Y.; An, S.; Gianchandani, Y.B. Arrayed architectures for multi-stage Si-micromachined high-flow Knudsen pumps. *J. Micromech. Microeng.* **2015**, *25*, 115026. [CrossRef]

21.	Qin, Y.; Gianchandani, Y.B. A fully electronic microfabricated gas chromatograph with complementary capacitive detectors for indoor pollutants. *Microsyst. Nanoeng.* **2016**, *2*, 15049. [CrossRef]

22.	Cheng, Q.; Qin, Y.; Gianchandani, Y.B. A Bidirectional Knudsen Pump with Superior Thermal Management for Micro-Gas Chromatography Applications. In Proceedings of the IEEE 30th International Conference on Micro Electro Mechanical Systems (MEMS), Las Vegas, NV, USA, 22–26 January 2017; pp. 167–170.

23.	Courson, R.; Cargou, S.; Conédéra, V.; Fouet, M.; Blatché, M.-C.; Serpentini, C.L.; Gué, A.-M. Low-cost multilevel microchannel lab on chip: DF-1000 series dry film photoresist as a promising enabler. *RSC Adv.* **2014**, *4*, 54847–54853. [CrossRef]

24.	Sharipov, F. Rarefied gas flow through a long tube at any temperature ratio. *J. Vac. Sci. Technol. A* **1996**, *14*, 2627–2635. [CrossRef]

25.	Sharipov, F.; Seleznev, V. Data on Internal Rarefied Gas Flows. *J. Phys. Chem. Ref. Data* **1998**, *27*, 657–706. [CrossRef]

26.	Sharipov, F. Non-isothermal gas flow through rectangular microchannels. *J. Micromech. Microeng.* **1999**, *9*, 394–401. [CrossRef]

27.	Sharipov, F.; Bertoldo, G. Rarefied gas flow through a long tube of variable radius. *J. Vac. Sci. Technol. A* **2005**, *23*, 531–533. [CrossRef]

28.	Graur, I.; Sharipov, F. Non-isothermal flow of rarefied gas through a long pipe with elliptic cross section. *Microfluid. Nanofluid.* **2009**, *6*, 267–275. [CrossRef]

29. Ritos, K.; Lihnaropoulos, Y.; Naris, S.; Valougeorgis, D. Pressure- and temperature-driven flow through triangular and trapezoidal microchannels. *Heat Transf. Eng.* **2011**, *32*, 1101–1107. [CrossRef]

30. Tatsios, G.; Lopez Quesada, G.; Rojas-Cardenas, M.; Baldas, L.; Colin, S.; Valougeorgis, D. Computational investigation and parametrization of the pumping effect in temperature-driven flows through long tapered channels. *Microfluid. Nanofluid.* **2017**, *21*, 99. [CrossRef]

31. Pantazis, S.; Valougeorgis, D.; Sharipov, F. End corrections for rarefied gas flows through circular tubes of finite length. *Vacuum* **2014**, *101*, 306–312. [CrossRef]

32. Valougeorgis, D.; Vasileiadis, N.; Titarev, V. Validity range of linear kinetic modeling in rarefied pressure driven single gas flows through circular capillaries. *Eur. J. Mech. B Fluids* **2017**, *64*, 2–7. [CrossRef]

33. Naris, S.; Valougeorgis, D.; Kalempa, D.; Sharipov, F. Flow of gaseous mixtures through rectangular microchannels driven by pressure, temperature, and concentration gradients. *Phys. Fluids* **2005**, *17*, 100607. [CrossRef]

34. Rojas Cardenas, M.; Graur, I.; Perrier, P.; Meolans, J.G. Thermal transpiration flow: A circular cross-section microtube submitted to a temperature gradient. *Phys. Fluids* **2011**, *23*, 031702. [CrossRef]

35. Rojas-Cárdenas, M.; Perrier, P.; Graur, I.; Méolans, J.G. Time-dependent experimental analysis of a thermal transpiration rarefied gas flow. *Phys. Fluids* **2013**, *25*, 72001. [CrossRef]

36. Sharipov, F. Data on the velocity slip and temperature jump on a gas-solid interface. *J. Phys. Chem. Ref. Data* **2011**, *40*, 023101. [CrossRef]

micromachines

MDPI

Article

Study of Flow Characteristics of Gas Mixtures in a Rectangular Knudsen Pump

Zhijun Zhang [1],*, Xiaowei Wang [1], Lili Zhao [2], Shiwei Zhang [1] and Fan Zhao [1]

[1]   School of Mechanical Engineering and Automation, Northeastern University, Shenyang 110819, China;
     xiaowwang812@163.com (X.W.); shwzhang@mail.neu.edu.cn (S.Z.); nvacuum_zhaofan@163.com (F.Z.)
[2]   School of Mechanical Engineering, Shenyang University, Shenyang 110044, China; zhaolili0214@163.com
*   Correspondence: Zhjzhang@mail.neu.edu.cn; Tel.: +86-2483679926

Received: 27 November 2018; Accepted: 22 January 2019; Published: 24 January 2019

Abstract: A Knudsen pump operates under the thermal transpiration effect or the thermal edge effect on the micro-scale. Due to the uneven temperature distribution of the walls in the channel axis direction or the constant temperature of the tips on the walls, directional thermally-induced flow is generated. In this paper the Direct Simulation Monte Carlo (DSMC) method is applied for N_2–O_2 gas mixtures in the ratios of 4:1, 1:1, and 1:4 with different Knudsen numbers in a classic rectangular Knudsen pump to study the flow characteristics of the gas mixtures in the pump. The results show that the changing in the gas physical properties does not affect the distribution of the velocity field, temperature fields, or other fields in the Knudsen pump. The thermal creep effect is related to the molecular mass of the gas. Even in N_2 and O_2 gas mixtures with similar molecular masses, N_2 can be also found to have a stronger thermal creep effect. Moreover, the lighter molecular weight gas (N_2) can effectively promote the motion of the heavier gas (O_2).

Keywords: Knudsen pump; thermally induced flow; gas mixtures; direct simulation Monte Carlo (DSMC); microfluidic

1. Introduction

It is well known that the thermally-induced flow of rarefied gas is generated by the temperature gradient along the walls of the Knudsen pump and that the gas is driven to flow from the low-temperature side to the high-temperature side. That is the basic mechanism of the Knudsen pump which was first put forward by Danish physicist Martin Knudsen in 1909 [1]. The Knudsen pump can provide consistent gas flow and has the advantage of having of no moving parts, a simple structure, ease of operation, long life span, low energy consumption, and wide energy sources. It is widely applied in Micro Electro Mechanical Systems (MEMS) such as gas separators [2,3], gas analysis [4–6], micro combustors [7,8], and micro-air vehicle systems [9,10].

The classic rectangular Knudsen pump is composed of a series of alternately connected wide and narrow micro-channels [1]. A tangential temperature gradient appears by imposing high-temperature and low-temperature heat resources for the two ends of the wide channels respectively. This generates a thermal creep effect for the gas flow. In recent years, with the development of materials technology and micro-machining technology, the pump structure can now be produced by using poly-silicon material, and using the inter-molecular gaps in porous materials such as aerogel membranes [10–12], mixed cellulose ester (MCE) [13,14], zeolite [15,16], porous ceramics [17,18] and Bi_2Te_3 [19,20] to construct the flowing channel of the Knudsen pump. Since the rectangular Knudsen pump has been proposed, many structures for the channel were successively designed and studied (Figure 1), including the sinusoidal micro-channel [21], matrix micro-channel [21], curved micro-channel with different curvature radii [22], alternately connected curved and straight micro-channel [23,24], tapered micro-channel [25], and ratchet micro-channel [26–29].

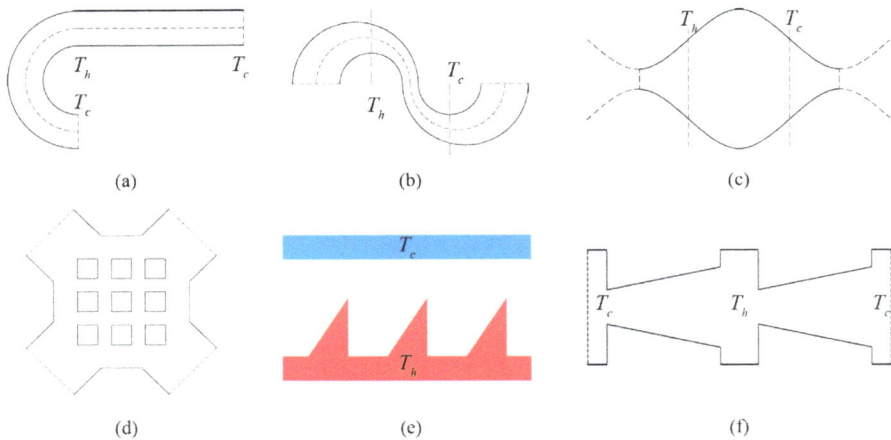

Figure 1. Common Knudsen pump channel structures: (**a**) Curve-straight channel, (**b**) double-curves channel, (**c**) sinusoidal channel, (**d**) matrix channel, (**e**) ratchet channel, (**f**) taper channel.

It is well known that the Boltzmann equation is the basic equation for solving the continuous, transition, and free-molecular regimes. It has been developed in mathematical methods such as the moment method and model equation in recent years. Chapman-Enskog solution is the most important representative of the moment method, whose first-order solution is the Navier–Stokes–Fourier (NSF) equation [30,31]. By adding velocity slip and temperature jump conditions, the Chapman-Enskog solution can be applied to the rarefied gas flow within a small Knudsen number. However, the regularized 13-moment equations [32] can be used for the rarefied gas flows with a large Knudsen number. The model equation simplifies the collision integral in the Boltzmann equation. The most famous model equations are the BGK model introduced by Bhatnagar, Gross and Krook [33] and McCormack model [34]. The BGK model is well applied to the study of transport characteristics of gas mixtures in micro-channels [35,36]. The McCormack model also shows good consistency with experimental results in studying the flow state of gas mixtures [37]. The Direct Simulation Monte Carlo (DSMC) method is a direct numerical solution of the Boltzmann equation, which eliminates the disadvantages in the mathematical solution of the Boltzmann equation. Although the DSMC method requires a large amount of internal storage space and long calculation time, its calculation results are highly consistent with experimental results [38–40].

Compared with other study methods [21–24,41–46], the DSMC method is widely used for heat and mass transfer in micro-channels [47–50]. There are many studies that apply the DSMC method in the flow of gas mixtures [51–53]. It is found that Knudsen pump shows good capability in gas separation by the simulation of DSMC [54]. While in the studies of gas flow in Knudsen pumps, the method of DSMC is widely employed [28,29,55]. For example, DSMC is used to study and simulate flow patterns of the gas in rectangular channels [55] and ratchet channels [28,29]. The studies of Knudsen pumps have generally been focused on innovation of the structure, optimization of performance, and practical application. The gas used in simulations is mostly monatomic noble gas. However, gas mixtures have been more widely applied than single gases, and the proportions of the noble gas in the air are very small. Moreover, the size of the micro-channel has already reached the nanometer level, ensuring that the Knudsen pump operates normally under atmospheric pressure.

In present study, the flow characteristics of gas mixture of N_2 and O_2 in Knudsen pump are simulated with the DSMC method. A classic rectangular channel is applied, that is more common and convenient to machine. The problem statement and the numerical method are presented in Section 2. The simulation results for the gas mixtures of N_2 and O_2 in three different ratios are discussed in Section 3, and the conclusions are in Section 4.

2. Problem Statement and Numerical Method

2.1. Problem Statement

The configuration consists of the alternately connected narrow and wide micro-channels shown in Figure 2. A periodic structure is well-established in the x-direction. For decreasing the calculation amount and improving the simulation efficiency, a basic unit presented by the green dash-dotted line in Figure 2 is extracted. The periodic boundary conditions on the inlet and outlet are used for the recurrence of the physical structure. When the simulation particles pass through these two boundaries, except for the change of the position in x-direction ($x_{\text{Inlet}} = x_{\text{Outlet}} - L$), all the other parameters such as distribution function, velocity, temperature, acting force Φ are definitely equal, that is $\Phi_{\text{Inlet}} = \Phi_{\text{Outlet}}$.

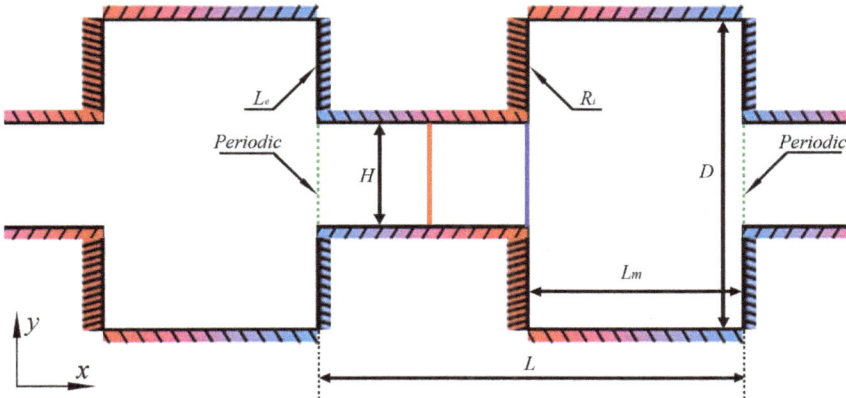

Figure 2. Configuration and geometric parameters.

The distance between the two walls of the narrow channel is indicated by H, and the distance (D) between the two walls of the wide channel is triple that of the narrow channel. The length of the basic unit is L; the length of the narrow channel and the wide channel in the x-direction are both equally L_m, which is half of L.

The configuration, temperature, are geometric parameters of the rectangular channel are $H = 1$ µm, $D = 3H = 3$ µm, $L = 4$ µm, and $L_m = L/2 = 2$ µm, respectively. In consideration of the thermal resistance of the materials in practice and the high heat flux density in the micro-scale, the reference temperatures of the cold walls and hot walls are, respectively, $T_c = 225$ K and $T_h = 375$ K [27,28]. In the paper, all parameters are only for the two-dimensional surface. Assume that the third dimension is infinite (defining the property as "empty"), and there is no influence of the scale effect in the z-direction on the flow characteristics. However, the OpenFOAM software (version 4.1, OpenCFD Ltd, Bracknell, England, UK) is limited to three-dimensional models, and to enable comparison to the mass flow rate in [28], the width in the z-direction is also considered as 20 µm in modeling.

The Knudsen number [40,56]:

$$Kn = \frac{\lambda}{H} = \frac{2(5 - 2\omega)(5 - 2\omega)}{15H} \left(\frac{M}{2\pi k T_{\text{m}}} \right)^{1/2} \left(\frac{\mu}{\rho} \right) \tag{1}$$

is defined by the characteristic dimension H, the distance between the two walls of the narrow channel. In Equation 1, P is the pressure. The mean temperature is $T_{\text{m}} = (T_{\text{h}} + T_{\text{c}})/2$ and $M = CM_a + (1 - C)M_b$ is the average molar mass of the mixture. The number densities of the components are denoted by n_a and n_b. C is the molar fraction of the first gas component. The dynamic viscosity μ is calculated from the Chapman-Enskog theory. It can be written as $\mu = \mu_a + \mu_b$, where μ_a and μ_b are assumed to be dependent on the temperature T_m according to the law [35,51].

2.2. DSMC Method

The DSMC method is based on three basic dynamic theories of rarefied gas, the ergodic assumption, binary collision assumption, and the molecular chaos assumption. The flowing characteristics of the real gas molecules in the micro-channels are represented by a set of simulation particles in the process of simulation. With its movements, inter-molecular collisions and the interaction of the boundary walls, some information of the simulation particles can be stored in the computer, including position, velocity, internal energy, and so on. The simulating process for the gas flow is achieved by applying statistics to obtain the average information of the simulation particles in all cells to represent the macroscopic variation.

Additionally, the time step of DSMC method should be far lower than the mean collision time. Therefore, the real processes of the molecular free movement and the inter-molecular collisions are decoupled into two consecutive steps. It is assumed that molecules are in uniform rectilinear motion in the original direction within one time step size. If collisions with the walls occur in the process of the free movement, the collision will be calculated first, and the post-velocity will be used for the free movement and inter-molecular collisions. Finally, the microscopic variables of all simulation particles in the cells can be used to depict the macroscopic physical quantities using statistics. There are many molecular models depicting inter-molecular collisions. The most famous are the hard sphere (HS), variable hard sphere (VHS), variable sphere (VS) and variable soft sphere (VSS) models [57]. The variable hard sphere (VHS) model is widely used in studies of the Knudsen pump. The VHS model is adopted in this work. The post-collision velocities of a colliding pair of molecules can be found in [38–40]. In addition, the no time counter (NTC) scheme [40] is used to ensure the correct number of collisions, which is consistent with the analytical theory.

For the boundary conditions, the constant cold temperature T_c and the constant hot temperature T_h are respectively exerted on the left and right walls of the wide channel (L_e and R_i in Figure 2). The wall can be made of high thermal conductivity material. Therefore, because of the heat transfer property, the positive constant temperature gradient is applied for the walls of the narrow channels. Additionally, the negative constant temperature gradient is applied for the walls of the wide channels in the actual applications. The walls are adopted as to completely diffuse reflection. All the particles colliding with the walls are diffusely reflected according to Maxwellian velocity distribution.

2.3. Code Validation

In the present study, an open source Direct Simulation Monte Carlo (DSMC) code, dsmcFoam [56,58] was employed. This solver has been tested in a lot of cases, such as 2D flow over a flat plate and a cylinder, and 3D supersonic flows over complex geometries. The dsmcFoam shows very good agreement with data provided by both analytical solutions and other contemporary DSMC codes. Furthermore, Shahabi et al. [29] applied this solver to study the physical mechanism of the thermally induced flow in ratchet Knudsen pump. In order to verify the feasibility of dsmcFoam in dealing with the thermally induced flow problems more clearly, thermally induced flow in square cavity was simulated in this paper. The same parameters were used, and the results were compared with the discrete unified gas kinetic method, dugksFoam [43]. It can be seen that both results agree well with each other, as shown in Figure 3 (the results of dugksFoam is not shown in Figure 3a).

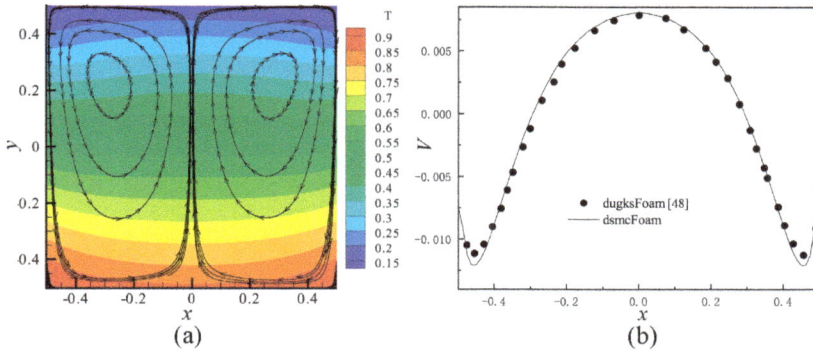

Figure 3. The simulation results of square cavity with dsmcFoam. (**a**) Temperature contours and velocity streamlines at $Kn = 0.1$ by dsmcFoam. (**b**) Profile of the V component of the velocity on horizontal line, passing through the center of the left primary vortex at $Kn = 0.1$. In (**b**), black points show the results extracted from [43], and the curve shows the results of dsmcFoam.

2.4. Grid, Particle and Time Step

Grid (or cell) size and the number of simulation particles in the cell are two main factors influencing the calculating efficiency and accuracy. Bird [38–40] pointed out that the cell size Δx should not exceed 1/3 of the mean free path λ, and the number of simulation particles N' in every cell should range from 20 to 30 which assumes that the number is averagely distributed.

$$N' = \frac{F_N}{N_C} \tag{2}$$

where N_C is the number in the cell and F_N is the sample number, which relies on the ratio of the number of real molecules N to equivalent particle number E_p:

$$F_N = \frac{N}{E_p} = \frac{P/kT_m}{E_p} \tag{3}$$

In terms of the gas mixtures, the cell size Δx should be smaller than 1/3 of the minimum mean free path ($\lambda_{min} = \min(\lambda_a, \lambda_b, \ldots)$). The relation between the time step Δt and the mean collision time (MCT) or the mean transit time (MTT) can be represented as follows [59]:

$$0.005 \le \Delta t/t_0 \le 0.5 \tag{4}$$

where, $t_0 = \lambda_{min}/v_m$ is the MCT, $v_m = \sqrt{2kT_m/m}$ is the most molecular probable speed. The effect of different time steps on the velocity and temperature distributions on the surface of the narrow channel outlet (indicated by the blue full line in Figure 2) compared in Figure 4a,b. It is demonstrated that a time step of $\Delta t < \lambda_{min}/(3v_m)$ could provide the time-step independent solutions.

Besides, in the actual simulation, when the number of simulation particles in the cell is over 15 and the cell size is $\lambda_{min}/3$, the results of the calculation do not have significant differences, as shown in Table 1. After comprehensive consideration, in this study, the number of simulation particles in the cell is around 15 and the cell size is about $\lambda_{min}/3$. With the rarefied degree of the rarefied gas increasing, 1/3 of the mean free path will probably exceed the size of the geometric model. Therefore, to avoid this circumstance, the oversize cell size hampers the normal division of the mesh. When the Knudsen number is over 0.387, the cell size remains (equal to the size for $Kn = 0.387$).

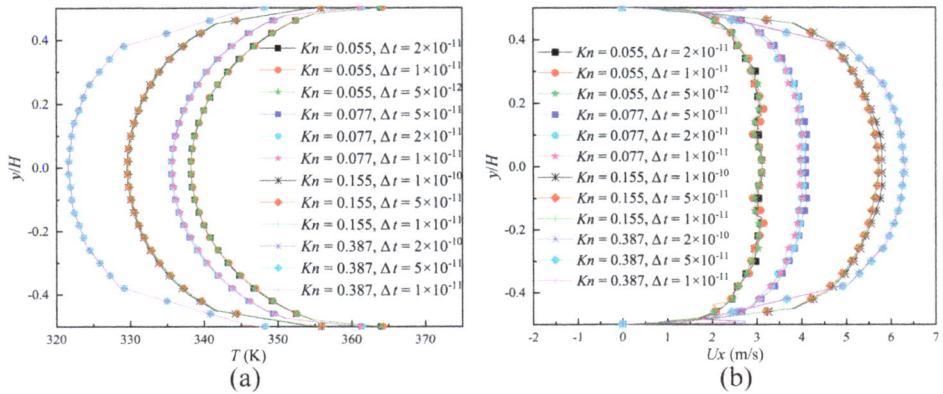

Figure 4. Calculation results of different time steps. (**a**) Velocity distributions on the outlet of the narrow channel; (**b**) temperature distributions on the outlet of the narrow channel.

Table 1. The mass flow rate for different cell sizes and the numbers of the simulator in the cell, at the reference for $H = 1$ μm, $D = 3$ μm, $L = 4$ μm, $L_m = 2$ μm, $T_c = 225$ K, $T_h = 375$ K and $Kn = 0.155$.

Parameter	Case 1	Case 2	Case 3	Case 4	Case 5	Case 6
Number of the cell N_C	9248	5100	3200	3200	3200	3200
$\Delta x / \lambda_{min}$	10/63	5/24	1/3	1/3	1/3	1/3
The number of the simulator in the cell n	15	15	10	15	20	30
Time step Δt(s)	4×10^{-11}	5×10^{-11}	1×10^{-10}	1×10^{-10}	1×10^{-10}	1×10^{-10}
Mass flow rate (10^{-11}) (kg/s)	3.8294	3.8328	3.8681	3.8367	3.8359	3.8384

3. Results and Discussion

This section mainly discusses the flowing characteristics of the gas mixtures of N_2 and O_2 in the ratios of 4:1, 1:1 and 1:4 in the rectangular channel Knudsen pump. The flow field, distribution of temperature gradient, distribution of velocity, and mass flow rate are thoroughly studied. The physical properties of the gases for the simulation particles of N_2 and O_2 in the VHS model are listed in Table 2.

Table 2. Physical properties of N_2 and O_2 for $T_0 = 273$ K.

Gas	Property			
	Molecular Mass m (10^{-27}) (kg)	Molecular Diameter d (10^{-10}) (m)	Internal Degree of Freedom	ω
N_2	46.50	4.17	2	0.74
O_2	53.12	4.07	2	0.77

3.1. Velocity and Temperature Distribution

For N_2 and O_2 in the three different mixed ratios, the representative velocity and temperature distribution in the rectangular channel for $Kn = 0.055$, $Kn = 0.387$, and $Kn = 3.87$ are illustrated respectively in Figure 5.

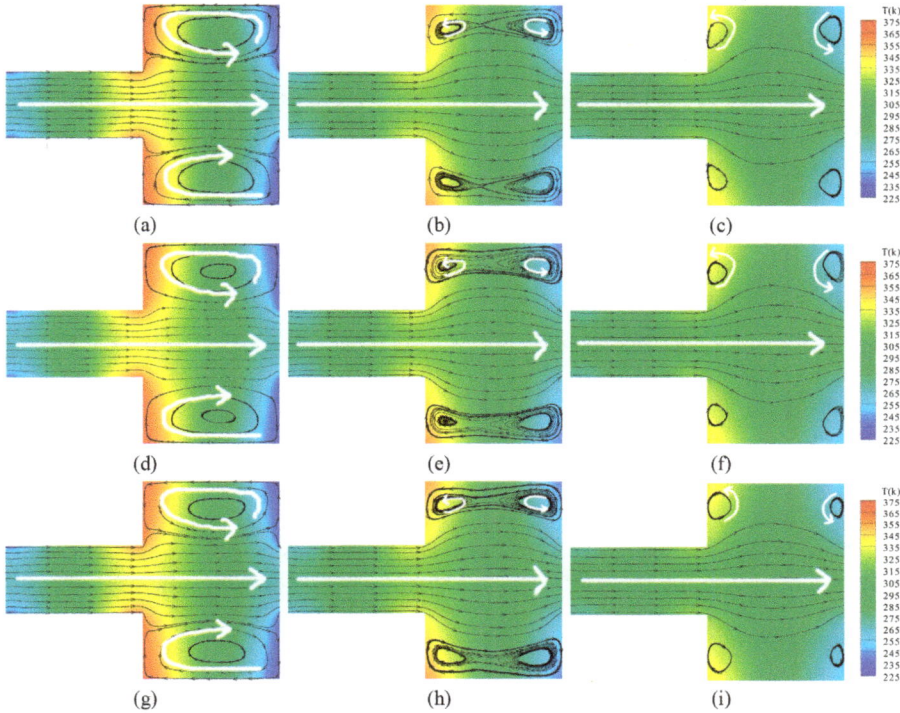

Figure 5. Velocity streamlines and temperature contours for different ratios of N_2 and O_2 and different Knudsen numbers. (**a**) N_2 and O_2 = 4:1, Kn = 0.055; (**b**) N_2 and O_2 = 4:1, Kn = 0.387; (**c**) N_2 and O_2 = 4:1, Kn = 3.87; (**d**) N_2 and O_2 = 1:1, Kn = 0.055; (**e**) N_2 and O_2 = 1:1, Kn = 0.387; (**f**) N_2 and O_2 = 1:1, Kn = 3.87; (**g**) N_2 and O_2 = 1:4, Kn = 0.055; (**h**) N_2 and O_2 = 1:4, Kn = 0.387; (**i**) N_2 and O_2 = 1:4, Kn = 3.87.

By observing the figures, it can be seen that gases flow forward from left to right in the narrow channels without differences, though the Knudsen number increases. However, large differences are triggered in the wide channels, and the rarefied degree of the gas is enhanced. A larger circular-flow vortex appears respectively in the upper side and the lower side of the wide channel for Kn = 0.055. When the Knudsen number increases, the velocity stream of gases near the central axis expands towards the upper side and the lower side. Within the larger vortex, two anticlockwise secondary vortexes are generated, illustrated by white arrows for Kn = 0.387. With the higher rarefied degree of the gas, the secondary vortexes completely replace the main vortexes. The secondary vortexes individually exist near the corners of the wide channel, shown by white arrows for Kn = 3.87.

In terms of the temperature field, as the rarefied degree of the gas increases, the total number of gas molecules decreases. The thermal conductivity is weakened. Thus, the energy transmitting from the walls to the field decreases dramatically, as does the temperature difference. For different ratios of the gas mixtures with the same Knudsen number, the general distributions of the temperature field and the velocity field do not change dramatically with the variations of the ratio.

3.2. Temperature Gradient

In the DSMC method, the overall temperature of the gas T is as follows [38–40],

$$T = \sum_{p=1}^{q} \left(\zeta_p T_p N'' \right) / \sum_{p}^{q} \left(\zeta_p N'' \right) \tag{5}$$

where p is the species of the simulation particle, q is the amount of the species of the simulation particle, $\sum N''$ is the weighted number of simulated molecules, and the temperature of species p is

$$T_p = \left(3T_{tr,p} + \zeta_{rot,p}T_{rot,p} + \zeta_{vib,p}T_{vib,p} + \zeta_{el,p}T_{el,p}\right)/\zeta_p \tag{6}$$

where, $T_{tr,p}$, $T_{rot,p}$, $T_{vib,p}$, and $T_{el,p}$ are translational temperature, rotational temperature, vibrational temperature and electronic temperature, respectively. $\zeta_{rot,p}$, $\zeta_{vib,p}$, and $\zeta_{el,p}$ are the corresponding degrees of freedom. The effective number of degrees of freedom ζ_p of species p is

$$\zeta_p = 3 + \zeta_{rot,p} + \zeta_{vib,p} + \zeta_{el,p} \tag{7}$$

Due to no consideration of the vibrational energy and the electronic energy ($\zeta_{vib,p} = \zeta_{el,p} = 0$), the temperature of species p is

$$T_p = \left(3T_{tr,p} + \zeta_{rot,p}T_{rot,p}\right)/\left(3 + \zeta_{rot,p}\right) \tag{8}$$

where,

$$T_{tr,p} = m_p\left\{\Sigma\left(u_p^2\right)'' + \Sigma\left(v_p^2\right)'' + \Sigma\left(w_p^2\right)'' - u_0^2 - v_0^2 - w_0^2\right\}/(3k) \tag{9}$$

$$T_{rot,p} = (2/k)\left(\Sigma\varepsilon''_{rot,p}/\zeta_p\right) \tag{10}$$

where, $\Sigma\varepsilon''_{rot,p}$ is the weighted sum of the rotational energy of the simulated molecules of species p.

By substituting Equations (9) and (10) into Equation (5), the overall temperature is obtained by the summation of all species of the simulation particles. Similarly, in the dsmcFoam, the overall temperature of the gas T is obtained by the mean value of the relevant macroscopic physical properties in the cell. The representation is as follows by dsmcFoam code [46],

$$T = \frac{2\left(\rho_{lKE} - 0.5\rho_m U^2 + \rho_{ilE}\right)}{k(3\rho_N + \rho_{iDF})} \tag{11}$$

where ρ_{lKE} is mean value of the linear kinetic energy density, ρ_m is the mean value of the mass density, U is the mean value of gas velocity, ρ_{ilE} is the mean value of the internal energy density, and ρ_N is the mean value of the real gas-molecular number density in the cell. ρ_{iDF} is the mean value of the internal degree of the freedom density, but the influence of internal degree of freedom is not considered in the VHS model; here, it is assumed that $\rho_{iDF} = 0$. To simulate the molecular motion, the random numbers are generated in the DSMC method. The average values of macroscopic physical quantities are applied to maximize the simulation accuracy.

Figure 6 illustrates the distribution of the temperature gradient for three gas mixtures and three Knudsen numbers on the central axis of the channel (indicated by the red full line in Figure 6) within one structure unit. It can be seen that the changes in the composition of the gas mixtures do not make a difference in the distribution of the temperature gradient on the central axis. The distribution patterns are similar to the asymmetric sinusoid shown in Figure 6a. This is because the size of the narrow channel is much smaller, and more energy is transferred to the central axis.

Moreover, it was found that for different Knudsen numbers, these distribution patterns always remain as asymmetric sinusoidal patterns. The maximum temperature gradients respectively reach the medium points of the narrow channel and the wide channel (shown as Figure 6b). For the reason that the rarefied degree of the gas increases and thermal conductivity weakens, the maximal values of temperature gradient decrease with increasing Knudsen numbers. However, the velocity for *Kn* = 0.387 is larger than the velocity for *Kn* = 0.155 (Discussion later). This is because the thermal creep effect hardly appears near the slip regime. It is not obvious, though the temperature gradient is larger. Therefore, we conclude that the value of the temperature gradient does not correspond to the performance of the thermal creep effect near the slip regime.

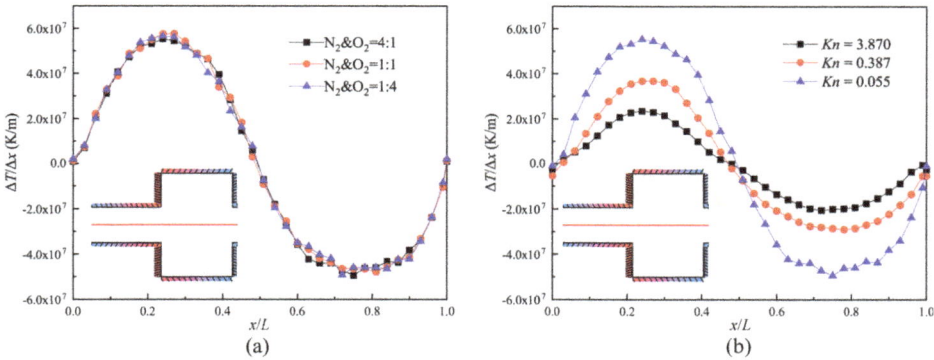

Figure 6. Temperature gradient on the central axis of the channel. (**a**) *Kn* = 0.055; (**b**) N_2 and O_2 = 4:1.

3.3. Mean Velocity

Figure 7 presents the mean velocity of each composition with the different gas mixture in a cross-section at $x = 1$ μm (indicated by the red full line in Figure 2) for different Knudsen numbers. Even in a mixture of N_2 and O_2 with a similar molecular mass, the velocity of N_2 is larger than that of O_2. The close relation between the thermal creep effect and the gas-molecular mass is clearly proved [60]. When the ratio of N_2 rises in the gas mixtures, the thermal creep effect is enhanced. This not only causes *n* increase in the velocity of N_2 itself, but also contributes to an increase in the velocity of O_2. That is, N_2 can promote the movement of O_2.

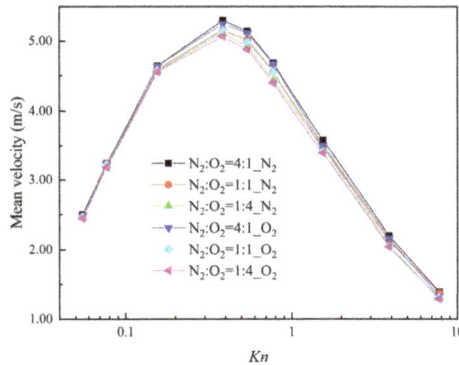

Figure 7. Mean velocities for different Knudsen numbers.

It was also found that the mean velocity of each composition in different gas mixtures reaches the maximum for *Kn* = 0.387. When the gas mixtures are in the slip regime (*Kn* < 0.077) and free-molecular regime (*Kn* > 7.75), no matter how changeable the compositions of the gas mixtures are, the mean velocities of each composition are almost the same. That is because the thermal creep effect cannot be effectively induced by the temperature gradient in the slip regime, while in the free-molecular regime the thermal creep effect is weaker. Therefore, regarding systems that process gas separation using the thermal creep effect, it should be guaranteed that gas mixtures are in a transition regime in order to improve the efficiency and quality of gas separation.

3.4. Mass Flow Rate

During one physical time t_{avg} which used for calculating the average of the macroscopic physical quantities, the mass flow rate is obtained by calculating the total mass of all particles that pass through the cross section in $x = 1$ μm (indicated by a red full line in Figure 2). This is demonstrated by the following equations:

$$M_f = \frac{\sum\limits_{i}^{\Delta N} E_p m_i sgn(\vec{c}_i \cdot \vec{n}_f)}{t_{avg}} \tag{12}$$

$$sgn(\vec{c}_i \cdot \vec{n}_f) = \begin{cases} 1 & if \ \vec{c}_i \cdot \vec{n}_f > 0 \\ 0 & if \ \vec{c}_i \cdot \vec{n}_f = 0 \\ -1 & if \ \vec{c}_i \cdot \vec{n}_f < 0 \end{cases} \tag{13}$$

where ΔN is the total number of the simulating particles passing through the cross section within t_{avg}, m_i is the mass of the molecules; $\vec{c}_i$ is the velocity of particles; $\vec{n}_f$ is the unit normal vector of the cross-section, and the positive and negative directions are the same as the x-axis. Likewise, in order to improve the accuracy of the simulation in the DSMC, the paper decreases the numerical error through the mean value of the mass flow rate for a large number of cycle numbers. The sampling interval is usually the same as the physical time, t_{avg}.

Therefore, the total mass flow rates of these three different N_2–O_2 gas mixes can be obtained. The mass flow rate of each composition for the different Knudsen numbers is illustrated in Figure 8. Mass flow rate decreases with the increase of the Knudsen number, and the maximum and minimum values occur respectively at $Kn = 0.055$ and $Kn = 7.75$. This is mainly because the enhancement of the gas rarefied degree leads to a decrease in the number of molecules. The variation of the mass flow rate is the largest in the Knudsen number range from 0.155 to 0.387. It is determined by the stronger thermal creep effect within this range.

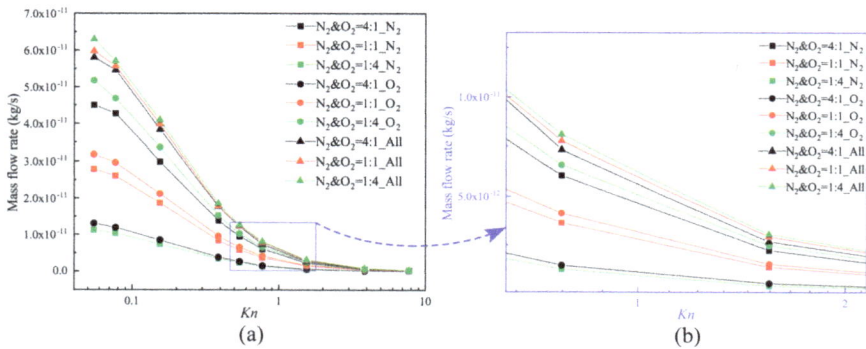

Figure 8. Mass flow rates for different Knudsen numbers.

Furthermore, though O_2 molecules are heavier and have higher densities in each of the gas mixtures, N_2 has the larger flowing velocity, and the number of N_2 gas molecules passing through the channel per unit time increases. Thus, the increase of the ratio of N_2 weakens the influence of the differences in the value of the densities. This increases the mass flow rate to some degree. The thermal creep effect is stronger, especially when the value of the Knudsen number is in the range of 0.155–0.387. The total mass flow rates of these three N_2–O_2 gas mixtures are much closer to each other.

4. Conclusions

The flow characteristics of N_2–O_2 gas mixtures in the rectangular Knudsen pump are studied by using the DSMC method. By exerting temperature gradient boundary conditions on the walls, a thermally induced flow (thermal creep flow) is successfully generated. The influences of N_2–O_2 gas mixtures in different ratios, and different gas rarefied degrees (different Knudsen numbers) on the flow characteristics of gases are well studied. The following conclusions can be drawn:

(a) Under the same Knudsen number, the flow fields of the three different gas mixtures in the Knudsen pump channel are highly similar. The distribution of the temperature gradient is all asymmetric sinusoid in nature on the central axis of the channel. That is, the variations of the gas compositions do not make a difference in the distribution of the flow field in the Knudsen pump channels.

(b) Even in N_2 and O_2 gas mixtures with similar molecular masses, N_2 is found to have a stronger thermal creep effect. It was successfully verified that the thermal creep effect has a relationship with the weight of the gas molecules.

(c) In gas mixtures, N_2 has a larger velocity than O_2. If the proportion of N_2 increases, the overall velocity also increases. The lighter gas can promote the movement of the heavier gas.

(d) The lighter gas and heavier gas respectively correspond to a larger volume flow rate and a larger mass flow rate. With the ratio of the lighter gas increasing, the incremental volume flow rate somewhat weakens the difference of the mass flow rate resulting from the difference of densities. Even though the ratios of each composition of the gas mixtures differ greatly, the total mass flow rates are almost equal, especially when the thermal creep effect is the strongest.

Author Contributions: Conceptualization, Z.Z.; Data curation, X.W. and L.Z.; Investigation, F.Z.; Methodology, X.W. and Z.Z.; Resources, X.W. and L.Z.; Software, X.W.; Visualization, L.Z.; Writing—original draft, X.W., Z.Z. and L.Z.; Writing—review & editing, Z.Z. and S.Z.

Funding: This research was supported by the National Natural Science Foundation of China (Grant Nos. 31371873, 31000665, 51176027, and 31300408) and Special Program for Applied Research on Super Computation of the NSFC-Guangdong Joint Fund (the second phase) of China.

Acknowledgments: Zhijun Zhang and Xiaowei Wang would like to thank Zhiguo Zhang (Northeastern University, China) for the technology support of computer code.

Conflicts of Interest: The authors declare no conflict of interest.

References

1. Knudsen, M. Eine revision der gleichgewichtsbedingung der gase. Thermische Molekularströmung. *Ann. Phys.* **1909**, *336*, 205–229. [CrossRef]

2. Nakaye, S.; Sugimoto, H. Demonstration of a gas separator composed of Knudsen pumps. *Vacuum* **2016**, *125*, 154–164. [CrossRef]

3. Nakaye, S.; Sugimoto, H.; Gupta, N.K.; Gianchandani, Y.B. Thermal method of gas separation with micro-pores. In Proceedings of the 2014 IEEE SENSORS, Valencia, Spain, 2–5 November 2014; pp. 815–818.

4. Terry, S.C.; Jerman, J.H.; Angell, J.B. A gas chromatographic air analyzer fabricated on a silicon wafer. *IEEE Trans. Electron. Devices* **1979**, *26*, 1880–1886. [CrossRef]

5. Ferran, R.J.; Boumsellek, S. High-pressure effects in miniature arrays of quadrupole analyzers for residual gas analysis from 10-9 to 10-2 Torr. *J. Vac. Sci. Technol. A-Vac. Surf. Films* **1996**, *14*, 1258–1265. [CrossRef]

6. Blomberg, M.; Rusanen, O.; Keranen, K.; Lehto, A. A silicon microsystem-miniaturised infrared spectrometer. In Proceedings of the International Conference on Solid State Sensors and Actuators, Chicago, IL, USA, 19 June 1997; pp. 1257–1258.

7. Fu, K.; Knobloch, A.J.; Cooley, B.A.; Walther, D.C.; Fernandez-Pello, C.; Liepmann, D.; Miyasaka, K. Microscale Combustion Research for Applications to MEMS Rotary IC Engine. In Proceedings of the ASME 35th National Heat Transfer Conference, Anaheim, CA, USA, 10–12 June 2001.

8. Yang, W.M.; Chou, S.K.; Shu, C.; Li, Z.W.; Xue, H. Combustion in micro-cylindrical combustors with and without a backward facing step. *Appl. Therm. Eng.* **2002**, *22*, 1777–1787. [CrossRef]

9.   Wei, S.; Berg, M.; Ljungqvist, D. Flapping and flexible wings for biological and micro air vehicles. *Prog. Aeosp. Sci.* **1999**, *35*, 455–505.

10. Ellington, C.P. The novel aerodynamics of insect flight: Applications to micro-air vehicles. *J. Exp. Biol.* **1999**, *202*, 3439–3448.

11. Zhao, S.; Jiang, B.; Maeder, T.; Muralt, P.; Kim, N.; Matam, S.K.; Jeong, E.; Han, Y.L.; Koebel, M.M. Dimensional and Structural Control of Silica Aerogel Membranes for Miniaturized Motionless Gas Pumps. *ACS Appl. Mater. Interfaces* **2015**, *7*, 18803–18814. [CrossRef] [PubMed]

12. Vargo, S.E.; Muntz, E.P. Initial results from the first MEMS fabricated thermal transpiration-driven vacuum pump. *Am. Inst. Phys.* **2001**, *585*, 502–509.

13. Young, M.; Han, Y.L.; Muntz, E.P.; Shiflett, G. Characterization and Optimization of a Radiantly Driven Multi-Stage Knudsen Compressor. In Proceedings of the RAREFIED GAS DYNAMICS: 24th International Symposium on Rarefied Gas Dynamics, Bari, Italy, 10–16 July 2004; American Institute of Physics: College Park, MD, USA, 2005; pp. 174–179.

14. Nakaye, S.; Sugimoto, H.; Gupta, N.K.; Gianchandani, Y.B. Thermally enhanced membrane gas separation. *Eur. J. Mech. B-Fluids* **2015**, *49*, 36–49. [CrossRef]

15. Gupta, N.K.; Gianchandani, Y.B. Thermal transpiration in zeolites: A mechanism for motionless gas pumps. *Appl. Phys. Lett.* **2008**, *93*, 193511. [CrossRef]

16. Gupta, N.K.; Gianchandani, Y.B. A knudsen pump using nanoporous zeolite for atmospheric pressure operation. In Proceedings of the IEEE International Conference on Micro Electro Mechanical Systems (MEMS), Wuhan, China, 13–17 January 2008; pp. 38–41.

17. Han, Y.L.; Young, M.; Muntz, E.P. Performance of Micro/Meso-Scale Thermal Transpiration Pumps at Low Pressures. In Proceedings of the ASME 2004 International Mechanical Engineering Congress and Exposition, Anaheim, CA, USA, 13–19 November 2004; pp. 257–265.

18. Han, Y.L.; Young, M.; Muntz, E.P.; Shiflett, G. Knudsen compressor performance at low pressures. In Proceedings of the AIP Conference, Rio de Janeiro, Brazil, 23–25 May 2005; pp. 162–167.

19. Faiz, A.; McNamara, S.; Bell, A.D.; Sumanasekera, G. Nanoporous Bi2Te3 thermoelectric based Knudsen gas pump. *J. Micromech. Microeng.* **2014**, *24*, 035002. [CrossRef]

20. Goldsmid, H.J.; Douglas, R.W. The use of semiconductors in thermoelectric refrigeration. *Br. J. Appl. Phys.* **1954**, *5*, 386. [CrossRef]

21. Bond, D.M.; Wheatley, V.; Goldsworthy, M. Numerical investigation into the performance of alternative Knudsen pump designs. *Int. J. Heat Mass Transf.* **2016**, *93*, 1038–1058. [CrossRef]

22. Bond, D.M.; Wheatley, V.; Goldsworthy, M. Numerical investigation of curved channel Knudsen pump performance. *Int. J. Heat Mass Transf.* **2014**, *76*, 1–15. [CrossRef]

23. Aoki, K.; Degond, P.; Mieussens, L.; Takata, S.; Yoshida, H. A diffusion model for rarefied flows in curved channels. Multiscale Model. *Multiscale Model. Simul.* **2008**, *6*, 1281–1316. [CrossRef]

24. Aoki, K.; Degond, P.; Mieussens, L. Numerical simulations of rarefied gases in curved channels: Thermal creep, circulating flow, and pumping effect. *Commun. Comput. Phys.* **2009**, *6*, 919–954. [CrossRef]

25. Tatsios, G.; Quesada, G.L.; Rojas-Cardenas, M.; Baldas, L.; Colin, S.; Valougeorgis, D. Computational investigation and parametrization of the pumping effect in temperature-driven flows through long tapered channels. *Microfluid. Nanofluid.* **2017**, *21*, 99. [CrossRef]

26. Würger, A. Leidenfrost gas ratchets driven by thermal creep. *Phys. Rev. Lett.* **2011**, *107*, 164502. [CrossRef]

27. Chen, J.; Baldas, L.; Colin, S. Numerical study of thermal creep flow between two ratchet surfaces. *Vacuum* **2014**, *109*, 294–301. [CrossRef]

28. Chen, J.; Stefanov, S.K.; Baldas, L.; Colin, S. Analysis of flow induced by temperature fields in ratchet-like microchannels by Direct Simulation Monte Carlo. *Int. J. Heat Mass Transf.* **2016**, *99*, 672–680. [CrossRef]

29. Shahabi, V.; Baier, T.; Roohi, E.; Hardt, S. Thermally induced gas flows in ratchet channels with diffuse and specular boundaries. *Sci Rep* **2017**, *7*, 41412. [CrossRef] [PubMed]

30. Kuddusi, L.; Çetegen, E. Thermal and hydrodynamic analysis of gaseous flow in trapezoidal silicon microchannels. *Int. J. Therm. Sci.* **2009**, *48*, 353–362. [CrossRef]

31. Gatignol, R.; Croizet, C. Asymptotic modeling of thermal binary monatomic gas flows in plane microchannels—Comparison with DSMC simulations. *Phys. Fluids* **2017**, *29*, 042001. [CrossRef]

32. Taheri, P.; Torrilhon, M.; Struchtrup, H. Couette and Poiseuille microflows: Analytical solutions for regularized 13-moment equations. *Phys. Fluids* **2009**, *21*, 7593. [CrossRef]

33. Bhatnagar, P.L.; Gross, E.P.; Krook, M. A model for collision processes in gases. I. Small amplitude processes in charged and neutral one-component systems. *Phys. Rev.* **1954**, *94*, 511–525. [CrossRef]

34. McCormack, F.J. Construction of linearized kinetic models for gaseous mixtures and molecular gases. *Phys. Fluids* **1973**, *16*, 2095–2105. [CrossRef]

35. Sharipov, F.; Kalempa, D. Gaseous mixture flow through a long tube at arbitrary Knudsen numbers. *J. Vac. Sci. Technol. A-Vac. Surf. Films* **2002**, *20*, 814–822. [CrossRef]

36. Naris, S.; Valougeorgis, D.; Kalempa, D.; Sharipov, F. Flow of gaseous mixtures through rectangular microchannels driven by pressure, temperature, and concentration gradients. *Phys. Fluids* **2005**, *17*, 100607. [CrossRef]

37. Szalmás, L. Flows of rarefied gaseous mixtures in networks of long channels. *Microfluid. Nanofluid.* **2013**, *15*, 817–827. [CrossRef]

38. Bird, G.A. *Molecular Gas Dynamics*; Clarendon Press: Gloucestershire, UK, 1976.

39. Bird, G.A. Monte Carlo simulation of gas flows. *Annu. Rev. Fluid Mech.* **1978**, *10*, 11–31. [CrossRef]

40. Bird, G.A. *Molecular Gas Dynamics and the Direct Simulation Monte Carlo of Gas Flows*; Clarendon Press: Gloucestershire, UK, 1994; Volume 508, p. 128.

41. Sharipov, F. Gaseous mixtures in vacuum systems and microfluidics. *J. Vac. Sci. Technol. A-Vac. Surf. Films* **2013**, *31*, 050806. [CrossRef]

42. Wang, R.; Xu, X.; Xu, K.; Qian, T. Onsager's cross coupling effects in gas flows confined to micro-channels. *Phys. Rev. Fluids* **2016**, *1*, 044102. [CrossRef]

43. Zhu, L.; Chen, S.; Guo, Z. dugksFoam: An open source OpenFOAM solver for the Boltzmann model equation. *Comput. Phys. Commun.* **2017**, *213*, 155–164. [CrossRef]

44. Zhu, L.; Yang, X.; Guo, Z. Thermally induced rarefied gas flow in a three-dimensional enclosure with square cross-section. *Phys. Rev. Fluids* **2017**, *2*, 123402. [CrossRef]

45. Zhu, L.; Guo, Z. Numerical study of nonequilibrium gas flow in a microchannel with a ratchet surface. *Phys. Rev. E* **2017**, *95*, 023113. [CrossRef] [PubMed]

46. Lotfian, A.; Roohi, E. Radiometric flow in periodically patterned channels: fluid physics and improved configurations. *J. Fluid Mech.* **2019**, *860*, 544–576. [CrossRef]

47. Prasanth, P.S.; Kakkassery, J.K. Direct simulation Monte Carlo (DSMC): A numerical method for transition-regime flows-A review. *J. Indian Inst. Sci.* **2013**, *86*, 169.

48. Liou, W.W.; Fang, Y. Heat transfer in microchannel devices using dsmc. *J. Microelectromech. Syst.* **2001**, *10*, 274–279. [CrossRef]

49. Ye, J.; Yang, J.; Zheng, J.; Xu, P.; Lam, C.; Wong, I.; Ma, Y. Effects of wall temperature on the heat and mass transfer in microchannels using the DSMC method. In Proceedings of the International Conference on Nano/Micro Engineered and Molecular Systems, Shenzhen, China, 5–8 January 2009; pp. 666–671.

50. Tantos, C.; Valougeorgis, D.; Pannuzzo, M.; Frezzotti, A.; Morini, G.L. Conductive heat transfer in a rarefied polyatomic gas confined between coaxial cylinders. *Int. J. Heat Mass Transf.* **2014**, *79*, 378–389. [CrossRef]

51. Szalmas, L.; Valougeorgis, D.; Colin, S. DSMC simulation of pressure driven binary rarefied gas flows through short microtubes. In Proceedings of the ASME 2011 9th International Conference on Nanochannels, Microchannels, and Minichannels, Edmonton, AB, Canada, 19–22 June 2011; pp. 279–288.

52. Vargas, M.; Stefanov, S.; Roussinov, V. Transient heat transfer flow through a binary gaseous mixture confined between coaxial cylinders. *Int. J. Heat Mass Transf.* **2013**, *59*, 302–315. [CrossRef]

53. Zade, A.Q.; Ahmadzadegan, A.; Renksizbulut, M. A detailed comparison between navier-stokes and dsmc simulations of multicomponent gaseous flow in microchannels. *Int. J. Heat Mass Transf.* **2012**, *55*, 46734681.

54. Sugimoto, H.; Shinotou, A. Gas separator with the thermal transpiration in a rarefied gas. In Proceedings of the AIP Conference Proceedings, Pacific Grove, CA, USA, 10–15 July 2010; pp. 784–789.

55. Han, Y.L. Thermal-creep-driven flows in knudsen compressors and related nano/microscale gas transport channels. *J. Microelectromech. Syst.* **2008**, *17*, 984–997.

56. Scanlon, T.J.; Roohi, E.; White, C.; Darbandi, M.; Reese, J.M. An open source, parallel DSMC code for rarefied gas flows in arbitrary geometries. *Comput. Fluids* **2010**, *39*, 2078–2089. [CrossRef]

57. Prasanth, P.S.; Kakkassery, J.K. Molecular models for simulation of rarefied gas flows using direct simulation monte carlo method. *Fluid Dyn. Res.* **2008**, *40*, 233–252. [CrossRef]

58. White, C.; Borg, M.K.; Scanlon, T.J.; Longshaw, S.M.; John, B.; Emerson, D.R.; Reese, J.M. dsmcFoam+: An OpenFOAM based direct simulation Monte Carlo solver. *Comput. Phys. Commun.* **2018**, *224*, 22–43. [CrossRef]

59. Gallis, M.A.; Torczynski, J.R.; Rader, D.J.; Bird, G.A. Convergence behavior of a new DSMC algorithm. *J. Comput. Phys.* **2009**, *228*, 4532–4548. [CrossRef]

60. Kosuge, S.; Takata, S. Database for flows of binary gas mixtures through a plane microchannel. *Eur. J. Mech. B-Fluids* **2008**, *27*, 444–465. [CrossRef]

micromachines

MDPI

Article

Gas Mixing and Final Mixture Composition Control in Simple Geometry Micro-mixers via DSMC Analysis

Stavros Meskos [1,*], **Stefan Stefanov** [1] **and Dimitris Valougeorgis** [2]

[1] Institute of Mechanics, Bulgarian Academy of Sciences, Acad. G. Bontchev St. bl. 4, 1113 Sofia, Bulgaria; stefanov@imbm.bas.bg

[2] Department of Mechanical Engineering, University of Thessaly—Pedion Areos, 38334 Volos, Greece; diva@mie.uth.gr

* Correspondence: stameskos@imbm.bas.bg

Received: 30 January 2019; Accepted: 1 March 2019; Published: 7 March 2019

Abstract: The mixing process of two pressure driven steady-state rarefied gas streams flowing between two parallel plates was investigated via DSMC (Direct Simulation Monte Carlo) for different combinations of gases. The distance from the inlet, where the associated relative density difference of each species is minimized and the associated mixture homogeneity is optimized, is the so-called mixing length. In general, gas mixing progressed very rapidly. The type of gas surface interaction was clearly the most important parameter affecting gas mixing. As the reflection became more specular, the mixing length significantly increased. The mixing lengths of the HS (hard sphere) and VHS (variable hard sphere) collision models were higher than those of the VSS (variable soft sphere) model, while the corresponding relative density differences were negligible. In addition, the molecular mass ratio of the two components had a minor effect on the mixing length and a more important effect on the relative density difference. The mixture became less homogenous as the molecular mass ratio reduced. Finally, varying the channel length and/or the wall temperature had a minor effect. Furthermore, it was proposed to control the output mixture composition by adding in the mixing zone, the so-called splitter, separating the downstream flow into two outlet mainstreams. Based on intensive simulation data with the splitter, simple approximate expressions were derived, capable of providing, once the desired outlet mixture composition was specified, the correct position of the splitter, without performing time consuming simulations. The mixing analysis performed and the proposed approach for controlling gas mixing may support corresponding experimental work, as well as the design of gas micro-mixers.

Keywords: binary gas mixing; micro-mixer; DSMC; splitter; mixing length; control mixture composition

---

1. Introduction

Micro-scale gas mixing is of major theoretical and industrial interest in the development and optimal design of gaseous MEMS/NEMS (micro/nano-electro-mechanical systems) devices [1]. Micro-mixers can be considered as the miniaturized counterparts of macro-mixers. When chemical reactions take place, they are referred to as micro-reactors. Most commonly, they are encountered as part of a larger assembly of micro-channels and micro-devices in various applications such as micro-pumps, micro-turbines, micro-engines and micro-sensors. The detailed investigation of the effects of the geometry, boundary conditions and operating conditions on mixing is essential in the design of these devices.

The computational analysis of gaseous flows in MEMS devices operating under different conditions in non-equilibrium flow regimes cannot be based on classical continuum models of fluid motion because the continuum assumption that the flow is locally in the near-equilibrium state is no longer valid. Thus, numerical methods that solve the Boltzmann equation such as the Discrete Velocity and the Lattice Boltzmann methods, as well as particle-based methods such as the Molecular Dynamics and the Direct Simulation Monte Carlo (DSMC) methods are used instead. The DSMC method [2,3] is arguably the most common and is employed in this work.

Most of the available work has focused on the investigation of the flow characteristics of already fully mixed rarefied gases inside micro-channels of various cross-sections. The flow of monatomic binary gas mixtures in the whole range of the Knudsen number, through long capillaries of circular and rectangular cross sections has been studied in [4,5] and [6,7] respectively. A comparative study between computational and experimental results for flows in long microchannels has been presented in [8]. All the aforementioned works solve the Boltzmann kinetic equation, for fully developed flows, replacing the collision term with the McCormack model for gas mixtures [9]. Time-dependent flow and heat configurations of binary gas mixtures have also been investigated based on the DSMC method [10–13]. The flow rates of transient binary gas mixture flows through circular capillaries of finite length, including some comparison with corresponding experimental results are presented in [10,11]. The associated separation effects for flows though short and long capillaries were analyzed in [12] and [5] respectively. The transient heat flow response of a binary gaseous mixture confined between coaxial cylinders for various values of temperature difference was studied in [13].

Contrary to binary gas flows of premixed gases, the process of gas mixing of two gases entering the mixing chamber from different inlets has attracted less attention. One of the first studies performed was in [14], where the dependence of the mixing length on the pressure ratio between the inlets and the outlets, as well as on the inlet velocities was investigated. The tested configuration consisted of two parallel gas streams of H_2 and O_2 entering the mixing chamber, with the outlet pressure constantly kept at 50 kPa. It was found that the mixing length was increased by increasing the pressure ratio between inlets and outlets, and between the two inlets. The same flow setup has also been examined in [15] testing the mixing of CO and N_2 while the outlet pressure was kept at zero (no back-flow). In this work, the terms, relative density difference and mixing coefficient were introduced in order to better describe the mixing process. It has been found that the mixing length is inversely proportional to the gas temperature and the Knudsen number, while it is proportional to the Mach number. It has been also shown that the wall characteristics have little effect on the mixing length. The mixing of CO and N_2 in a T-shaped micro-mixer by keeping the outlet pressure low enough to prevent back-flow is presented in [16]. It was found that at a higher Knudsen number the mixing length is reduced, while increasing the inlet pressure resulted in an increased mixing length. Also, increasing the flow and the wall temperatures resulted in reduced mixing lengths, with the wall temperature effect being the more significant one. In [17] an effort was made to improve the mixing efficiency by attaching two bumps at the upper and lower walls, just after the end of the inlet plate separating the two gas streams. In [18] the mixing process of N_2 and CO in a microchannel was investigated and the results showed that the mixing length increased when the inlet-outlet pressure difference was increased, while it reduced when the pressure ratio of the two species at the inlet was increased. More recently, CO and N_2 mixing at an angle by applying Y-shaped inlets was investigated [19]. In the same work, the effect of replacing a "larger" micro-mixer with many "smaller" ones was also studied. Finally, the influence of gas rarefaction on the diffusive mass transfer in slip and transitional regimes has been examined [20]. All the works related to gas mixing employ the DSMC method.

The present work focused on gas mixing. More specifically, the mixing of two parallel gas streams entering a micro-channel was simulated using the DSMC method. The influence of the wall accommodation coefficients, the temperature, the molecular mass and diameter ratios of the two species and the implemented intermolecular collision model on the mixing process was investigated. Various gas mixtures were simulated and in all cases the mixing length and the flow field in the mixing micro-channel were computed. Then, another plate parallel to the micro-channel walls was positioned downstream at a lesser distance than the mixing length in order to split the partially mixed stream into two outlet streams with specified compositions. In this way it was possible to control the output mixture composition by properly positioning the splitter while keeping the inlet conditions fixed. The relationship between the splitter position and the outlet mixture composition was investigated. The micro-mixer geometry with the associated flow setup is given in Section 2, followed by a brief description of the DSMC scheme and the numerical parameters involved in Section 3. Then, the detailed mixing length analysis is presented in Section 4, while the proposed methodology for controlling the output mixture composition is given in Section 5. The manuscript concludes with some final remarks in Section 6.

2. Micro-Mixer Geometry and Flow Setup

Consider the pressure driven two-dimensional gas mixing flow of two rarefied gases into a rather simple micro-mixer, as shown in Figure 1. It consisted of two parallel plates of finite length L, located at $y = 0$ and $y = 2H$, and of a third parallel plate of length $d < L/3$, located at $y = H$. The origin of all three plates was at $x = 0$. The light gas, denoted by the subscript "1" entered from above ($H \leq y \leq 2H$), while the heavy gas, denoted by the subscript "2" entered from below ($0 \leq y \leq H$). Furthermore, the inlet pressures P_{in}^1 and P_{in}^2 at $x = 0$ were fixed, while the exit pressure at $x = L$ was set equal to zero (expansion into vacuum). The inlet and wall temperatures are T_0 and T_w respectively, with the latter one being the same for all plates. Maxwell diffuse-specular boundary conditions were considered at the walls. To cover a wide range of mixing with regard to the molecular mass of the involved gases the following combinations were considered: CO-N_2, Ne-Ar, He-Ne, He-Ar, and He-Xe. The molecular mass ratio of the light over the heavy species ranged from 1 for CO- N_2 down to 0.03 for He-Xe.

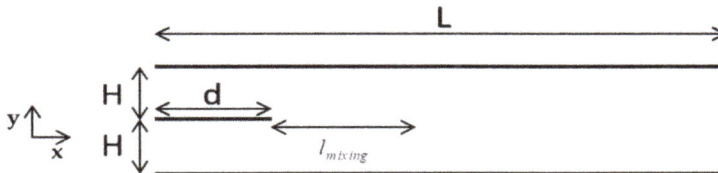

Figure 1. Schematic of gas micro-mixer.

Due to the imposed pressure difference there was a steady-state flow of the two species along the channel, while in parallel, due to molecular diffusion, the two species were mixed. The areas $x \leq d$ and $x > d$ were considered as the inlet and mixing zones respectively and the corresponding chambers were defined as the inlet and mixing chambers. The two species start mixing, upon entering the mixing chamber. The mixing process progressed gradually, mainly in the flow direction x up to some distance, denoted by L_{mix}, where the two species were considered as mixed, provided that some mixing criteria related to the homogeneity of the mixture was fulfilled. The length $l_{mix} = L_{mix} - d$ defined as the mixing length, characterized each mixing process and is of major importance in the present work.

Next, in order to be able to control the composition of the binary mixture produced, in terms of its two species, another plate, the so-called splitter, was added parallel to all other plates, with its origin at some point $(x, y) = (d + w, h)$, where $w < l_{mix}$ and $0 < h < 2H$, as shown in Figure 2. All other flow conditions remained the same. Positioning the splitter within the specified zone of the flow domain, where the mixing process was still in progress but not yet completed, resulted in having two outlet mixture streams of specific compositions, different from the fully mixed one without the splitter. Therefore, by locating the splitter at different points inside the mixing zone, and by fixing w and h accordingly, the mixing process could be controlled and the desired mixture composition at the outlet could be recovered. The areas $x \leq d$, $d < x \leq d + w$ and $x > d + w$ were considered as the inlet, premixing and mixing zones respectively.

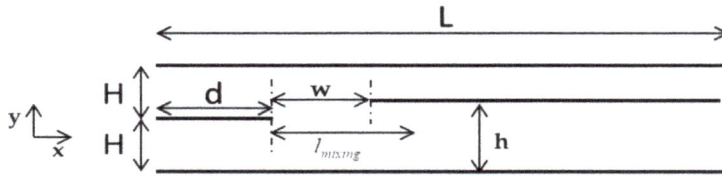

Figure 2. Schematic of gas micro-mixer with splitter.

As noted above, a mixture is considered as fully mixed, provided that some mixing condition, which is directly connected to the number densities of the mixture components, is fulfilled. The orthogonal flow domain $0 \leq x \leq L$, $0 \leq y \leq 2H$ is discretized into $i = 1,2,\dots,I$ rows and $j = 1,2,\dots,J$ columns, while the number density of each species at the (i,j) cell is denoted as $n^a_{i,j}$, with $a = 1,2$. Then, the relative number density difference is defined as:

$$\xi^a_j = \frac{1}{I}\sum_{i=1}^{I}\frac{\left|n^a_{i,j} - \overline{n}^a_j\right|}{\overline{n}^a_j} \tag{1}$$

where $\overline{n}^a_j$ is the average number density of species $a = 1,2$ of all cells in column j, given by:

$$\overline{n}^a_j = \frac{1}{I}\sum_{i=1}^{I}n^a_{i,j} \tag{2}$$

In this work the relative number density difference ξ^a_j, will be referred to from now on as "RDD^a". As the species number density at each cell of a channel column tends to coincide with the species average number density of the column, the RDD^a tends to zero. It is evident that the relative density difference is a measurement of the homogeneity of the mixture. Ideally, perfect mixing implies that the relative density difference of both components is equal to zero. Since this is not practically achievable, the mixtures were considered fully mixed provided that $RDD^a < \varepsilon$, where, depending upon the flow setup, ε took values from 0.005 up to 0.025 or on a percentage basis from 0.5% up to 0.25%. The mixing length l^a_{mix} of each mixture component $a = 1,2$, is the distance, where the mixing criteria is fulfilled, while if the fully mixed criterion, for several reasons, cannot be satisfied, the mixing length l^a_{mix} is the distance, where the corresponding minimum values of the RDD^a are observed. In general, $l^1_{\text{mix}} \neq l^2_{\text{mix}}$ and therefore, the mixing length of the mixture is specified as $l_{\text{mix}} = \max(l^a_{\text{mix}})$. Similarly, the mixture number density difference is defined as $RDD = \max(RDD^a)$. It is noted that the definition of the RDD according to Equations (1) and (2) is an extension of the one presented by Wang and Li [15], with the present definition providing, as discussed in Appendix A, more reliable results free of computational noise.

An associated quantity of interest in the present work is the cell molar fraction of the mixture defined as:

$$C^a_{i,j} = n^a_{i,j}/n_{i,j} \tag{3}$$

where $n_{i,j} = n^1_{i,j} + n^2_{i,j}$ is the cell total number density and obviously, $C^1_{i,j} + C^2_{i,j} = 1$. The average column molar fraction $\overline{C}^a_{i,j}$ is defined similar to the corresponding number density in Equation (2).

Following the description of the numerical scheme in Section 3, the flow configurations of Figures 1 and 2 were simulated and results are presented in Sections 4 and 5, respectively. The results in Section 4 relate to the computation of the mixing length in terms of all flow parameters, while the results in Section 5 relate to the control of the mixing process and to the production of gas mixtures of specific compositions.

3. Computational Scheme

The DSMC method, which is used in the present study, is a well-known particle-based stochastic method that simulates gas flows very efficiently across a wide range of the Knudsen number. The main principle of the method is decoupling the collisions and the motion of particles. This is achieved by considering a time step smaller than the amount of time a particle travels a mean free path with the most probable velocity. Each simulated particle represents a huge number of real molecules, usually greater than 10^{18}. The computational domain is divided into cells. Particles are moving through the cells, collisions are occurring only between particles within the same cell and finally, molecular speeds and number densities are sampled at each cell. This process is repeated at each time step and the simulation continues in time until steady-state conditions are reached. There are two possibilities for averaging. The first is done by time averaging a large number of time steps after reaching the steady-state. This averaging method is sufficient when only the steady-state solution is of interest. However, in order to derive accurate results for the transient period, an ensemble averaging method is required as well, where many independent simulations are performed and results are additionally averaged at the same time steps by ensemble averaging over all simulations. An in-house DSMC code was developed and employed in this work, which combines both time and ensemble averaging methods. For a detailed description of the method, the reader is referred to [21–23].

In Table 1, the computational properties of the implemented DSMC method, applied in all simulations unless otherwise stated, are presented. The grid consists of square cells with $\Delta x = \Delta y = 1/30$. For the sampling procedure, a coarser grid was used which was exactly half of the original one. For the collision procedure, the fine grid was used. The time step was always less than one-third of the mean free time and, in particular, was set approximately equal to the time needed for a particle to travel the shortest length of a cell with the most probable velocity. For example, for the mixing flow case of CO-N_2, it was set at 0.2 ns. The weight factor is defined as the ratio of real number density to the simulated number of particles and was set in a way that would result in fluxes approximately equal to 200 particles per cell per time step. Note that this number represents only the calculated inlet flux and not the net flux.

Table 1. Direct Simulation Monte Carlo (DSMC) properties.

Property	Value
Collision scheme	No Time Counter (NTC)
Molecular model	VSS (monoatomic)
Time averaging	20 kinetic steps
Total samples	500 (1 sample = 20 kin. steps)
Ensemble averaging	50 (simulations)

The pressure and temperature at the inlets ($x = 0$) were kept constant. The number density is derived from the equation of state and then, the fluxes are calculated as [3]:

$$\dot{N} = \left[n \exp(-s_n^2) \pm \sqrt{\pi} s_n \{1 \pm erf(s_n)\} \right] / (2\sqrt{\pi}\beta) \qquad (4)$$

where n is the weighted number density, $s_n = u_0 \beta$ is the dimensionless inlet velocity, with u_0 being the inlet dimensional velocity and $\beta = \sqrt{m/(2k_B T_0)}$ is the inverse of the most probable speed (k_B is the Boltzmann constant). In this study, the inlet bulk velocity is zero and Equation (3) is reduced to:

$$\dot{N} = n / (2\sqrt{\pi}\beta) \qquad (5)$$

Initially, vacuum conditions are imposed inside the computational domain and the initial molecular velocities are sampled from the Maxwellian distribution.

4. Mixing Length Analysis

The mixing length in terms of various parameters affecting the mixing process was computed for the flow configuration shown in Figure 1. More specifically, the parameters involved included the effect of the channel length, the accommodation coefficient, the wall temperature, the intermolecular collision model and the molecular mass ratio of the mixture components. In the base case scenario, the half distance between the plates is $H = 1$ μm, the channel length $L = 8$ μm and the length of the inlet middle plate $d = 2$ μm. The working gases are CO and N_2 having a molecular mass ratio equal to one. Also, the inlet pressure and temperature are $P_{in}^1 = P_{in}^2 = P_{in} = 0.2$ atm and $T_0 = 300$ K, while the temperature of all plates is taken as equal to the inlet temperature ($T_w = T_0$). Finally, purely diffuse reflection is considered at the walls (the accommodation coefficient is equal to one) and the intermolecular collision model is the variable soft sphere (VSS) model. In the base case scenario, the reference Knudsen number is approximately 0.3 and the flow is in the transition regime.

To obtain an initial view of the gas mixing in the micro-mixer, the flow of He and Xe, keeping all other parameters as in the base case scenario, was considered. The corresponding contours of the molar fraction of He and the variation of the relative density difference of the two species (RDD^a), are plotted in Figure 3a,b respectively. At $x = 0$, the molar fraction of He is very close to one in the upper inlet and close to zero in the lower one. Then, the two species start to mix, even in the inlet zone, where the molar fraction of He is gradually decreased in the upper part, and increased in the lower part reaching, at the end of the inlet zone ($x = 2$ μm), values of approximately 0.7 and 0.3, respectively. On entering the mixing zone, gas mixing is rapid and in a distance of about $x = 4$ μm the two species were fully mixed. The corresponding variation of the relative density difference of He and Xe in the mixing zone is also shown. Both RDD^a rapidly decreased reaching a minimum value at approximately $x = 4.2$ μm, and then they slightly increased up to the channel end. The molar fraction of He in the resulting fully mixed He-Xe mixture was approximately 0.57.

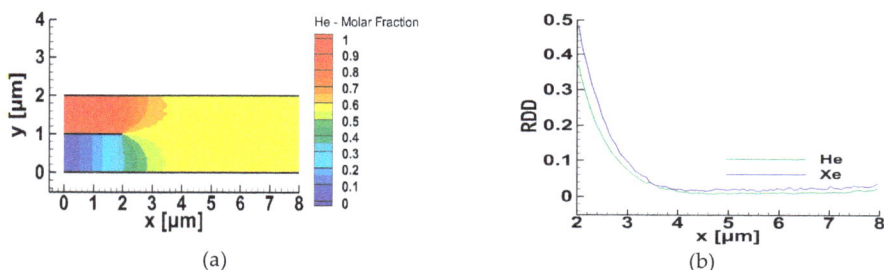

Figure 3. Gas mixing of He and Xe in the micro-mixer: (**a**) molar fraction contours of He, and (**b**) evolution of the relative density difference of He and Xe in the mixing zone.

A more detailed view of the behavior of the species' relative density difference in the downstream part of the mixing chamber is shown in Figure 4, where the RDD^a are plotted for the flow of He and Xe through micro-mixers with $L = 8$ μm and $L = 12$ μm, keeping all other parameters according to the base scenario. In both cases, the relative density differences rapidly reduced as the gases entered the mixing chamber ($x > 2$), reaching some minimum values at $4 < x < 5$. The distances where the minimum RDD^a are observed correspond to the mixing lengths l_{mix}^a. Then, as x is further increased, the relative density differences slightly increase up to the end part of the channels, where they rapidly increase, due to gas separation plus end-effects, which strongly affect the homogeneity of the mixture. The RDD^a curves of He and Xe, for both channel lengths, never cross each other. Overall, it is noted that for both channel lengths, $L = 8$ μm and $L = 12$ μm, the evolution of the relative density difference was similar and more importantly, the minimum values of RDD^a and the corresponding mixing lengths l_{mix}^a were identical. This observation clarifies that the mixing process is independent of the channel length, provided that the channel is longer than some critical value, allowing the evolution of the mixing process. It is also clarified that taking $L = 8$ μm in the base case scenario is adequate in order to investigate the effect of all other parameters.

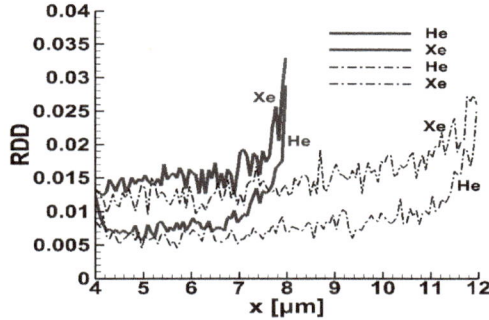

Figure 4. Detailed view of the evolution of the relative density difference of He and Xe in the downstream part of the mixing zone of micro-mixers with a total length of 8 μm (solid grey) and 12 μm (dash-dot black).

Next, the effect of the gas–surface interaction on the mixing process was examined for the base case flow scenario. In Figure 5 the mixing length l_{mix} in terms of the accommodation coefficient $\alpha \in [0, 1]$ is plotted. The limiting cases of $\alpha = 0$ and $\alpha = 1$ correspond to purely specular and diffuse reflection, respectively. In Figure 5a, the same accommodation coefficient is applied to all micro-mixer walls, while in Figure 5b the accommodation varies only in the upper wall ($y = 2H$) and is kept equal to one in all other walls. Also, it is assumed that both gases have the same α. Results are presented for two values of the mixture relative density difference, RDD, namely 0.5% and 1%. It is clearly seen that α has a significant effect and more specifically, the mixing length increases as the reflection becomes more specular. For $RDD < 1\%$ the mixing length approximately doubled when the accommodation coefficient was reduced from $\alpha = 1$ to $\alpha = 0$. As expected, the mixing length also increased as the homogeneity criterion was reduced from 1% to 0.5%. This reduction increased the mixing length about 1 μm in Figure 5a and even more in Figure 5b. Also, in Figure 5b, the required threshold values of $RDD < 1\%$ and $RDD < 0.5\%$ were not recovered at all for $\alpha < 0.3$ and $\alpha < 0.7$, respectively and therefore are not shown in the figure. It may be clearly stated that in the transition regime the gas–surface interaction plays a significant role in the mixing process, which may be more important than the interaction between particles.

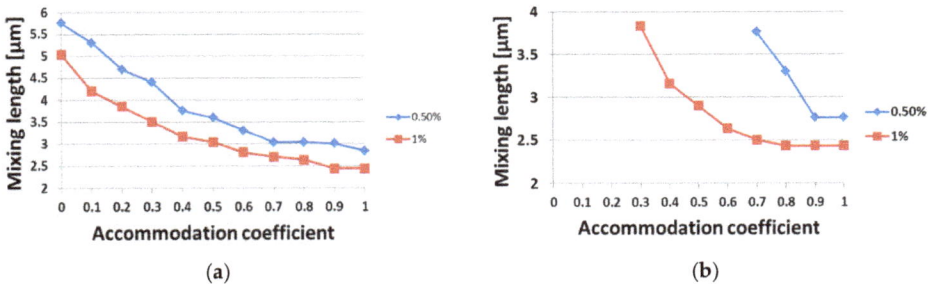

Figure 5. Mixing length variation over different accommodation coefficients α at walls: (**a**) all walls have the same $\alpha \in [0, 1]$ and (**b**) the upper wall has $\alpha \in [0, 1]$, while all other walls have $\alpha = 1$.

Next, the effect of the temperature of the walls was investigated. In Table 2, the mixing length l_{mix} of the mixture is tabulated with the temperature of the walls being uniform and equal to $T_w = 300, 450$ and 600 K. The inlet temperature is always kept at $T_0 = 300$ K. In addition to the mixing length, the corresponding C^1 and RDD^a values are provided. The largest RDD^a is also the RDD of the mixture. In the last column of Table 2, a mean value of the relative density difference, denoted by $\langle RDD \rangle$, is also given. It has been computed by averaging all RDD^a in the mixing chamber from $x = 4.5$ μm up to $x = 7$ μm, i.e., approximately from the fully mixed distance up to the region just before the channel end. This mean value provides an estimate of the overall mixture homogeneity downstream of the mixing length. It is seen that a significant increase in the walls temperature yields a rather small decrease of the mixing length and the molar fraction of CO remains almost constant at 50%. The corresponding minimum RDD of the mixture increased slightly from 0.59% at 300 K to 0.75% at 450 K and 600 K, indicating that the mixture became less homogenous. This is further strengthened by the mean value of the relative density difference $\langle RDD \rangle$, which increased from 0.76% to 1.0%. Similar conclusions were drawn based on the corresponding results for flow mixing of He-Xe presented in Table 3. Overall, it may be stated that the temperature of the walls had a minor effect on the mixing process.

Table 2. Effect of the wall temperature on the mixing length for CO-N_2 flow.

T_w (K)	l_{mix} (μm)	C^1	RDD^1	RDD^2	$\langle RDD \rangle$
300	2.8	50.1%	0.59%	0.55%	0.76%
450	2.7	49.9%	0.75%	0.59%	0.98%
600	2.5	50.1%	0.64%	0.75%	1.0%

Table 3. Effect of the wall temperature on the mixing length for He-Xe flow.

T_w (K)	l_{mix} (μm)	C^1	RDD^1	RDD^2	$\langle RDD \rangle$
300	2.2	57.7%	0.83%	1.16%	0.93%
450	2.3	57.9%	0.87%	0.95%	0.98%
600	2.2	57.6%	0.85%	1.02%	1.04%

Computations were performed with several intermolecular collision models and their effect on the gas mixing results is reported next. In Table 4, the mixing length of the mixture is tabulated employing the hard sphere (HS) and the variable hard sphere (VHS) models, proposed by Bird [24], as well as the variable soft sphere (VSS) molecular model, proposed by Koura and Matsumoto [25]. The corresponding C^1 and RDD^a values are also provided. The temperature of the walls was kept equal to the inlet temperature at 300 K. It is interesting to see that the mixing length varied significantly between the models. Compared to the VSS model, which is the model in the base case scenario, the mixing lengths with the VHS and HS models increased 12.4% and 17.4%, respectively, with the RDD remaining the same at approximately 0.60%. This is a significant increase in the mixing length considering that all wall and inlet/outlet boundaries were at isothermal conditions.

Table 4. Effect of the molecular model on the mixing length for CO-N_2 flow.

Molecular Model	l_{mix} (μm)	C^1	RDD^1	RDD^2	Relative Difference of the HS and VHS with Respect to the VSS
HS	3.3	49.8%	0.60%	0.55%	17.4%
VHS	3.15	50.1%	0.59%	0.53%	12.4%
VSS	2.8	50.1%	0.59%	0.55%	-

In order to further investigate the effect of the molecular model, a comparison between the corresponding velocity profiles was conducted. The comparison was done along the symmetry axis at the beginning ($x = 2$ μm), the middle ($x = 5$ μm) and the end ($x = 8$ μm) of the mixing chamber for each model and the computed axial velocities are tabulated in Table 5. It can be seen that the associated reported relative velocity differences, also presented in Table 5, are very small up to approximately 3%. Thus, although the effect of the intermolecular model on the bulk velocities was small, its effect on the mixing process was significant. The authors in [15,16] arrived at the same conclusion by investigating different inlet bulk velocities. These findings, along with the conclusion by Koura and Matsumoto that the VSS should be preferred in gas mixture flows as a more reliable model [25], explains the reasoning behind considering the VSS model in the basic case flow scenario.

Table 5. Effect of the molecular model on the mixture velocities at several locations along the symmetry axis for CO-N_2 flow.

x (μm)	Velocity (× 422.08 m/s)			Relative Velocity Difference		
	VSS	VHS	HS	$\frac{\|VSS-VHS\|}{VSS}$	$\frac{\|VSS-HS\|}{VSS}$	$\frac{\|VHS-HS\|}{VHS}$
2	0.1837	0.1885	0.1858	2.61%	1.14%	1.43%
5	0.3133	0.3232	0.3172	3.16%	1.24%	1.88%
8	0.7692	0.7785	0.7682	1.21%	0.13%	1.32%

Finally, the effect of the molecular mass ratio of the two components of the mixture is considered. In Table 6, the mixing lengths l_{mix}^a are provided for the following combinations: CO-N_2, Ne-Ar, He-Ne, He-Ar, and He-Xe. The corresponding molecular mass and diameter ratios of the light over the heavy species are depicted in the second and third columns respectively. In addition, the range of the RDD^a in the region downstream of the corresponding mixing length is tabulated in order to demonstrate the mixture homogeneity, with the minimum values referring to the mixing length. It is seen that the mixing length of each component depends weakly on the molecular mass, with the mixing length of the light and heavy species varying by no more than 10%. Also, no conclusive comment can be made concerning the behavior of the mixing length with regard to the molecular mass, since, as the molecular mass ratio is decreased the mixing lengths of the species may either increase or decrease. It is also seen however, that the range of variation of RDD^2 is always larger than the corresponding RDD^1, except of course for the CO-N_2 mixture, where the two molecular masses are the same. The values of RDD^1 and RDD^2 ranged from 0.5% to 1.0% and 0.5% to 2.4%, respectively, with the largest inhomogeneity referring to Xe. Actually, in the specific flow scenarios, if the threshold value is set to $\varepsilon = 0.5\%$, the mixture will never be considered fully mixed, because of the inhomogeneity of the heavy species of the mixtures, and the best possible mixing is the one demonstrated in Table 6. It is evident that the heavy species of the mixture are less uniformly distributed compared to the light species. In general, it can be stated that, independent of the mixture composition, the mixing lengths were approximately the same but with different homogeneity.

Table 6. Effect of the molecular masses of the mixture components on the mixing length.

Species	m_1/m_2	D_1/D_2	RDD^1 Range Downstream of the Mixing Length	l_{mix}^1 (μm)	RDD^2 Range Downstream of the Mixing Length	l_{mix}^2 (μm)
CO-N_2	1.000	1.002	0.5–0.7%	2.83	0.5–0.7%	2.83
Ne-Ar	0.506	0.659	0.7–1.1%	2.23	1.0%	2.16
He-Ne	0.200	0.845	0.7–1.0%	2.30	1.0–1.5%	2.50
He-Ar	0.100	0.559	0.5–1.1%	2.30	1.1–1.6%	2.56
He-Xe	0.030	0.404	0.6–1.0%	2.23	1.4–2.4%	2.16

5. Micro-Mixer with Splitter

The analysis presented above provides a detailed view of the mixing process including the downstream composition of the mixture, based on all input parameters and conditions. However, having a solid knowledge of the output molar fraction of the mixture and furthermore controlling the output mixture composition, before the simulations are performed and concluded, is not a trivial task, even for a micro-mixer of specific geometry. This is due to the large number and range of the input parameters, which is typical in binary gas mixture simulations in the transition regime, making a complete parametrization study computationally very expensive. Obviously, controlling the mixing output is of major importance in technological applications.

Arguing that the output composition of the mixture may be somehow correlated to the ratios of the input pressures or number fluxes of the species is, in general, erroneous. Such an argument is valid only if the components have equal molecular masses (e.g., CO-N_2). Otherwise, due to the different molecular masses of the two species, this approach cannot be applied in the transition regime. This is clearly demonstrated in the following numerical experiment. Consider the mixing of He-Xe in the micro-mixer of Figure 1 with $P_{in}^{He} = 0.2$ atm, $\dot{N}_{in}^{He} = 623$ particles per cell area per time step, $T_0 = T_w = 300$ K, $\alpha = 1$ and the VSS model. Two different inlet conditions are considered:

(a) the inlet pressure of the two species is the same, i.e., $P_{in}^{He} = P_{in}^{Xe}$, resulting in an inlet flux ratio equal to $\dot{N}_{in}^{He} / \dot{N}_{in}^{Xe} = 6$;

(b) the inlet number fluxes of the two species are the same, i.e., $\dot{N}_{in}^{He} = \dot{N}_{in}^{Xe}$, resulting in a pressure ratio equal to $P_{in}^{He} / P_{in}^{Xe} = 0.175$.

The corresponding mixing results are presented in Figure 6a,b respectively, where the contours of the molar fraction of He in the micro-mixer are plotted. The downstream molar fractions of He, where the mixing criterion of *RDD* <0.5% has been fulfilled, are also provided and they are $C^{He} = 57.3\%$ and $C^{He} = 19.6\%$ in the cases of equal inlet pressures and fluxes, respectively. Then, the ratios of the molar fractions C^{He} / C^{Xe} are found to be 1.35 in the case of equal inlet pressures (Figure 6a) and 0.243 in the case of equal inlet fluxes (Figure 6b). Both molar fraction ratios are completely different compared to the corresponding inlet flux and pressure ratios, which are equal to 6 and 0.175 respectively.

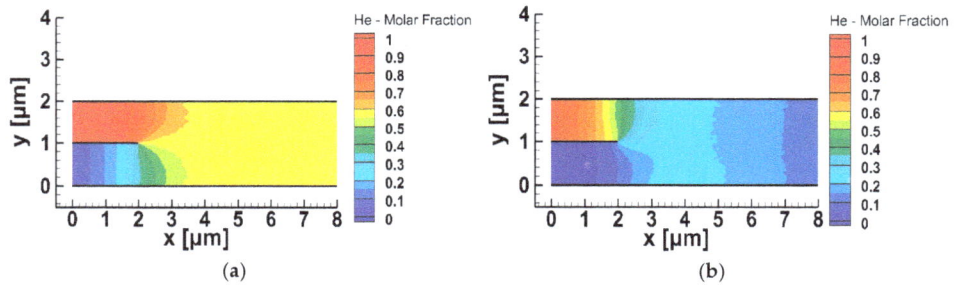

Figure 6. Molar fraction of He in the micro-mixer imposing: (**a**) equal inlet pressures of He and Xe, and (**b**) equal inlet fluxes of He and Xe.

Based on all the above, a rather simple approach was proposed to control the output mixture composition, by modifying the device slightly (hardware part) and implementing a proposed algorithm (software part). As shown in Figure 2, a plate parallel to the others was positioned in the mixing chamber with its origin at some point $(x, y) = (d + w, h)$, where $d < w < d + l_{mix}$ and $0 < h < 2H$. Since the flow is split into two mainstreams this plate was named the "splitter". The mixture composition above and below the splitter differed from each other, as well as from the final fully mixed one without the splitter. Therefore, locating the splitter in various positions in the mixing chamber would result in different mixture compositions. It is noted that the splitter was considered without thickness and therefore, the proposed design, as depicted in Figure 2 and used in the simulations, serves mainly as a proof of concept.

The next step was the cartography of the mixture molar fraction in the mixing chamber with the splitter in various positions, and then, the identification of the proper position of the splitter in order to provide a prescribed downstream mixture composition in a computationally efficient manner.

The proposed algorithm was demonstrated by considering mixing He and Xe in the flow configuration shown in Figure 2, positioning the splitter in 25 different locations. More specifically, the origin of the splitter was positioned at all the combinations of the following x, y coordinates:

$x = 3.50, 4.50, 4.75, 5.0, 5.25; y = 0.2, 0.5, 1.0, 1.5, 1.8$.

All other parameters remained the same as in the base case scenario. All 25 flow cases were simulated and the corresponding downstream molar fractions of He (C^{He}) are tabulated, on a percentage basis, in Table 7 for the mixture above the splitter (upper outlet), and in Table 8 for the mixture below the splitter (lower outlet). Analyzing the tabulated results, it is seen that, depending on the splitter location, C^{He} varies from 27–52% and from 54–85% in the lower and upper streams, respectively. This is easily justified, since He is entering the micro-mixer from above and for the same reason, as y is increased, i.e., as the splitter is located closer to the upper plate, C^{He} is also increased. Also, as x is increased, C^{He} is increased, while the corresponding upper and lower values of C^{He} are closer to each other. Both observations are explained by the fact that as the splitter is located further downstream, the homogeneity of the mixture just before the splitter is increased. More importantly, it becomes clear that based on this concept, mixtures with a wide range of compositions may be produced.

Table 7. Downstream molar fraction (%) of He at the lower outlet for 25 different positions of the splitter.

y (µm) \ x (µm)	2.5	2.75	3	3.25	3.5
0.2	27.84	32.92	37.16	40.68	43.4
0.5	24.81	31.18	36.31	40.45	43.47
1	31.11	37.35	41.6	44.49	46.55
1.5	44.87	46.75	48.13	49.12	49.79
1.8	50.05	50.89	51.18	51.37	51.47

Table 8. Downstream molar fraction (%) of He at the upper outlet for 25 different positions of the splitter.

y (µm) \ x (µm)	2.5	2.75	3	3.25	3.5
0.2	54.22	53.62	53.13	52.76	52.47
0.5	58.82	56.81	55.3	54.13	53.28
1	72.15	65.49	61.64	58.26	56.01
1.5	81.66	75.11	69.33	64.62	61
1.8	84.92	79.85	74.73	70.08	66.18

Furthermore, in order to have a detailed view of the molar fraction filed in the micro-mixer with the splitter, in Figure 7 the contours of C^{He} are plotted for two indicative (out of the 25) flow cases, with the splitter origin located at (3.25, 0.2) and (2.5, 1.8). In both cases, the inlet, premixing and the two mixing areas resulting in mixtures of the same components but different molar fractions are clearly defined. When the splitter was positioned closer to the lower plate the mole fractions of the He-Xe mixture at the upper and lower outlets were 52.7% and 40.7% respectively, while when the splitter was positioned closer to the upper plate the corresponding values were 84.9% and 50%, respectively. When the splitter was positioned closer to the lower plate the mole fractions of the He-Xe mixture at the upper and lower outlets were 52.7% and 40.7%, respectively, while when the splitter was positioned closer to the upper plate, the corresponding values were 84.9% and 50%, respectively. These results were qualitatively expected since He is entering from the upper inlet. It is clearly seen that He-Xe mixtures in a wide range of mole fractions may be deduced.

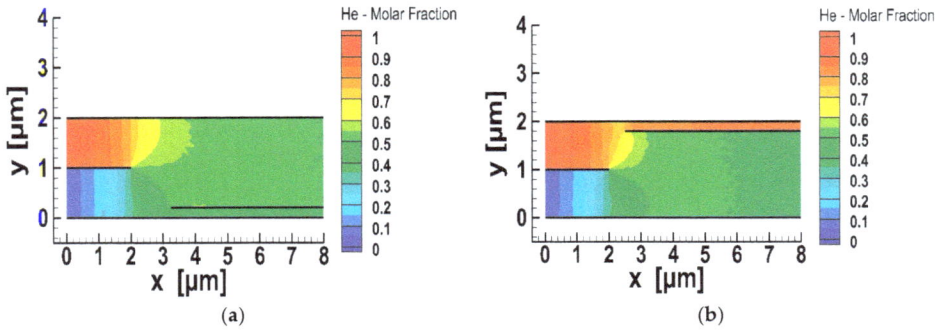

Figure 7. Gas mixing of He-Xe in the micro-mixer with splitter: Molar fraction contours of He with the splitter positioned at points (**a**) (3.25, 0.2) and (**b**) (2.5, 1.8).

Next, the objective was to develop a simple computational tool capable of providing the correct position of the splitter once the desired outlet mixture composition was specified, without performing the whole simulation each time, which requires computing resources and is time consuming. Based on the data in Tables 7 and 8, a least square approach in the x and y coordinates was implemented, to derive two third order polynomials, one for above and one for below the splitter, to approximate the molar fraction in the corresponding streams. The general form of the two polynomials is:

$$C^{\text{He}}(x,y) = a + bx + cy + dx^2 + ey^2 + fx^3 + gy^3 + hxy + ix^2y + jxy^2 \tag{6}$$

where the coefficients of the two polynomials are given in Table 9. The relative error between the molar fractions provided by Equation (6) and the corresponding ones in Tables 7 and 8 was less than $\pm 3\%$, while the Euclidean norm of the error vector was approximately 2.8. The accuracy is considered as very good, and certainly much better than that obtained by other tested approximation formulas.

Table 9. Coefficients of the third order polynomials for the upper and lower outlet.

Coefficient	Lower Outlet	Upper Outlet
a	$-1.4699150657035133 \times 10^2$	$6.5921177492441657 \times 10^1$
b	$1.2957775301457653 \times 10^2$	$-1.5812706830864261 \times 10^1$
c	$-2.0134328955701239 \times 10^1$	$1.1039559381541720 \times 10^2$
d	$-2.8591563402844741 \times 10^1$	$4.2562568218902328 \times 10^0$
e	$6.1478013059979929 \times 10^1$	$-1.4992529652493928 \times 10^1$
f	$2.0906666666617326 \times 10^0$	$-1.8133333333999069 \times 10^{-1}$
g	$-8.6358974358975775 \times 10^0$	$-2.5153846153845123 \times 10^0$
h	$-1.1499560838143527 \times 10^1$	$-4.3247339251155680 \times 10^1$
i	$3.5492776886033468 \times 10^0$	$2.4340288924554212 \times 10^0$
j	$-1.0025917815277760 \times 10^1$	$8.3286615998969680 \times 10^0$

The contours of the molar fractions provided by the third order polynomials are provided for the lower and upper outlets in Figure 8a,b respectively, helping visualize the cartography of the molar fraction maps. The contours have been plotted in the same two dimensional space with the results in Tables 7 and 8, i.e., for $2.5 \leq x \leq 3.5$ and $0.2 \leq y \leq 1.8$. The molar fraction step between the contours is small enough to allow an accurate estimation.

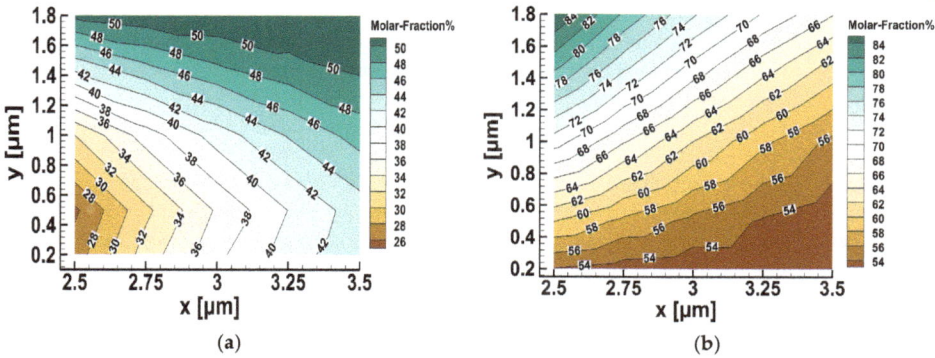

Figure 8. Molar fraction contours of He in the lower (**a**) and upper (**b**) outlets based on positioning the splitter in 25 different positions.

Once the desired mixture composition was prescribed, then based on Equation (3) for the molar fraction and the associated contours of Figure 8, it was possible to estimate the proper location of the splitter in order to obtain the specified composition, as follows:

Assume that the specified downstream molar fraction of He is $C^{He} = 40\%$. Observing the contours in Figure 8, it is easily seen that the specified composition may be obtained only in the lower outlet stream and for (x, y) coordinates deducing a molar fraction at contour line 40%. In order to ensure the validity of the approach, one of the two coordinates was fixed, e.g., $x = 3$ μm and Equation (7) for the lower outlet was solved by an iterative method to find $y = 0.93$. This result was next validated by running the whole simulation for the flow field in the micro-mixer with the origin of the splitter located at (3, 0.93). The computed C^{He} field is shown in Figure 9. At the lower outlet, the downstream molar fraction of He was computed to be 40.7%, which is only slightly different than the specified one. The introduced discrepancy is due to the small error introduced in the regression model.

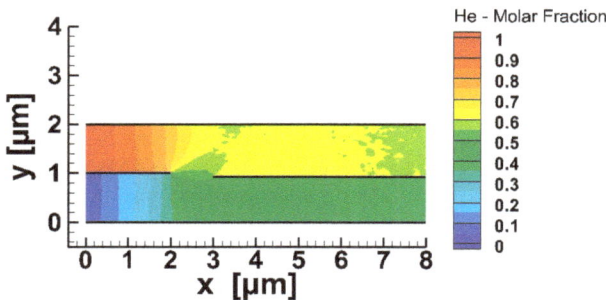

Figure 9. Molar fraction contours of He in the micro-mixer with the splitter positioned at (3, 0.93).

Obviously, the approximate expression (6) is valid only for He-Xe mixing but the proposed methodology can be also applied in other gas mixtures, depending upon the technological application.

6. Summary and Concluding Remarks

The mixing process of two pressure driven steady-state rarefied gas streams flowing between two parallel plates was investigated. The micro-mixer assembly was simple and consisted of the inlet and mixing zones. Upon entering the mixing zone, gas mixing progressed very rapidly and then it slowed down, reaching in an asymptotic manner, at some distance from the inlet zone, the optimum mixture homogeneity. This distance was defined as the mixing length. Then, further downstream, the homogeneity of the mixture slightly deteriorated up to the end part of the channel, where the homogeneity decreased more rapidly, due to gas separation and end effects. The homogeneity of the mixture was evaluated based on the relative difference between the species number density at each cell of a channel cross section and the corresponding average number density of the cross section. In general, gas mixing is upgraded as the mixing length is decreased and the relative density difference is decreased, otherwise the mixture homogeneity is increased.

The mixing length, with the associated relative density difference, was computed for different combinations of gases, in terms of the parameters affecting the mixing process, in a micro-mixer assembly having length and height equal to 8 μm and 2 μm, respectively. The mixing length, in most flow scenarios, varied between 2 μm and 4 μm. It has been shown that increasing the channel length does not affect the mixing length and the associated relative density difference, i.e., the grade of the mixture homogeneity. Varying the wall temperature also has a minor effect on the gas mixing. The type of gas surface interaction is the most important parameter affecting gas mixing and more specifically, as the reflection becomes more specular, the mixing length significantly increases. In cases with high specular reflection, the mixing length may be doubled or even tripled. The intermolecular collision models of HS, VHS and VSS were implemented and it was found that the computed mixing length of the HS and VHS models were about 20% and 10% higher than for VSS, while the corresponding relative density differences were negligible. It is noted that in gas mixture modeling the VSS model is considered the most reliable one. Concerning the mixture composition, it was deduced that the ratio of the molecular masses of the two components ranged, in the present study, from 1 down to 0.03, and had a minor effect on the mixing length and a more important one on the relative density difference. More specifically, the mixture became less homogenous as the molecular mass ratio reduced, i.e., the molecular masses of the two species are quite different.

Next, a rather simple approach has been proposed to control the output mixture composition. A plate, parallel to the others, the so-called splitter, was added in the mixing zone of the micro-mixer assembly, separating the downstream flow into two outlet mainstreams. It has been shown that locating the splitter in various positions in the mixing zone will result in different downstream mixture compositions. Intensive simulations were performed for 25 different positions of the splitter and the molar fractions of the produced binary mixtures above and below the splitter have been deduced. Based on these data, simple approximate expressions have been derived capable of providing, once the desired outlet mixture composition is specified, the correct position of the splitter, without performing time consuming simulations. The analysis has been performed and validated for He-Xe flow but can also be easily applied to other gas mixtures.

It is hoped that the mixing analysis performed and the proposed approach for controlling gas mixing in order to obtain binary gas mixtures with specific compositions, will support corresponding experimental work, as well as the design and optimization of gas mixing micro devices.

Author Contributions: Conceptualization, S.S. and S.M.; software, S.M.; validation, S.M., S.S. and D.V.; writing—original draft preparation, S.M.; writing—review and editing, S.S. and D.V.; visualization, S.M.; supervision, S.S. and D.V.; project administration, S.S.

Funding: This project received funding from the European Union's Framework Programme for Research and Innovation Horizon 2020 (2014–2020) under the Marie Sklodowska-Curie Grant Agreement No. 643095.

Conflicts of Interest: The authors declare no conflicts of interest.

Appendix A

The *RDD* expression introduced by Wang and Li [15] is given by:

$$\xi_j^a = \frac{\left| n_{\text{up},j}^a - n_{\text{low},j}^a \right|}{\max \left(n_{\text{up},j}^a, n_{\text{low},j}^a \right)} \tag{A1}$$

where, ζ_a^j is the relative density difference RDD^a, of species $a = 1, 2$ at column j, while $n_{\text{up},j}^a$ and $n_{\text{low},j}^a$ correspond to the number densities of the species in the same column at the two cells adjacent to the upper and lower walls, respectively. The RDD definition, implemented in the present work (see Equation (1)), instead of only the two cells adjacent to the upper and lower walls, considers all column cells. A comparison between the two definitions in terms of the corresponding distribution of the RDD along the channel was performed.

Consider the steady-state mixing flow of CO-N$_2$ in the micromixer of Figure 1, with $P_{\text{in}}^1 = P_{\text{in}}^2 = 0.2$ atm, $T_0 = T_w = 300$ K and purely diffuse reflection at the walls. The computed distributions of the relative density differences by the present expressions and the one in [15] along the channel are plotted in Figure A1. Simulations were performed by setting $RDD^a < \varepsilon$, with $\varepsilon = 0.5\%$. The RDD^a distributions of both species, $a = 1, 2$, obtained by the two definitions were qualitatively similar and more important the computed mixing length was the same and equal to $l_{\text{mix}} = 2.8$ µm. However, the introduced DSMC statistical noise, which is a very significant quantity in probabilistic computations, was much higher in the simulations via the expression in [15]. This is more clearly shown in the detail of Figure A1, where the evolution of RDD from a length near the mixing point until the end of the channel, as calculated in [15] and by Equation (1) respectively, are plotted.

Furthermore, the fact that the variation of RDD^1 is identical to RDD^2 is due to both having the same molecular masses of CO and N$_2$, while the sudden increase of both RDD^a at the channel end is due to gas separation and end effects, an effect that is only captured by the proposed method. Therefore, in the present work, the RDD is computed according to Equation (1). These issues are further discussed in Section 4.

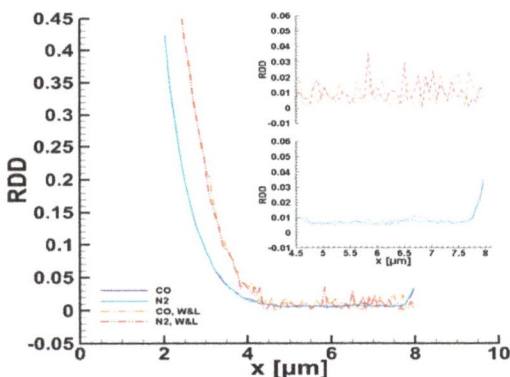

Figure A1. Distribution of relative density difference of CO and N$_2$ calculated with the present definition and Wang and Li's definition in [15]; the results are zoomed for a specific channel length at the upper right.

References

1. Lee, C.-Y.; Chang, C.-L.; Wang, Y.-N.; Fu, L.-M. Microfluidic Mixing: A Review. *Int. J. Mol. Sci.* **2011**, *12*, 3263–3287. [CrossRef] [PubMed]
2. Bird, G.A. *Molecular Gas Dynamics and the Direct Simulation of Gas Flows*; Clarendon Press: Oxford, UK, 1994; ISBN 0-19-856195-4.
3. Bird, G.A. *The DSMC Method Version 1.2*; Amazon Distribution GmbH: Leipzig, Germany, 2013; pp. 49–54, ISBN 978-1492112907.
4. Sharipov, F.; Kalempa, D. Gaseous mixture flow through a long tube at arbitrary Knudsen numbers. *J. Vac. Sci. Technol.* **2002**, *20*, 814. [CrossRef]
5. Sharipov, F.; Kalempa, D. Separation phenomena for gaseous mixture flowing through a long tube into vacuum. *Phys Fluids* **2005**, *17*, 127102. [CrossRef]

6. Valougeorgis, D.; Naris, S. Shear driven micro-flows of gaseous mixtures. *Sens. Lett.* **2006**, *4*, 46–52.
7. Naris, S.; Valougeorgis, D.; Kalempa, D.; Sharipov, F. Flow of gaseous mixtures through rectangular microchannels driven by pressure, temperature, and concentration gradients. *Phys. Fluids* **2005**, *17*, 100607. [CrossRef]
8. Szalmas, L.; Pitakarnnop, J.; Geoffroy, S.; Colin, S.; Valougeorgis, D. Comparative study between computational results for binary rarefied gas flows through long microchannels. *Microfluid Nanofluid* **2010**, *9*, 1103–1114. [CrossRef]
9. McCormack, F.J. Construction of linearized kinetic models for gaseous mixtures and molecular gases. *Phys. Fluids* **1973**, *16*, 2095–2105. [CrossRef]
10. Vargas, M.; Naris, S.; Valougeorgis, D.; Pantazis, S.; Jousten, K. Hybrid modeling of time-dependent rarefied gas expansion. *J. Vac. Sci. Technol. A* **2014**, *32*, 021602. [CrossRef]
11. Vargas, M.; Naris, S.; Valougeorgis, D.; Pantazis, S.; Jousten, K. Time-dependent rarefied gas flow of single gases and binary gas mixtures into vacuum. *Vacuum* **2014**, *109*, 385–396. [CrossRef]
12. Valougeorgis, D.; Vargas, M.; Naris, S. Analysis of gas separation, conductance and equivalent single gas approach for binary gas mixture flow expansion through tubes of various lengths into vacuum. *Vacuum* **2016**, *128*, 1–8. [CrossRef]
13. Vargas, M.; Stefanov, S.; Roussinov, V. Transient heat transfer flow through a binary gaseous mixture confined between coaxial cylinders. *Int. J. Heat Mass Trans.* **2013**, *59*, 302–315. [CrossRef]
14. Yan, F.; Farouk, B. Numerical simulation of gas flow and mixing in a microchannel using the direct simulation montecarlo method. *J. Microscale Therm. Eng.* **2002**, *6*, 235–251. [CrossRef]
15. Wang, M.; Li, Z. Gas mixing in microchannels using the direct simulation Monte Carlo method. *Int. J. Heat Mass Trans.* **2006**, *49*, 1696–1702. [CrossRef]
16. Le, M.; Hassan, L. DSMC simulation of gas mixing in T-shape micromixer. *Appl. Therm. Eng.* **2007**, *27*, 2370–2377. [CrossRef]
17. Reyhanian, M.; Croizet, C.; Gatignol, R. Numerical analysis of the mixing of two gases in a microchannel. *Mech. Ind.* **2013**, *14*, 453–460. [CrossRef]
18. Darbandi, M.; Lakzian, E. Mixing Enhancement of two Gases in a Microchannel Using DSMC. *Appl. Mech. Mater.* **2013**, *307*, 166–169. [CrossRef]
19. Darbandi, M.; Sabouri, M. Detail study on improving micro/nano gas mixer performances in slip and transitional flow regimes. *Sensor. Actuators B Chem.* **2015**, *218*, 78–88. [CrossRef]
20. Darbandi, M.; Sabouri, M. Rarefaction effects on gas mixing in micro- and nanoscales. In Proceedings of the ASME 2016 5th International Conference on Micro/Nanoscale Heat and Mass Transfer, Biopolis, Singapore, 4–6 January 2016; ASME: New York City, NY, USA, 2016.
21. Stefanov, S. On DSMC Calculations of Rarefied Gas Flows with Small Number of Particles in Cells. *J. Sci. Comput.* **2011**, *33*, 677–702. [CrossRef]
22. Roohi, E.; Stefanov, S. Collsionpartrner selection schemes in DSMC: From micro/nano flows to hypersonic flows. *Phys. Rep.* **2016**, *656*, 1–38. [CrossRef]
23. Chen, J.; Stefanov, S.; Baldas, L.; Colin, S. Analysis of flow induced by temperature fields in ratchet-like microchannels by Direct Simulation Monte Carlo. *Int. J. Heat Mass Trans.* **2016**, *99*, 672–680. [CrossRef]
24. Bird, G.A. Monte-carlo simulation in an engineering context. *Prog. Astronaut. Aeronaut.* **1981**, *74*, 239–255.
25. Koura, K.; Matsumoto, H. Variable soft sphere molecular model for inverse-power-law or Lennard-Jones potential. *Phys. Fluids A Fluid* **1991**, *3*, 2459. [CrossRef]

micromachines

MDPI

Article

Design of a Novel Axial Gas Pulses Micromixer and Simulations of its Mixing Abilities via Computational Fluid Dynamics

Florian Noël [1,2,3], Christophe A. Serra [3] and Stéphane Le Calvé [1,2,*]

1   ICPEES UMR 7515, Université de Strasbourg/CNRS, F-67000 Strasbourg, France; florian.noel@etu.unistra.fr
2   In'Air Solutions, 25 rue Becquerel, 67087 Strasbourg, France
3   Institut Charles Sadron (ICS) UPR 22, Université de Strasbourg/CNRS, F-67000 Strasbourg, France; ca.serra@unistra.fr
*   Correspondence: slecalve@unistra.fr

Received: 31 January 2019; Accepted: 19 March 2019; Published: 23 March 2019

Abstract: Following the fast development of microfluidics over the last decade, the need for methods for mixing two gases in flow at an overall flow rate ranging from 1 to 100 NmL·min^{-1} with programmable mixing ratios has been quickly increasing in many fields of application, especially in the calibration of analytical devices such as air pollution sensors. This work investigates numerically the mixing of pure gas pulses at flow rates in the range 1–100 NmL·min^{-1} in a newly designed multi-stage and modular micromixer composed of 4 buffer tanks of 300 µL each per stage. Results indicate that, for a 1 s pulse of pure gas (formaldehyde) followed by a 9 s pulse of pure carrier gas (air), that is a pulses ratio of 1/10, an effective mixing up to 94–96% can be readily obtained at the exit of the micromixer. This is achieved in less than 20 s for any flow rate ranging from 1 to 100 NmL·min^{-1} simply by adjusting the number of stages, 1 to 16 respectively. By using an already diluted gas bottle containing 100 ppm of a given compound in an inert gas same as the carrier gas, concentrations ranging from 10 to 90 ppm should be obtained by adjusting the pulses ratio between 1/10 and 9/10 respectively.

Keywords: gas mixing; pulsed flow; modular micromixer; multi-stage micromixer; modelling

---

1. Introduction

The homogenization of a gaseous chemical mixture is of particular interest in many, sometimes complex processes with multiple and varied applications [1–6]. This is particularly the case for the generation of gas mixtures of known concentrations for supplying chemical reactors [7,8] or analytical devices for their calibration [9,10]. Most of the time, the applied flow rates vary from a few litres per minute for gas calibration generators [11–13] to several m^3 per hour for industrial processes [14].

Today, analytical devices become more and more miniaturized and industrial processes are increasingly using microfluidic devices to improve energy [15] and chemical reactions [16] yields. Thus, low gas flow rates, typically less than 1 NmL·min^{-1}, start to be considered. However, some applications still require flow rates over 1 NmL·min^{-1} but lower than few hundreds of NmL·min^{-1}. Then, manipulating gas in microchannels has become a way of achieving the miniaturization of many devices in several fields of application. Fluids manipulation implies, *inter alia*, droplets generation [17–19], multi-phase flows [20,21] and fluids mixing which is a crucial aspect in several fields of application. Various ways of mixing gas flows presenting a radial heterogeneity (Figure 1a) already exist, going from passive to electronics- or sound-driven mixing techniques [22] but none have been already experimentally used to solve for axial heterogeneity (Figure 1b).

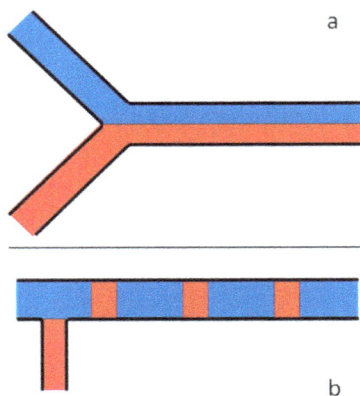

Figure 1. Schematic representation of a two-fluid flow with radial (**a**) or axial (**b**) heterogeneity before diffusion mixing operates.

At total flow rates around or lower than a few NmL·min^{-1} (e.g., for air pollution sensors), low mixing ratios become an issue for the usual mixers since they require to control a very small flow rate of the fluid to be diluted (typically below 1 NmL·min^{-1}). This implies to rely on highly accurate flow controllers which is a real issue for gas flows technologies. However, this issue is negated by the pulsed flow mixing, where only the generation time of each fluid matters in the mixing process. In this case, the flow pattern is characterized by sequences of alternated slugs of two different gases flowing in a channel which are expected to be fully mixed at the exit of the mixer. Such flow pattern is obtained by generating alternate pulses of the two individual gas flows while keeping the overall flow rate constant. In this case, the mixing ratio depends on the ratio between the generation time of the first fluid and that of the second fluid.

Numerous studies in the literature have focused on the design, manufacture and validation of microfluidic chips for the homogenization of liquid solutions as reported in a recent review paper [23]. Conversely, far fewer studies have focused on obtaining homogeneous gas mixtures using microfluidic devices [24–29], with only one paper to date proposing a way of mixing axially heterogeneous gaseous flows [10] (see Table 1).

Table 1. Technological solutions used to obtain homogeneous gas mixtures in microfluidic devices

Short Description/Technology Used	Approach	Applications	Type of Heterogeneity	Total Flow Rate ($NmL \cdot min^{-1}$)	Mixing Time (s)	Microchip Design	Reference
Fluids collision inducing oscillations for mixing liquids or gases	Experimental (liquids)	Fuel technology	Radial	-	-	Fixed	Tesař et al., 2000 [24]
Multilamination by using V-shaped microstructures	Experimental	Chemical reaction engineering	Radial	1000–10,000	6×10^{-4}	Modular	Haas-Santo et al., 2005 [25]
Microchannels network generating discrete concentrations of O_2 in N_2	Experimental	Biotechnology, cell culture	Radial	16.2	4	Fixed	Polinkovsky et al., 2009 [26]
Mixing of 9 gas flows at different O_2 concentrations to create O_2 concentration gradients	Experimental	Cell culture	Radial	108	24	Fixed	Adler et al., 2009 [27]
Diffusion between parallel flow channels through PDMS layer to create an O_2 concentration gradient	Experimental	Cell culture	Radial	80	20	Modular	Lo et al., 2010 [28]
Splitting of the flow between inlet and outlet chambers, followed by a buffer tank	Modelling	Calibration gases generation	Axial	-	120	Fixed	Martin et al., 2012 [10]
Basic T-shaped mixer	Experimental	Microcombuster, fuel technology	Radial	5–250	3.7×10^{-3}	Modular	Huang et al., 2017 [29]
Multistage mixing microchips for pulsed gas flow	Modelling	Gas mixture generation	Axial	1–100	20	Modular	This work

Here, Martin et al. reported a way of mixing such gaseous flows but in a non-flexible approach as the mixing device was designed for a given flow rate and a given mixing ratio. However, many processes require to either use several mixing ratios (chemical reactions) or varying flow rates (industrial processes, calibration of analytical devices) and often both of them. Therefore, another approach is needed for these applications. Table 1 summarizes the main technological solutions envisaged for obtaining homogeneous gas mixtures using microfluidic circuits at modular flow rates and mixing ratios. All these methods except for one have been used to mix continuous gas flows resulting in radial heterogeneity. Haas-Santo et al. [24] presented a very interesting way of mixing at flow rates up to 10,000 NmL·min^{-1}, with a mixing time lower than 600 μs by using V-shaped microstructures that are splitting the flow into numerous sub-flows before recombining them. Another approach to create various concentrations of gases consists in injecting in two different inlet microchannels discrete concentrations of gases (pure N_2 and pure O_2) that are physically mixed in a three split and recombined levels of a tree-shaped micromixer [25]. In the nine outlet microchannels, the resulting concentrations of O_2 varied from 0 to 100%. In another study, a gradient of O_2 in N_2 (0–100%) was obtained by using nine parallel inlet microchannels of three different O_2 concentrations (0%, 50% or 100%). The gradient of concentration was achieved by diffusion through the PDMS walls in between the microchannels [26,27]. Here, the flow rates are smaller, ranging from 1 to 80 NmL·min^{-1}, which corresponds to the investigated flow rates range. However, these methods cannot be used to generate a single chosen concentration. The gradient method makes it impossible to pick up one concentration inside the flow and the discrete concentrations method is limited by the number of concentrations being generated. This latter would also imply heavy wastes of gases, as one would pick only one channel for the chosen concentration and leave the others to exhaust. Finally, the basic T-shaped mixer studied by Huang et al. [28] shows homogeneity after only a few millimetres at 200 NmL·min^{-1} in a 550 μm × 125 μm (width × height) channel, showing the effectiveness of the diffusion process while mixing gases. However, this process is not as effective to balance the concentration of species in an axially heterogeneous flow, because the movement induced by the flow is opposed to diffusion between the two gases.

The objective of this work is then to develop a flexible mixing device made of microfluidic chips able to mix and homogenize pulsed gaseous flows at flow rates ranging from 1 to 100 NmL·min^{-1}. To meet this challenge, the strategy is based on the creation of 4 parallel sub-flows entering buffer tanks patterned on a microchip in order to decrease the gas linear velocity and promote gas diffusion and mixing. The gases mixture being recombined passed the buffer tanks. The innovative part concerns the flexibility of the device with respect to the targeted gas flow rate. Indeed, to adapt to the needs of the user, a novel multi-stage modular system has been imagined and validated by numerical simulations in the range 1–100 NmL·min^{-1}.

2. Materials and Methods

2.1. Design of the Microfluidic Chip

The design of all the chips presented in this work have been made using the Autodesk Inventor software.

2.1.1. Technical Constraints and Objectives

The objective of this study is to develop a design to mix pulses of a standard gas A delivered during $t_A = 1$ s with pulses of pure air B delivering during $t_B = 9$ s both at the inlet flow rate Q = 5 NmL·min^{-1}, this case being considered as the extreme case in terms of pulses ratio $(t_A/(t_A+t_B))=1/10)$ and axial heterogeneity. Afterwards, several copies of this design may be used in series for mixing at higher flow rates up to 100 NmL·min^{-1} with the same values of t_A and t_B. The pressure at the device's outlet and the temperature of the gas mixture are set to 1 atm and 23 °C respectively. Under these conditions at Q = 5 NmL·min^{-1}, the volume of a pulse of standard gas A is $V_A = 0.0832$ NmL and the volume of a pulse of pure air B is $V_B = 0.7488$ NmL. Finally, the depth of the design is set to 1 mm, so that the mixer fits into a 5 cm × 10 cm rectangular chip.

2.1.2. Strategy used to Design and Define the Mixing Microchip Pattern and Size

In the case of gases, diffusion coefficients are very high, in the order of 0.1 cm^2·s^{-1} [30], compared to those of liquid, generally in the order of 1.10^{-5} cm^2·s^{-1} [7,8]. This makes diffusion very effective at low flow rates, where the mass transfer by diffusion is significantly higher than that from advection. This is illustrated by the Peclet number P_e, which is the ratio of the rate of advection to rate of diffusion and is given by the following equation [31]:

$$P_e = \frac{L_c \times v}{D},$$ (1)

where L_c is the characteristic length (in m), v is the average fluid's velocity (in m·s^{-1}) and D is the diffusion coefficient (in m^2·s^{-1}).

Increasing the flow rate increases the value of P_e and then makes mass transfer by diffusion less efficient. Therefore, it is harder to achieve a fully homogeneous mixture at high flow rates by relying mainly on diffusion. The value of P_e can be reduced by decreasing the fluid velocity. This can be done either by increasing the cross-section area through which the fluid passes or by splitting the flow.

However, increasing the cross-section is even more interesting because it also increases the contact area between two mixture layers [32], enhancing the effectiveness of the axial diffusion process. Splitting the flow into several channels of same cross-section decreases the velocity but it does not increase this contact area between layers. It is why it has been chosen to focus mainly on increasing the cross-section and creating an optimized design using buffer tanks.

The volume of this buffer tank must be at least equal to $V_{AB} = V_A + V_B = 0.8333$ NmL in order to properly mix every pulse of standard gas A with a pulse of pure air B at $Q = 5$ NmL·min^{-1}. Nevertheless, this volume should be kept close to the volume V_{AB} to minimize the response time of the mixer. Therefore, the choice was arbitrarily focused on a buffer tank volume that was approximately $1.5 \times V_{AB}$, that is, 1.2 mL.

Given this latter volume and an imposed depth of 1 mm for the buffer tank, key variables become its width and its length. At a constant volume, increasing its length implies a decrease in its width and a reduction in the cross-section area, which would reduce the effectiveness of axial diffusion. On the other hand, increasing the width creates important variations in fluid velocity, leaving dead volumes at the edges of the buffer tank. Indeed, Figure 2a shows that there is a corridor with a stable mean velocity (in green) in the middle of the buffer tank. The velocity then decreases quickly down to almost 0 away from this corridor.

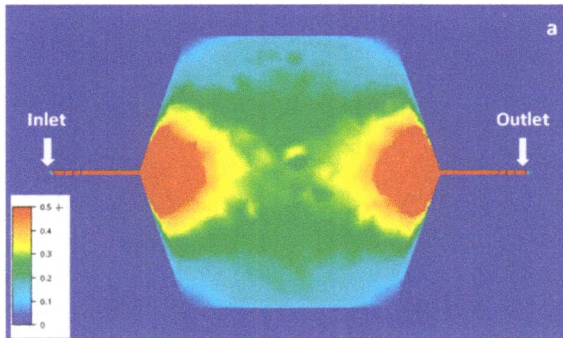

Figure 2. *Cont.*

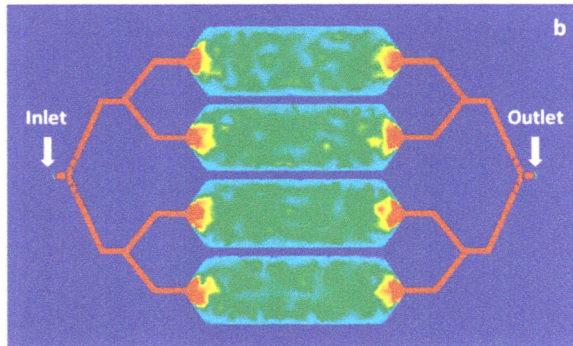

Figure 2. Velocity amplitude of a 3 cm long and 4 cm wide buffer tank (**a**) and of a set of 4 buffer tanks in parallel of 3 cm long and 1 cm wide each (**b**). The depth is fixed to 0.1 cm, the flow rate at the inlet is set to 5 NmL·min^{-1} and the pressure at the outlet is maintained at atmospheric pressure. The velocity scale is in cm·s^{-1} and is the same for insets. In red are represented areas with a velocity superior or equal to 0.5 cm·s^{-1}. Both chips are 86 cm long and 49 cm wide. These simulations were performed using the Autodesk CFD (Computational Fluid Dynamics) software.

In order to increase the cross-section area without creating large dead volumes, it has been chosen to combine both aforementioned solutions and split the flow into numerous parallel buffer tanks before collecting all the fluid back into one single outlet. Figure 2b shows an example with 4 buffer tanks. The total cross-section area and the length of each of them are equal to those of the buffer tank presented in Figure 2a but the width of each buffer tank is 4 times smaller. In this way, the dead volumes have been greatly minimized. Of course, this final geometry may still be slightly improved in absence of size constraints in order to make the fluid velocity gradient even smoother.

Even though doubling several times the number of buffer tanks would have improved the mixer's effectiveness, it has to be fitted into the 10 cm × 5 cm maximum size of the chip. In addition to these maximum dimensions, some convenience distances must be respected between the different elements for fabrication and assembly constraints, as illustrated in Figure 3:

- 5 mm between the inlet/outlet and the edges of the chip, to leave space for the fluidic connectors;
- 8 mm for every flow division: 4 mm for splitting and 4 mm for redirecting the flow into the right direction;
- 3 mm between channels and the walls of the chip;
- 1 mm between channels.

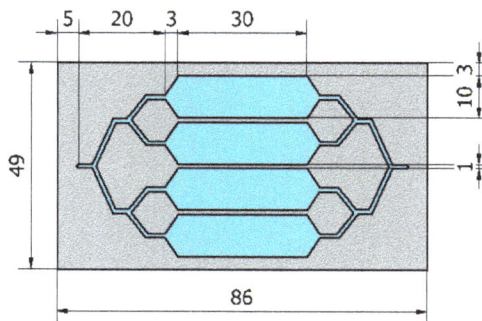

Figure 3. Dimensions of the optimized chip with 4 buffer tanks (given in mm).

Because of these parameters, it would not be possible to increase the number of buffer tanks to 8, as it would exceed both target dimensions of 10 cm × 5 cm. Therefore, the 4 buffer tanks mixer (Figure 2b) is considered as the optimal system for the targeted application.

As illustrated in Figure 3, each buffer tank is 3 cm long and 1 cm wide which, considering the depth of 0.1 cm, makes the total volume of the 4 buffer tanks equal to 1.2 mL. Finally, the total volume of the single stage chip including channels and buffer tanks is 1.686 mL.

2.1.3. Elaboration of the Multi-Stage Micromixer

In order to homogenize gases mixtures with the same $t_A = 1$ s and $t_B = 9$ s at higher flow rates, several stages presenting the same pattern of 4 buffer tanks are connected in series. Thus, a flexible multi-stage micromixer composed of many identical chips can be easily obtained. As an example, Figure 4 represents a 4 stages micromixer. The first chip (1 on Figure 4), at the top and bottom of the assembly, is equipped with inlet and outlet fluidic connectors for 1/16" outer diameter tubings. The second chip (2 on Figure 4) presents a pattern of 4 buffer tanks on both sides. The third chip (3 on Figure 4) consists in a single hole to connect the outlet of one chip n°2 and the inlet of another chip n°2 and simultaneously to bond the lower part and the upper part of these two chips. Since only holes and two-dimensional patterns on each side of the stage are required, all these chips can be manufactured easily using a micro-milling machine. With this flexible design, it is possible to connect many mixing stages in series, with only one inlet and one outlet.

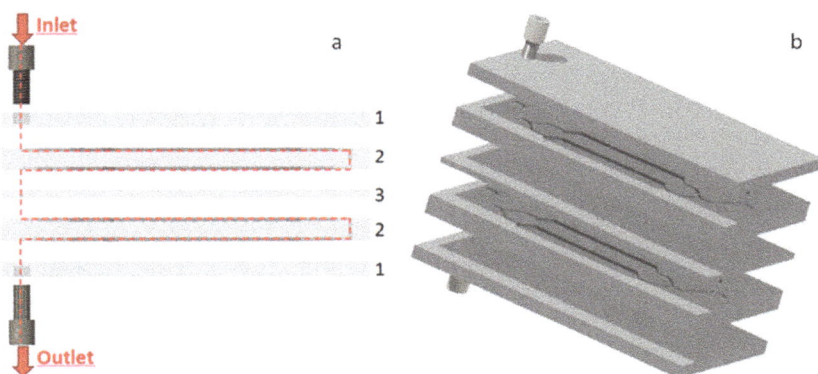

Figure 4. (a) Principle of the multi-stage design. The red line displays the path of the fluid flow. The design is made of 3 different chips: **1.** This chip is used to fix a 1/16" outer diameter connector serving as inlet or outlet ports. It is also the rooftop for microchannels; **2.** The chip presents 2 patterns containing 4 buffer zones each, one on each side of the chip. The outlet of one pattern is connected to the inlet of the other via a hole. **3.** An intermediate chip used as a rooftop for microchannels, separates the channels of the different chips number 2. (b) 3D schematics of the 4-stage mixer.

Considering that a single chip is made of 2 stages, the total volume of one double side chip (2 on Figure 4) including channels and buffer tanks is 3.372 mL, increasing the mixing capacity to mix pulses of gas A of $t_A = 1$ s with pulses of gas B of $t_B = 9$ s at a theoretical flow rate Q = 10 NmL·min^{-1}.

2.2. Methodology for Simulation of the Gas Flow and Mixing

Several simulations have been made to determine the impact of the number of stages and the flow rate variations over the homogeneity and the time needed to achieve a steady state at the outlet. These simulations have been conducted for a compressible gas phase using the Autodesk CFD (Computational Fluid Dynamics) software and all the parameters used, and their values are listed in Table S1. The gases used for these simulations were air and formaldehyde. The temperature was considered constant all over the micromixer and equal to 23 °C. Reynolds number was calculated to be 110 at the inlet of the micromixer, which has the lowest section and highest flow rate (100 NmL·min^{-1}) in the entire micromixer (1 mm^2). This indicates that the flow is laminar in the whole system. In practice, A may represent a mixture of gas pollutant already very diluted in air (typically commercially available mixtures with concentrations in the range 0.1–100 ppm). However, the diffusion coefficient between formaldehyde and air (0.176 cm^2·s^{-1}) is very close to the self-diffusion coefficient of air (0.178 cm^2·s^{-1}), so the results would be identical.

The mixing of both gases at different points along the flow pathway was monitored using a scalar variable denoted S which is representative of gas A concentration in the mixture. For pure formaldehyde, the scalar was 1, while for pure air the scalar was 0. Therefore, a fully homogeneous mixture made of half formaldehyde and half air would have then got an average scalar of 0.5. In this study, mixing 1 s of formaldehyde at scalar 1 with 9 s of pure air at scalar 0 led to an average scalar of:

$$S = \frac{t_{airA}}{t_{airA} + t_{airB}} = 0.1. \tag{2}$$

The properties of the microchips' material are those of Polyether Ether Ketone (PEEK), which is a non-reactive material used in many microfluidic applications, especially for tubings and fittings [33] and easily manufactured by micro-milling machines. An average reference mesh density of 757 nodes cm^{-3} was used for all the simulations. A comparison, for a flow rate of 25 NmL·min^{-1} and a pulses ratio of 1/10, with densities twice higher and lower has been made in order to confirm that the results achieved a satisfactory precision (Figure S1). Doubling the mesh density led to variations of the scalar value at the exit of the 4th mixing stage in the range of 2% compared to the reference mesh density. This was assumed to be a reliable test to validate the mesh density used for the simulations.

The diffusion coefficient between formaldehyde and air was calculated using Chapman-Enskog equation with a precision of about 8% according to Cussler [34]:

$$D = \frac{1.86 \times 10^{-3} \times T^{\frac{3}{2}} \times \left(\frac{1}{M_A} + \frac{1}{M_B}\right)^{\frac{1}{2}}}{p \times \sigma_{AB}^2 \times \Omega}, \tag{3}$$

where D is the diffusion coefficient (in cm^2·s^{-1}), T is the temperature (in K), M_A and M_B are the molecular weights of gas A and gas B respectively (in g·mol^{-1}) and p is the pressure (in atm) σ_{AB} is the average kinetic diameter between gases A and B, given by:

$$\sigma_{AB} = \frac{\sigma_A + \sigma_B}{2}, \tag{4}$$

where σ_A and σ_B are the kinetic diameters of gases A and B respectively. The collision integral Ω is a dimensionless quantity in the order of 1, which represents the interaction between both gases at a given temperature. Its value can be more precisely determined by calculating the Lennard-Jones potential between the two species [34]. Considering Ω = 1 and given the other parameters for gases A and B having the same properties than those of pure air, the diffusion coefficient used in these simulations was D = 0.178 cm^2·s^{-1}.

The values of both molecular weight M and kinetic diameter σ of air and several Volatile Organic Compounds (VOCs) are given in Table 2 [35–37], as well as their diffusion coefficient in air, resulting from these 2 parameters according to Equation (3). The calculated diffusion coefficients of VOCs in air vary in the range 0.071–0.176 cm^2·s^{-1}, the value of 0.071 cm^2·s^{-1} being 2.5 times lower than air's self-diffusion coefficient. Because this factor could have an impact on the effectiveness of the mixing device, a comparison was done between an air-formaldehyde mixture (D = 0.176 cm^2·s^{-1}) and an air-toluene mixture (D = 0.087 cm^2·s^{-1}). Even though other compounds such as o-Xylene and Naphthalene have lower diffusion coefficient in air, toluene was chosen for this comparison because it is one of the major indoor air pollutants along with formaldehyde.

Table 2. Diffusion coefficient of several gases in air at 1 atm and 23 °C.

Gas	Molecular Weight (g·mol^{-1})	Kinetic Diameter σ (Å)	Diffusion Coefficient in Air (cm^2·s^{-1})
Air	28.97	3.71 [8]	0.178
Formaldehyde (HCHO)	30.03	3.73 [35]	0.176
Acetaldehyde (CH$_3$CHO)	44.05	7.27 [36]	0.074
Benzene (C$_6$H$_6$)	78.11	5.85 [37]	0.089
Toluene (C$_7$H$_8$)	92.14	5.85 [37]	0.087
Ethylbenzene	106.17	6.00 [37]	0.083
p-Xylene	106.16	5.85 [37]	0.086
m-Xylene	106.16	6.80 [37]	0.071
o-Xylene	106.16	6.80 [37]	0.071
Naphthalene	128.17	6.20 [37]	0.078

Regarding the simulations, there were 2 boundary conditions at the inlet, which were the flow rate and the scalar of the gas entering the channels. The individual gases flow rates were varied depending on the scenario, while the scalar of the gas at the inlet is 1 for 1 s, then 0 for 9 s alternatively, simulating pulses of 1 s of formaldehyde alternating with pulses of 9 s of pure air. The boundary condition at the outlet was 1 atmosphere, since the set-up is supposed to operate at the atmospheric pressure.

The time step of the simulations was always 1 s, with 5 iterations per time step which allowed attaining a satisfactory convergence criterium (residual value of 1.0×10^{-7}). This time step was verified to be low enough to guarantee reliable results by comparing the convergence criterium with that of a simulation performed with a time step of 0.1 s. The calculated scalar values were found to be the same within a difference lower than of 0.5%. However, for simulation time saving purposes, the time step of 1 s has been preferred for all simulations. For the same reasons, the results were saved only every 3 time' steps (3 s).

3. Results

For every simulation, the first pulse of gas A entered the first stage at $t_0 = 20$ s. Because of the compressibility of the gas, pure gas B was generated during the first 20 s in order to reach a steady-state flow. Table 3 summarizes all the simulations that have been run, depending on the flow rate of the gaseous mixture and the number of mixing stages. Flow rates of 1, 5 and 10 NmL·min^{-1} have not been studied for 8 and 16 stages since (i) a satisfactory mixing has been reached with only 4 stages at these low flow rates, (ii) their corresponding response time at the outlet would have been become too long. Similarly, the 16 stages running at 25 NmL·min^{-1} was not studied either.

Table 3. Presentation of the simulations realized at different flow rates varying between 1 and 100 NmL·min^{-1} and numbers of stages in the range 1–16. Configurations marked with a ✓ have been simulated, while configurations marked with a ✗ have not been studied. The total microfluidic circuit volume of the stages set is also calculated and mentioned on the right.

Flow Rate (NmL·min^{-1}) Number of Stages	1	5	10	25	50	100	Total Volume (mL)
1	✓	✓	✓	✓	✓	✓	1.686
2	✓	✓	✓	✓	✓	✓	3.372
3	✓	✓	✓	✓	✓	✓	5.058
4	✓	✓	✓	✓	✓	✓	6.744
8	✗	✗	✗	✓	✓	✓	13.488
16	✗	✗	✗	✗	✓	✓	26.976

The scalar value for pure formaldehyde is monitored at the cross-section's centre of the microchannel's exit of the studied stages, as shown in Figure 5 for the 4th stage. At flow rates of 5 and 10 NmL·min^{-1}, the scalar value slowly approaches the target value of 0.1 and stabilizes afterwards. At high flow rates, the scalar value oscillates around the targeted scalar value and the amplitude of these oscillations increases with the flow rate. For 50 NmL·min^{-1}, the percentage of oscillation around the targeted scalar value is close to 20% while for flow rates from 25 NmL·min^{-1} and below, it is under 5%. This witnesses that the flow rate of 50 NmL·min^{-1} is too high to achieve a good mixing and indicates that more stages are required as illustrated in the following (see Figure 6).

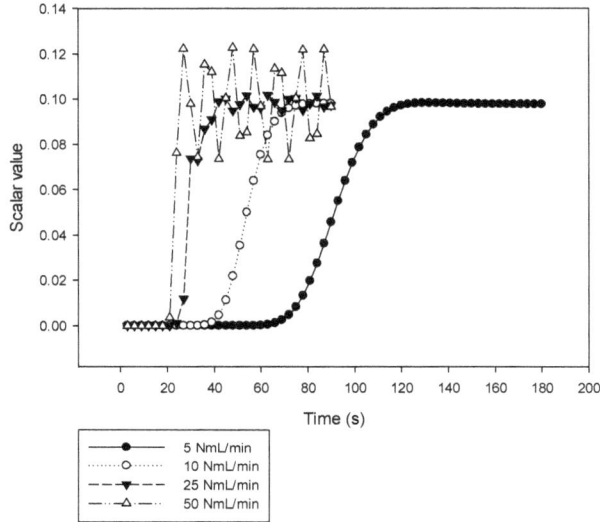

Figure 5. Effect of different flow rates on the variations of the scalar variable as a function of time at the exit of the last stage of a 4-stages micromixer.

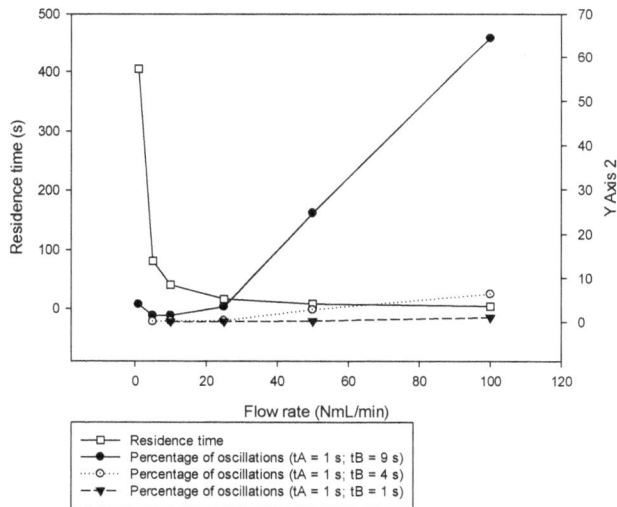

Figure 6. Residence time and percentage of oscillations around the targeted scalar 0.1 for flow rates ranging from 1 to 100 NmL·min^{-1} at the exit of the 4th mixing stage. Percentage of oscillations is given for $t_A = 1$ s and $t_B = 9$ s, 4 s and 1 s.

To investigate whether the gas mixing operates in the buffer tanks of a stage or in the channels connecting the stages, the variations of the scalar value has been plotted versus time for a flow rate of 25 NmL·min^{-1} between (i) the inlet and outlet of one of the four parallel tank of the first stage (Figure S2a) and (ii) the exit of one tank of the first stage and the inlet of one tank of the second stage (Figure S2b). It is clearly seen that the mixing takes place exclusively in the tanks.

Figure 6 shows for the 4 stages device both variations of the residence time and the percentage of oscillation around the targeted value as a function of the flow rate for 3 different pulses ratios (1/10; 1/5; 1/2). As expected, it is observed that the residence time decreases with the flow rate while the percentage of oscillation at a pulses ratio of 1/10 sharply increases for flow rates higher than 10 NmL·min^{-1}. Assuming that a satisfactory mixing is characterized by a percentage of oscillation lower than 5–6%, only flow rates up to 25 NmL·min^{-1} can be considered for the highest pulses ratio investigated, that is 1/10. For lower pulses ratios, since the axial diffusion pathway for gas mixture homogenization is reduced, the flow rate to reach the targeted percentage of oscillations is increased. In other terms, for a given flow rate, the lower the pulse flow rate, the lower is the percentage of oscillations and better is the mixing. For example, at 100 NmL·min^{-1}, the percentage of oscillations reaches 64.4%, 6.4% and 1.0% for a pulses' ratio of 1/10, 1/5 and 1/2 respectively and 24.7%, 2.8% and 0.2% at 50 NmL·min^{-1}.

The percentage of oscillations for different number of stages is plotted versus the flow rate in the range 1–100 NmL·min^{-1} (see Figure S3). For a given flow rate and a pulses ratio of 1/10, the percentage of oscillations decreases with the number of stages since a longer residence time promotes a better gas mixing.

The number of stages requested to achieve a percentage of oscillation lower than 5–6% as a function of the flow rate and the subsequent residence time is represented in Figure 7 for a pulses ratio of 1/10. Above 1 NmL·min^{-1}, the number of required stages increases linearly ($R^2 = 0.99$) from 1 to 16 for flow rate ranging between 5 and 100 NmL·min^{-1}, respectively whereas the residence time stays almost constant in the range 16.2–20.2 s. It means that whatever the flow rate desired, it is possible to adjust the number of stages to achieve a satisfactory mixing with a constant residence time of around 16–20 s. Furthermore, this number of stages will decrease in case of a higher pulses' ratio, for example, 1/5 or 1/2.

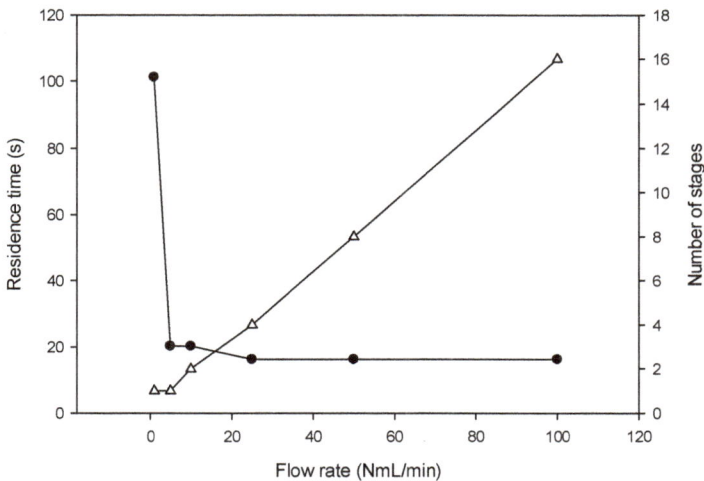

Figure 7. Number of stages (Δ) required for the oscillations of the scalar value at the micromixer's exit to be lower than 5% of the targeted value 0.1 at flow rates ranging from 1 to 100 NmL·min^{-1}. Variations of the residence time (●) with the flow rate for the corresponding number of stages required. At 100 NmL·min^{-1}, the 5% threshold is not reached: the percentage of oscillations is 6.1% after the 16th stage of mixing.

Finally, a comparison between an air-formaldehyde mixture, an air-toluene mixture and air-air has been made in order to investigate the effect of the diffusion coefficient on the mixing's efficiency.

In Figure 8a,b, the scalar value is monitored at the exit of a 2-stages and a 4-stages device respectively for different gas mixtures where B is pure air (t_B = 9 s) and A is either pure formaldehyde, pure toluene or pure air (t_A = 1 s) at a given flow rate of 25 NmL·min^{-1}. At the 2 stages device's exit, the amplitude of the oscillations is close to 20% of the targeted scalar value whatever the nature of gas B, while it reaches 5 % for the 4-stages device.

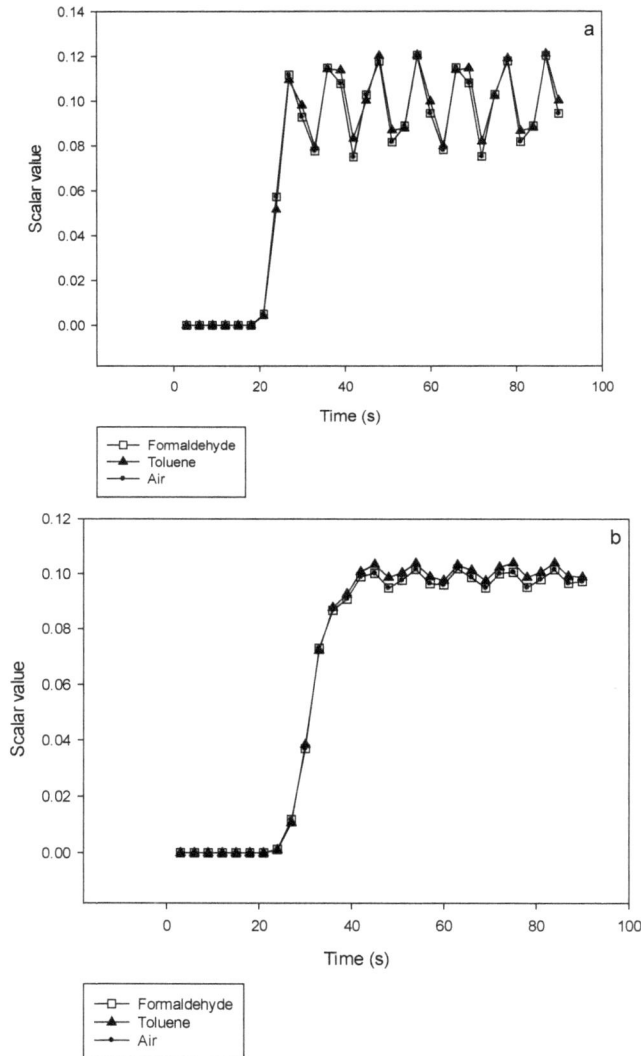

Figure 8. Comparison of the mixing's efficiency between formaldehyde, toluene and air being mixed with pure air at 25 NmL·min^{-1}. Variations of the scalar value with time at the exit of a 2-stages (**a**) or 4-stages (**b**) device, illustrating that there is almost no time response difference between formaldehyde, toluene and air.

One may have expected that the gas mixing efficiency would have been dependent on the diffusion coefficient in air for the different studied gases, that is $D_{HCHO} = 0.176$ cm$^2 \cdot$s^{-1}; $D_{toluene} = 0.087$ cm$^2 \cdot$s^{-1} and $D_{air} = 0.178$ cm$^2 \cdot$s^{-1}. However, the simulations shown in Figure 8 demonstrate that differences between those 3 gases are already very small after 2 stages and negligible for the 4 stages device. Such results indicate the strong efficiency of the mixing device developed in this work.

Because there is no difference, neither in stability nor in response time, between air, formaldehyde and toluene, it can be assumed that the simulations of this work can be used for all the VOCs listed in Table 2, with diffusion coefficients ranging from 0.071 to 0.176 cm$^2 \cdot$s^{-1}.

In case of gas A is already composed of a diluted gaseous component in air, the same behaviour is expected as that observed for a pure gas A, which permits to potentially generate extremely low gas concentrations of A.

Other alternative solution to dilute more the gas A consists in increasing the ratio between times t_A and $t_A + t_B$, for example, $t_A = 0.1$ s and $t_B = 9.9$ s. For a given flow rate of 5 NmL$\cdot$min^{-1}, the percentage of oscillation decreases with number of stages: 44.7% (1 stage); 10.1% (2 stages); 3.4% (3 stages) and 0.5% (4 stages). As illustrated in Figure S4, the corresponding residence time is increased from 16s to around 48 s and 64 s for 3 and 4 stages, respectively.

4. Discussion

As aforementioned, few experimental studies have investigated the mixing of 2 gases in microfluidic devices (see Table 1) by the radial approach. Only one study [10] refers to the axial approach and was a numerical investigation like in the present work.

As presented in Table 1, most of the microchips used for mixing gases have a fixed design [10,25,26,29], meaning that either the flow rate or the mixing ratios or both, cannot be changed using a single microchip. This is an issue for these devices where many applications require on line and rapid parameters changes. For instance, this is the case for analytical calibration instruments or during the generation of a protective gaseous mixture during welding. Conversely, the technical flexible solution presented in this work offers to the user the possibility to have a modular system where mixing ratios and total gas flow rates can easily vary in the ranges 1/10–9/10 and 1–100 NmL$\cdot$min^{-1}, respectively.

Most of gas mixing studies have been based on a radial approach where they required mixing times ranging between 4 and 24 s (see Table 1) [25–27]. The technical solution proposed by Polinkovski et al. [25] achieved homogeneity after 4 s using a microchannels network to generate discrete concentrations of O$_2$ in N$_2$. Two other microchips, used to generate O$_2$ concentrations for cell culture, exhibited mixing time of 20–24 s [26,27]. In addition, most of them [26,27] proposed a fixed design limiting any parameters changes. The proposed modular device allows the selection of the right number of stages needed for a given flow rate in order to obtain a response time of 20 s whatever the flow rate with oscillations of the targeted concentration lower than 5% (Figure 7).

Haas-Santo et al. [24] and Huang et al. [28] have developed very fast mixers for radial heterogeneity, with mixing times down to 0.6 and 3.7 ms, respectively. The former allows efficient gas mixing at very high flow rates up to 10,000 NmL$\cdot$min^{-1}, while the latter operates at a range of flow rates (5–250 NmL$\cdot$min^{-1}) close to the range of interest (1–100 NmL$\cdot$min^{-1}). However, these devices require very accurate and expensive techniques in clean rooms to manufacture microchips integrating small microchannels in the order of 100 μm × 100 μm and 550 μm × 125 μm, respectively [24,28]. On the contrary, the chip has larger channels of 1 mm × 1 mm and tanks of 10 mm × 1 mm, so that it can be thus easily manufactured using a micromilling machine. Furthermore, these two devices [24,28] could never have been used for mixing pulsed gas flows since the small microchannels cross section would have induced the generation of long plugs of gases to be mixed; which in turn would have required a long mixing time. This fully justifies the development of a specific micromixer for the axial mixing of gases.

Moreover, the device allows response times of a few tens of seconds, by accommodating the number of stages according to the total flow rate, conversely to the approximately 2 min reported by the other axial numerical diffusion study [10], not to mention its lack of flexibility.

5. Conclusions

A new axial gas pulses multi-stage micromixer has been designed to allow an efficient mixing between two pulses of different gases (A into B), easily allowing the generation of different gas concentrations for many applications such as cell culture or analytical calibration.

Once the dimensions and the geometry of a single stage determined and optimized, modelling was performed by CFD and has demonstrated that the device allows response times of few tens of seconds. This response time was obtained by adjusting the number of stages (1 to 16) according to the total gas flow rate (1 to 100 NmL·min^{-1}) to reach 94–95% of the theoretical mixing ratio calculated by the pulses ratio; and corresponds to the residence time of the micromixer. As an example, a single stage of 4 buffer tanks achieves homogeneity at a flow rate of 5 NmL·min^{-1} within 20 s for a pulses ratio of 10% ($t_A = 1$ s and $t_B = 9$ s), while 16s and 16 stages are required for a flow rate of 100 NmL·min^{-1}.

If the gas A is already diluted in gas B at a concentration of 100 ppm (commercial product), then the targeted concentrations of A at the device exit can vary in the range 10–90 ppm for a given flow rate by adjusting the A pulses ratio, 1/10 to 9/10 respectively. 1/100 dilution could be also achieved by combining pulses of $t_A = 0.1$ s and $t_B = 9.9$ s if it is acceptable to have a longer mixing time. In practice, pulses times can be easily controlled by software using electronic valves to become automatic and user friendly. Thus, the proposed device could serve for calibration purposes for instance.

Supplementary Materials: The following are available online at http://www.mdpi.com/2072-666X/10/3/205/s1, Table S1: Parameters of the simulations, Figure S1: Comparison of the results for mesh densities equal to 0.5, 1 and 2 times the reference mesh density with the same mixing parameters (Q = 25 NmL·min^{-1} and $t_A/(t_A + t_B) = 1/10$) at the exit of the 4th mixing stage, Figure S2: Scalar values variations with respect to time between the inlet and the outlet of a tank from the first stage (a) and between the outlet of this tank and the inlet of a tank from the second stage (b), Figure S3: Scalar variations at the exit of different numbers of mixing stages with respect to the flow rate. The variations are presented as a +/− percentage of the targeted scalar 0.1, Figure S4: Scalar value after 1 to 4 stages at a flow rate of 5 NmL·min^{-1} for a pulses ratio of 1/100.

Author Contributions: Conceptualization, F.N., S.L.C., C.A.S.; methodology, F.N., S.L.C.; software, F.N.; validation, F.N., S.L.C., C.A.S.; formal analysis, F.N.; investigation, F.N.; resources, S.L.C., C.A.S.; writing—original draft preparation, F.N., S.L.C., C.A.S.; writing—review and editing, F.N., S.L.C., C.A.S.; visualization, F.N.; supervision, S.L.C., C.A.S.; project administration, S.L.C.; funding acquisition, S.L.C.

Funding: This research was funded by "Association Nationale de la Recherche et de la Technologie," grant number 2016/1089, by European Commission through the LIFE programme, grant number LIFE17 ENV/FR/000330 and by SME In'Air Solutions. The APC was funded by LIFE17 European Commission through the LIFE programme.

Acknowledgments: This study was supported by the CIFRE programme between SME In'Air Solutions (Strasbourg, France) and the French National Association of Research and Technology (ANRT, grant number 2016/1089) and was also funded by the European Commission through the LIFE programme (SMART'IN AIR, LIFE17 ENV/FR/000330).

Conflicts of Interest: The authors declare no conflict of interest.

References

1. Robbins, P.A.; Swanson, G.D.; Micco, A.J.; Schubert, W.P. A fast gas-mixing system for breath-to-breath respiratory control studies. *J. Appl. Physiol. Respir. Environ. Exerc Physiol.* **1982**, *52*, 1358–1362. [CrossRef] [PubMed]

2. Dantas, H.V.; Barbosa, M.F.; Moreira, P.N.T.; Galvão, R.K.H.; Araújo, M.C.U. An automatic system for accurate preparation of gas mixtures. *Microchem. J.* **2015**, *119*, 123–127. [CrossRef]

3. Continuous Flow Type Gas Blending Facility Used for Autonomous and System Diving—ScienceDirect. Available online: https://www.sciencedirect.com/science/article/pii/S1876610217311748 (accessed on 17 December 2018).

4. Fletcher, G.C.; Summers, G.; Corrigan, V.K.; Johanson, M.R.; Hedderley, D. Optimizing Gas Mixtures for Modified Atmosphere Packaging of Fresh King Salmon (Oncorhynchus tshawytscha). *J. Aquat. Food Prod. Technol.* **2005**, *13*, 5–28. [CrossRef]
5. Hood, M.E. Gas Mixing Device for Draught Beer Dispensing. U.S. Patent 2,569,378, 25 Spetember 1951.
6. Mvola, B.; Kah, P. Effects of shielding gas control: Welded joint properties in GMAW process optimization. *Int. J. Adv. Manuf. Technol.* **2017**, *88*, 2369–2387. [CrossRef]
7. Shmelev, V.M.; Nikolaev, V. Propane conversion in a chemical compression reactor. *Russ. J. Phys. Chem. B* **2011**, *5*, 235–243. [CrossRef]
8. Zethræus, B.; Adams, C.; Berge, N. A simple model for turbulent gas mixing in CFB reactors. *Powder Technol.* **1992**, *69*, 101–105. [CrossRef]
9. Christensen, P.L.; Nielsen, J.; Kann, T. Methods to produce calibration mixtures for anesthetic gas monitors and how to perform volumetric calculations on anesthetic gases. *J. Clin. Monit. Comput.* **1992**, *8*, 279–284. [CrossRef]
10. Martin, N.A.; Goody, B.A.; Wang, J.; Milton, M.J.T. Accurate and adjustable calibration gas flow by switching permeation and diffusion devices. *Meas. Sci. Technol.* **2012**, *23*, 105005. [CrossRef]
11. Rosenberg, E.; Hallama, R.A.; Grasserbauer, M. Development and evaluation of a calibration gas generator for the analysis of volatile organic compounds in air based on the injection method. *Fresenius J. Anal. Chem* **2001**, *371*, 798–805. [CrossRef]
12. Monsé, C.; Broding, H.; Hoffmeyer, F.; Jettkant, B.; Berresheim, H.; Brüning, T.; Bünger, J.; Sucker, K. Use of a Calibration Gas Generator for Irritation Threshold Assessment and As Supplement of Dynamic Dilution Olfactometry. *Chem. Sens.* **2010**, *35*, 523–530. [CrossRef]
13. Pérez Ballesta, P.; Baldan, A.; Cancelinha, J. Atmosphere Generation System for the Preparation of Ambient Air Volatile Organic Compound Standard Mixtures. *Anal. Chem.* **1999**, *71*, 2241–2245. [CrossRef] [PubMed]
14. Ricker, N.L.; Muller, C.J.; Craig, I.K. Fuel gas blending benchmark for economic performance evaluation of advanced control and state estimation. *J. Process. Control.* **2012**, *22*, 968–974. [CrossRef]
15. Safdar, M.; Jänis, J.; Sánchez, S. Microfluidic fuel cells for energy generation. *Lab Chip* **2016**, *16*, 2754–2758. [CrossRef]
16. Seong, G.H.; Crooks, R.M. Efficient Mixing and Reactions within Microfluidic Channels Using Microbead-Supported Catalysts. *J. Am. Chem. Soc.* **2002**, *124*, 13360–13361. [CrossRef] [PubMed]
17. Shang, L.; Cheng, Y.; Zhao, Y. Emerging Droplet Microfluidics. *Chem. Rev.* **2017**, *117*, 7964–8040. [CrossRef] [PubMed]
18. Zhu, P.; Wang, L. Passive and active droplet generation with microfluidics: A review. *Lab. Chip* **2016**, *17*, 34–75. [CrossRef]
19. Christopher, G.F.; Anna, S.L. Microfluidic methods for generating continuous droplet streams. *J. Phys. D Appl. Phys.* **2007**, *40*, R319. [CrossRef]
20. Baroud, C.N.; Willaime, H. Multiphase flows in microfluidics. *C. R. Phys.* **2004**, *5*, 547–555. [CrossRef]
21. Zhao, C.-X.; Middelberg, A.P.J. Two-phase microfluidic flows. *Chem. Eng. Sci.* **2011**, *66*, 1394–1411. [CrossRef]
22. Lee, C.-Y.; Chang, C.-L.; Wang, Y.-N.; Fu, L.-M. Microfluidic Mixing: A Review. *Int. J. Mol. Sci.* **2011**, *12*, 3263–3287. [CrossRef]
23. Suh, Y.K.; Kang, S. A Review on Mixing in Microfluidics. *Micromachines* **2010**, *1*, 82–111. [CrossRef]
24. Haas-Santo, K.; Pfeifer, P.; Schubert, K.; Zech, T.; Hönicke, D. Experimental evaluation of gas mixing with a static microstructure mixer. *Chem. Eng. Sci.* **2005**, *60*, 2955–2962. [CrossRef]
25. Polinkovsky, M.; Gutierrez, E.; Levchenko, A.; Groisman, A. Fine temporal control of the medium gas content and acidity and on-chip generation of series of oxygen concentrations for cell cultures. *Lab. Chip* **2009**, *9*, 1073–1084. [CrossRef] [PubMed]
26. Adler, M.; Polinkovsky, M.; Gutierrez, E.; Groisman, A. Generation of oxygen gradients with arbitrary shapes in a microfluidic device. *Lab Chip* **2010**, *10*, 388–391. [CrossRef] [PubMed]
27. Lo, J.F.; Sinkala, E.; Eddington, D.T. Oxygen gradients for open well cellular cultures via microfluidic substrates. *Lab Chip* **2010**, *10*, 2394–2401. [CrossRef] [PubMed]
28. Huang, C.-Y.; Wan, S.-A.; Hu, Y.-H. Oxygen and nitrogen gases mixing in T-type micromixers visualized and quantitatively characterized using pressure-sensitive paint. *Int. J. Heat Mass Transf.* **2017**, *111*, 520–531. [CrossRef]

29. Tesař, V.R.; Tippetts, J.; Low, Y.-Y. Oscillator Mixer for Chemical Microreactors. In Proceedings of the 9th International Symposium on Flow Visualization, Edinburgh, UK, 22–25 August 2000.

30. Wilke, C.R.; Lee, C.Y. Estimation of Diffusion Coefficients for Gases and Vapors. *Ind. Eng. Chem.* **1955**, *47*, 1253–1257. [CrossRef]

31. Squires, T.M.; Quake, S.R. Microfluidics: Fluid physics at the nanoliter scale. *Rev. Mod. Phys.* **2005**, *77*, 977–1026. [CrossRef]

32. Faanes, A.; Skogestad, S. A systematic approach to the design of buffer tanks. *Comput. Chem. Eng.* **2000**, *24*, 1395–1401. [CrossRef]

33. Tsao, C.-W. Polymer Microfluidics: Simple, Low-Cost Fabrication Process Bridging Academic Lab Research to Commercialized Production. *Micromachines* **2016**, *7*, 225. [CrossRef]

34. Cussler, E.L. *Diffusion: Mass Transfer in Fluid Systems*, 2nd ed.; Cambridge University Press: New York, NY, USA, 1997; ISBN 978-0-521-45078-2.

35. (PDF) Adsorption of Low-Concentration Formaldehyde from Air by Silver and Copper Nano-Particles Attached on Bamboo-Based Activated Carbon. Available online: https://www.researchgate.net/publication/271305127_Adsorption_of_Low-Concentration_Formaldehyde_from_Air_by_Silver_and_Copper_Nano-Particles_Attached_on_Bamboo-Based_Activated_Carbon (accessed on 8 January 2019).

36. Gauf, A.; Navarro, C.; Balch, G.; Hargreaves, L.R.; Khakoo, M.A.; Winstead, C.; McKoy, V. Low-energy elastic electron scattering by acetaldehyde. *Phys. Rev. A* **2014**, *89*, 022708. [CrossRef]

37. Weng, Y.; Qiu, S.; Ma, L.; Liu, Q.; Ding, M.; Zhang, Q.; Zhang, Q.; Wang, T. Jet-Fuel Range Hydrocarbons from Biomass-Derived Sorbitol over Ni-HZSM-5/SBA-15 Catalyst. *Catalysts* **2015**, *5*, 2147–2160. [CrossRef]

micromachines

MDPI

Article

Development of a Toluene Detector Based on Deep UV Absorption Spectrophotometry Using Glass and Aluminum Capillary Tube Gas Cells with a LED Source

Sulaiman Khan [1,2,3], David Newport [1] and Stéphane Le Calvé [2,3,*]

1   School of Engineering, Bernal Institute, University of Limerick, V94 T9PX Limerick, Ireland;
    sulaiman.khan@ul.ie (S.K.); david.newport@ul.ie (D.N.)
2   Université de Strasbourg, CNRS, ICPEES UMR 7515, F-67000 Strasbourg, France
3   In'Air Solutions, 67087 Strasbourg, France
*   Correspondence: slecalve@unistra.fr

Received: 14 February 2019; Accepted: 11 March 2019; Published: 18 March 2019

Abstract: A simple deep-ultraviolet (UV) absorption spectrophotometer based on ultraviolet light-emitting diode (UV LED) was developed for the detection of air-borne toluene with a good sensitivity. A fiber-coupled deep UV-LED was employed as a light source, and a spectrometer was used as a detector with a gas cell in between. 3D printed opto-fluidics connectors were designed to integrate the gas flow with UV light. Two types of hollow core waveguides (HCW) were tested as gas cells: a glass capillary tube with aluminum-coated inner walls and an aluminum capillary tube. The setup was tested for different toluene concentrations (10–100 ppm), and a linear relationship was observed with sensitivities of 0.20 mA·U/ppm and 0.32 mA·U/ppm for the glass and aluminum HCWs, respectively. The corresponding limits of detection were found to be 8.1 ppm and 12.4 ppm, respectively.

Keywords: ultraviolet light-emitting diode (UV LED); spectrophotometry; UV absorption; gas sensors; Benzene, toluene, ethylbenzene and xylene (BTEX); toluene; hollow core waveguides; capillary tubes

1. Introduction

Monitoring of air quality in indoor spaces is critical for healthy living. Nowadays most of our daily activities are based in indoor spaces where exposure to various indoor air pollutants is inevitable [1]. Indoor air can contain various volatile organic compounds (VOCs) among other pollutants such as air-borne particles, microorganisms, household odours, and gases. VOCs are organic compounds with a high vapour pressure at room temperature; they readily evaporate into a gaseous phase at room temperature. Some of the common VOCs are acetaldehyde, acetone, benzene, carbon tetra chloride, ethyl acetate, heptane, hexane, isopropyl alcohol, formaldehyde, naphthalene, styrene, toluene, and xylenes [1,2]. Benzene, toluene, ethylbenzene and xylene (BTEX) are aromatic hydrocarbons and are some of the most hazardous pollutants among VOCs. Toluene is a colourless VOC with a sweet, pungent odour, density of 0.866 g·cm^{-3} at 20 °C, and boiling point of 110.7 °C [3]. Its sources of generation in indoor spaces are common household items, i.e., cleaning products, paint thinners, adhesives, synthetic fragrances, nail polish, and cigarette smoke. Automobile emissions are the main source of toluene in outdoor air environments [4]. Exposure to toluene can affect the central nervous system, liver, kidney, and skin [3]. The American Conference of Governmental Industrial Hygienists (ACGIH) have established the threshold limit value (TLV) of 50 ppm for toluene for 8 h exposure. In addition, the Occupational Safety and Health Administration (OSHA) recommends permissible exposure limits (PEL) for toluene of 200 ppm as the 8 h time-weighted average (TWA) concentration [5].

Detection of aromatics VOCs at ppm and sub-ppm ranges requires a sensitive and accurate method. Different techniques have been applied for the detection of different VOCs, for instance, electrochemical gas sensors [6], micro gas chromatography (μ-GC) [7], photoionization detectors [8], piezoelectric-based gas sensors, i.e., surface acoustic wave [9], quartz crystal microbalances [10], and tuning forks [11], gravimetric-based gas sensors [12], metal-oxide semiconductor gas sensors [13], and optical sensors such as colorimetric gas sensors [14], non-dispersive infrared gas sensors [11], and ultraviolet (UV) spectrophotometry gas sensors [15]. Among these, optical gas sensors are highly sensitive, they have minimal drift issues and rapid time responses, and they facilitate real time and in situ measurements without changing the chemical nature of gases.

UV spectrophotometry is a non-destructive, rapid time response with minimal cross-responses to other gases as long as its design is carefully considered. The technique involves direct measurement of a molecular absorption at a specific wavelength which offers an inherently reliable approach for gas sensing with excellent selectivity [16]. BTEX gases absorb strongly in the deep UV range, i.e., 250–270 nm [17], indicating they can be detected using deep UV spectrophotometry.

Recently, deep ultraviolet-light-emitting diodes (UV-LEDs) with narrow bandwidths (<30 nm) have been developed, which matches with the absorption band of many molecules. It provides a portable source without the need of monochromators or filters. LEDs have good stability, robustness, flexibility in output intensity, low power consumption and low heat generation [18]. In order to broaden the emission band, an array of LEDs of different wavelength can be used to cover a broad range of molecules [19]. UV-LEDs have been applied in the detection of different gases, for instance, NO_2 [20], O_3 [21], SO_2 [22], and BTEX [23].

In this work, we have applied UV spectrophotometry to detect toluene using a fiber-coupled deep UV-LED with aluminium- based hollow core waveguides (HCWs) coupled to mini spectrometer. We have assembled the fluid and optic parts using 3D connectors, which makes the alignment and sealing of the setup easier.

2. Materials and Methods

2.1. Spectrophotometry

UV absorption spectrophotometry is a direct optical gas detection method that is based on the unique absorption spectra (fingerprints) at a specific wavelength. In spectrophotometry, the molecular absorption level is measured according to the Beer–Lambert law:

$$A = \sigma c l = \log \frac{I_0}{I} \tag{1}$$

where A is absorbance, σ (cm^2/molecule) is the absorption cross-section, c (molecules/cm^3) is concentration of gas molecules, and l (cm) is length of gas cell. I_0 and I are the transmitted intensities recorded for the background gas (i.e., nitrogen) and toluene gas concentrations in the gas cell, respectively. The constant σ is the absorption cross-section, which is a molecule-specific property and has a constant value at a specific wavelength. It represents the effective area of a molecule that is needed for a photon to transverse the molecule.

The sensitivity depends on the optical path length, which is defined by the design of the absorption gas cell. The absorption cell can be single pass, multi-pass, or a resonant cavity. The single pass cell is relatively easy to manufacture, has a quick time response and is relatively easy to couple with light sources and detectors compared to multi-pass or resonant cells. HCW offer a compact and efficient alternative to gas cells. They provide a compact platform for the interaction of photons and gas molecules to realize a quantitative and molecular specific absorption spectroscopy. The HCWs guide the radiation in a leaky-mode and the radiation is propagated by metallic reflection inside the coaxial hollow core. The transmission of UV in HCW depends on material, size and geometry of HCW. HCW with smaller diameter have higher attenuation losses compared to the higher diameter. They have been applied in IR spectrometry and other sensing applications, e.g., biomedical and toxicology.

2.2. Instrumentation

A fiber-coupled deep UV-LED (Mightex Systems, Pleasanton, CA, USA) with a peak at 260 nm and power range from 45–80 μW was used. A mini-spectrometer (Hamamatsu mini-spectrometer C10082CH, Iwata, Japan) with a spectral detection range from 200–800 nm and integration time of 10–10,000 ms was used to record the intensity of the transmitted UV light. Two HCWs: one glass capillary tube with an inner wall coated with aluminium (Doku Engineering, Japan) and the other aluminium capillary tube as a waveguide (Advent, London, UK), were tested. The gas cell was coupled with the source and detector using optical fibers (Ocean Optics, Largo, FL, USA) for UV applications (range 200–1100 nm) with a core diameter of 400 μm. The optics and fluidics connections were 3D printed. The connectors were designed using Solidworks 2018 and 3D printed (Ultimaker 3, Geldermalsen, The Netherlands) using Acrylonitrile Butadiene Styrene (ABS). Toluene was delivered from a gas cylinder with concentration of 100 ppm ± 2% (Air Products, Aubervilliers, France). The gaseous flow was controlled via two mass flow controllers (Bronkhorst, Gelderland, The Netherlands) with a full-scale range of 20 mL/min ± 0.5% and 50 mL/min ± 0.5%.

2.3. Experimental Setup

The schematic of experimental setup is shown in Figure 1a. In order to ensure a stable intensity, a constant current was supplied to the LED and the emission intensity was varied by changing the input current (0–30 mA). The opto-mechanical components were aligned using 3D printed holders on the optical breadboard to avoid baseline shifts due to mechanical movement. The gas cell was thermally insulated to minimize thermal fluctuations.

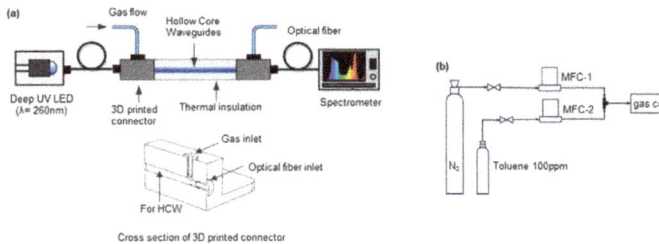

Figure 1. (**a**) Experimental setup for toluene detection. (**b**) Toluene concentration generation setup.

Different concentrations of toluene were generated in nitrogen using a configuration of mass flow controller (MFC) as shown in Figure 1b. The desired concentration of toluene was obtained by mixing toluene with nitrogen using MFC-2 and MFC-1, respectively. A total gas flow of 40 mL/min was injected into the waveguide gas cell. The sealing of the opto-fluidics connector was tested, and less than 3% leakage was found.

3. Results and Discussion

The glass HCW and aluminium HCW were investigated for gas sensing applications. The glass HCW was composed of a glass capillary tube with a thin inner coating of aluminium. The aluminium HCW was a capillary tube made of tempered aluminium with 99.7% purity. The lengths of the glass HCW (inner diameter, 1mm; internal gas cell volume, 27 μL) and aluminium HCW (inner dimeter, 2 mm; internal gas cell volume, 157 μL) installed were 34 cm and 50 cm, respectively. Aluminium was selected because it had good reflective and transmittance properties (i.e., attenuation loss, 0.2 dB/m) for deep UV application, compared to other reflective metals like silver [24]. It also has good chemical compatibility with toluene.

The absorption spectra of toluene and the emission band of UV-LED was measured and compared, as shown in Figure 2. Toluene had three peaks for a wavelength range of 250 to 270 nm. The bandwidth (full width at half maximum (FWHM)) of the LED is 10 nm, centred at 260 nm, covering the absorption spectra of toluene, which implies that the setup can be applied for measuring the absorbance of toluene.

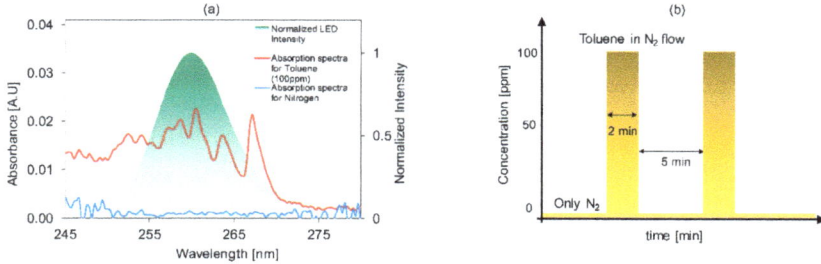

Figure 2. (**a**) Absorption spectra of toluene (100 ppm), background gas, i.e., nitrogen, and emission spectrum of deep UV-LED. (**b**) Schematics of flushing routine of nitrogen and toluene for measuring reference and measured intensity respectively.

The measurement sequence was started by flushing nitrogen through the gas cell, followed by injecting different concentrations of toluene for 1 to 2 min until a stable signal was obtained as shown in Figure 2b. The recorded intensities for nitrogen and toluene concentrations were used as a reference and measured intensity, respectively. A dark intensity was recorded and subtracted from both references, i.e., I_0 and the measured I intensities to calculate the absorbance according to Equation (1).

The setup with the glass HCW was tested for a concentration range of 10–100 ppm. The absorbance increased linearly with an increasing toluene concentration according to the Beer–Lambert law, as shown in Figure 3.

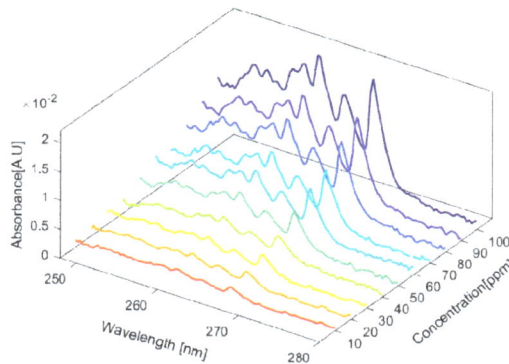

Figure 3. Absorbance of toluene at different wavelengths for different concentrations of toluene using the glass hollow core waveguide (HCW).

The absorbance peaks at $\lambda_1 = 260.3$ nm, $\lambda_2 = 263.1$ nm, and $\lambda_3 = 267.2$ nm were plotted for different concentrations of toluene, and a linear relation was obtained, as shown in Figure 4. The combined uncertainties in the toluene concentration were calculated according to BIPM guidelines [25], by taking into account the uncertainties of MFCs and gas cylinders. The sensitivity was calculated by taking the slope of the plot, representing absorbance vs. concentration. Sensitivities of 0.20 mA·U/ppm, 0.15 mA·U/ppm, and 0.19 mA·U/ppm were obtained for λ_1, λ_2, and λ_3 respectively. According to the absorption cross-section values at λ_1, λ_2, and λ_3, the absorbance values should be higher for λ_3, but relative lower values were observed due to the emission profile of LED and limited resolution of spectrometer. A limit of detection of 8.2 ppm was calculated for λ_1 from standard deviation of the calibration data assuming that the data is normally (Gaussian) distributed, using the equations [26].

$$x_{\text{LOD}} = \frac{S_y t}{r} \tag{2}$$

x_{LOD} is the limit of detection, where r and S_y represent sensitivity of the linear fit and average standard deviation respectively.

$$r = \frac{\Delta y}{\Delta x} = \frac{n \sum (x_i y_i) - \sum x_i \sum y_i}{D} \tag{3}$$

$$S_y = \sqrt{\frac{\sum (y_i - r x_i - b)^2}{n - 2}} \tag{4}$$

where b is intercept of the calibration curve i.e. the signal offset.

$$b = \frac{n \sum x_i^2 \sum y_i - \sum (x_i y_i) \sum x_i}{D} \tag{5}$$

$$D = n \sum x_i^2 - \left(\sum x_i \right)^2 \tag{6}$$

where t is Student t-function. x_i and y_i are the calibration curve points and n is the number of data points on calibration curve.

The same experiment was repeated for the aluminium HCW and a good linearity was found for different concentrations of toluene (20–100 ppm) as shown in Figure 5. Sensitivities of 0.32 mA·U/ppm, 0.23 mA·U/ppm and 0.30 mA·U/ppm were obtained for λ_1, λ_2 and λ_3 respectively. A limit of detection of 12.5 ppm was calculated for λ_1.

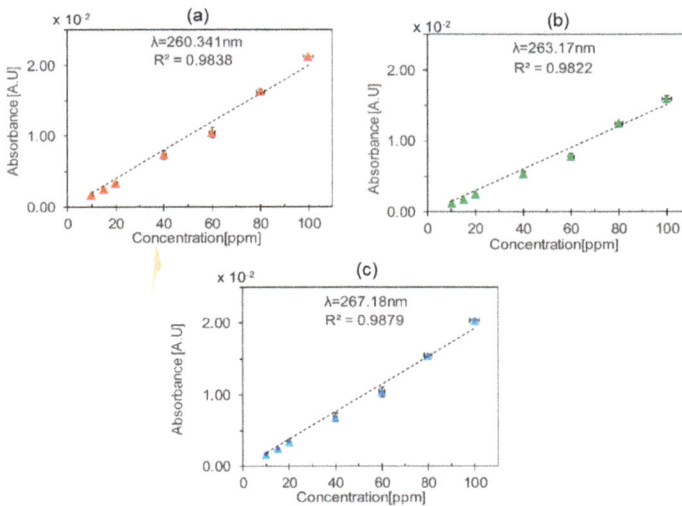

Figure 4. Absorbance vs. toluene concentration using the glass HCW at different wavelengths: (a) λ_1 = 260.34 nm, (b) λ_2 = 263.17 nm, (c) λ_3 = 267.18 nm. The vertical and the horizontal error bars represent standard deviations in absorbance values and combined uncertainties in the generated concentrations defined by the uncertainties of mass flow controllers (MFCs) and gas cylinder concentrations, respectively.

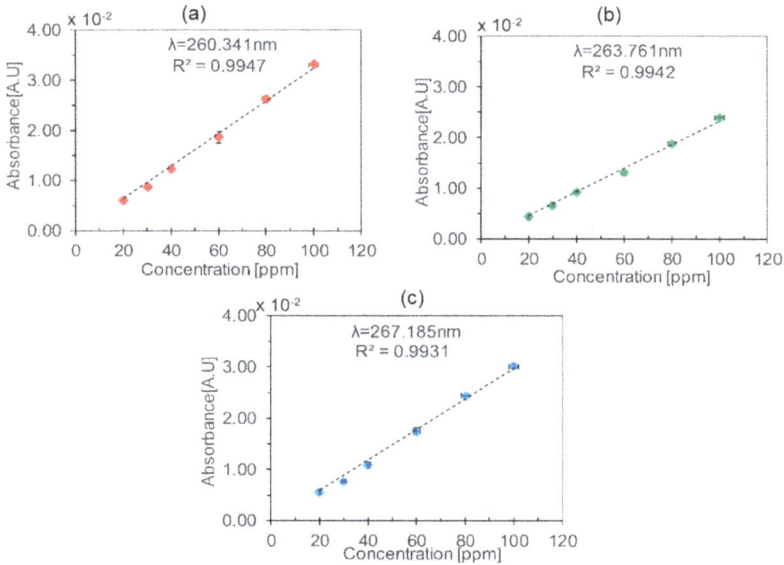

Figure 5. Absorbance vs. toluene concentration using the aluminium HCW at different wavelengths: (**a**) λ_1 = 260.34 nm, (**b**) λ_2 = 263.17 nm, and (**c**) λ_3 = 267.18 nm. The vertical and the horizontal error bars represent standard deviations in absorbance values and combined uncertainties in the generated concentrations defined by the uncertainties of MFCs and gas cylinders, respectively.

The repeatability and reproducibility of the system were tested for a toluene concentration of 30 ppm using aluminium HCW by varying the gas cell flow rate, LED current input, and the integration time of the spectrometer. A good repeatability (relative standard deviation, RSD = 2.5%) was found for five experiments using a flow rate of 20 mL/min with LED current input of 30 mA and integration time of 100 ms (experiment 2 in Figure 6). Reproducibility with RSD = 4.8% was observed for five different experiments at different conditions, as shown in Figure 6.

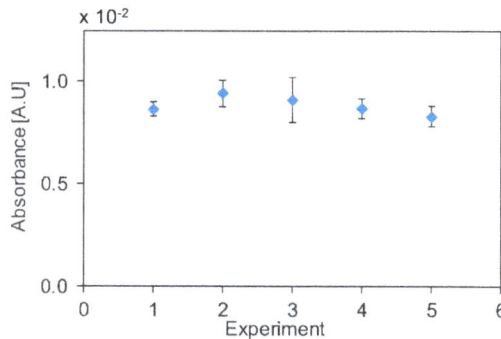

Figure 6. Absorbance for different experimental conditions for repeatability and reproducibility. (1) flowrate, 40 mL/min; (2) flowrate, 20 mL/min; (3) flowrate, 10 mL/min; (4) Detector integration time, 50 ms; and (5) LED current input, 15 mA.

The performance of the two HCWs for toluene detection applications is compared and summarized in Table 1. The aluminium HCW had a higher sensitivity compared to the glass HCW at the different peak wavelengths. Also, the aluminium HCW had good mechanical properties and can be easily assembled with fluidic and optical components. There were low coupling optical losses associated with the aluminium HCW, resulting in improved sensitivity. On the other hand, the glass HCW was fragile, and it was challenging to obtain a smooth end surface for alignment with the LED source and detector. The aluminium HCW offered a simple, cost-effective, and robust approach for absorption spectrometry with a good level of sensitivity. On the other hand, the glass HCW was a good candidate for application where a low gas cell volume is needed, for example integration of the detector with a μ-GC.

By comparing the performance of the sensor with other toluene sensing methods, the UV absorption detector based on HCW had a good sensitivity and linearity at room temperature. For instance, resistive gas sensors have a linearity range between 10–100 ppm, but they operate at high temperature [13]. Fiber optic toluene sensors have limited linearity and selectivity. Optical fiber sensors based on a long period grating for toluene has an operating range of 0–60 ppm [27], and Fabry Perot fiber sensors face the issue of non-linearity [28].

The setup has demonstrated a good sensitivity for toluene and has potential to be used for detecting BTEX molecules. The setup can be applied for ultra-low and high-selective detection of BTEX molecule mixtures by coupling it with a pre-concentration unit with/without a GC-column for separation purposes.

Table 1. Comparison of aluminium and glass HCWs for toluene detection.

HCW	Length (cm)	Wavelength (nm)	Sensitivity $(mA \cdot U \cdot ppm^{-1})$	Sensitivity/Length $(mA \cdot U \cdot ppm^{-1} \cdot cm^{-1})$
Aluminium	50	260.34	0.32	0.00640
		263.76	0.23	0.00460
		267.18	0.30	0.00600
Glass	34	260.34	0.20	0.00588
		263.76	0.15	0.00441
		267.18	0.19	0.00559

4. Conclusions and Future Scope

In this study we have demonstrated a simple and sensitive deep UV absorption spectrophotometry for detection of air-borne toluene. A fiber-coupled deep UV-LED was employed as a light source, and a spectrometer was used as detector with a gas cell in between. 3D printed opto-fluidics connectors were designed to integrate the gas flow with a UV source and detector. A glass HCW with aluminium coating and an aluminium HCW were tested as a gas cell. The setup was tested for different toluene concentrations (10–100 ppm), and a linear relationship was observed with sensitivities of 0.20 mA·U/ppm and 0.32 mA·U/ppm for glass HCW and aluminium HCW, respectively, at 260 nm. The limits of detection of 8.15 ppm and 12.45 ppm were calculated for glass HCW and aluminium HCW, respectively. The sensitivity and selectivity of the setup can be improved by coupling it with a pre-concentration unit and a micro GC column, respectively. This study provides a guide for the design of aluminium-based HCWs for UV spectrophotometry and can be applied to detect a number of molecules which show UV absorption for example, ozone, benzene, xylenes, NO_2 and SO_2.

Author Contributions: Conceptualization, experiments, data analysis and writing by S.K.; supervision and review by D.N. and S.L.C.

Funding: This work was funded by European Union's Horizon 2020 research and innovation programme under the Marie Skłodowska-Curie Innovative Training Network-MIGRATE (Miniaturized Gas flow foR Applications with Enhanced Thermal Effects), grant agreement No. 643095 [H2020-MSCA-ITN-2014]. This work has also received funding from the Clean Sky 2 Joint Undertaking under the European Union's Horizon 2020 research and innovation program under grant agreement No 687014 (MACAO Project).

Conflicts of Interest: The authors declare no conflict of interest.

References

1. Koppmann, R. Chemistry of Volatile Organic Compounds in the Atmosphere. In *Handbook of Hydrocarbon and Lipid Microbiology*; Springer: Berlin/Heidelberg, Germany, 2010; pp. 267–277.

2. Khan, F.I.; Ghoshal, A.K. Removal of Volatile Organic Compounds from polluted air. *J. Loss Prev. Process Ind.* **2000**, *13*, 527–545. [CrossRef]

3. Patnaik, P. *A Comprehensive Guide to the Hazardous Properties of Chemical Substances*; John Wiley: Hoboken, NJ, USA, 2007.

4. Agency for Toxic Substances and Disease Registry (ATSDR). *Toxicological Profile for Toluene*; U.S. Department of Health and Human Services, Public Health Service: Atlanta, GA, USA, 2017.

5. Barsan, M.E. *NIOSH-Pocket Guide to Chemical Hazards*; DHHS: Pittsburgh, PA, USA, 2005.

6. Sekhar, P.K.; Subramaniyam, K. Detection of Harmful Benzene, Toluene, Ethylbenzene, Xylenes (BTEX) Vapors Using Electrochemical Gas Sensors. *ECS Electrochem. Lett.* **2014**, *3*, B1–B4. [CrossRef]

7. Haghighi, F.; Talebpour, Z.; Sanati-Nezhad, A. Through the years with on-a-chip gas chromatography: A review. *Lab Chip* **2015**, *15*, 2559–2575. [CrossRef] [PubMed]

8. Rezende, G.C.; le Calvé, S.; Brandner, J.J.; Newport, D. Micro photoionization detectors. *Sens. Actuators B Chem.* **2019**, *287*, 86–94. [CrossRef]

9. Jakubik, W.P. Surface acoustic wave-based gas sensors. *Thin Solid Films* **2011**, *520*, 986–993. [CrossRef]

10. Kumar, A.; Brunet, J.; Varenne, C.; Ndiaye, A.; Pauly, A. Phthalocyanines based QCM sensors for aromatic hydrocarbons monitoring: Role of metal atoms and substituents on response to toluene. *Sens. Actuators B Chem.* **2016**, *230*, 320–329. [CrossRef]

11. Chen, C.; Campbell, K.D.; Negi, I.; Iglesias, R.A.; Owens, P.; Tao, N.; Tsow, F.; Forzani, E.S. A new sensor for the assessment of personal exposure to volatile organic compounds. *Atmos. Environ.* **2012**, *54*, 679–687. [CrossRef]

12. Fanget, S.; Hentz, S.; Puget, P.; Arcamone, J.; Matheron, M.; Colinet, E.D.; Andreucci, P.; Duraffourg, L.; Myers, E.; Roukes, M.L. Gas sensors based on gravimetric detection—A review. *Sens. Actuators B Chem.* **2011**, *160*, 804–821. [CrossRef]

13. Mirzaei, A.; Kim, J.H.; Kim, H.W.; Kim, S.S. Resistive-based gas sensors for detection of benzene, toluene and xylene (BTX) gases: A review. *J. Mater. Chem. C* **2018**, *6*, 4342–4370. [CrossRef]

14. Askim, J.R.; Mahmoudi, M.; Suslick, K.S. Optical sensor arrays for chemical sensing: The optoelectronic nose. *Chem. Soc. Rev.* **2013**, *42*, 8649–8682. [CrossRef]

15. Allouch, A.; le Calvé, S.; Serra, C.A. Portable, miniature, fast and high sensitive real-time analyzers: BTEX detection. *Sens. Actuators B Chem.* **2013**, *182*, 446–452. [CrossRef]

16. Hodgkinson, J.; Tatam, R.P. Optical gas sensing: A review. *Meas. Sci. Technol.* **2013**, *24*, 012004. [CrossRef]

17. Tunnicliff, D.; Brattain, R.; Zumwalt, L. Benzene, Toluene, Ethyl benzene, 0-Xylene, m-Xylene, and p-Xylene: Determination by ultravoilet spectrophotometry. *Anal. Chem.* **1949**, *21*, 890–894. [CrossRef]

18. Bui, D.A.; Hauser, P.C. Analytical devices based on light-emitting diodes—A review of the state-of-the-art. *Anal. Chim. Acta* **2015**, *853*, 46–58. [CrossRef] [PubMed]

19. Bui, D.A.; Kraiczek, K.G.; Hauser, P.C. Molecular absorption measurements with an optical fibre coupled array of ultra-violet light-emitting diodes. *Anal. Chim. Acta* **2017**, *986*, 95–100. [CrossRef] [PubMed]

20. Hawe, E.; Fitzpatrick, C.; Chambers, P.; Dooly, G.; Lewis, E. Hazardous gas detection using an integrating sphere as a multipass gas absorption cell. *Sens. Actuators A Phys.* **2008**, *141*, 414–421. [CrossRef]

21. Aoyagi, Y.; Takeuchi, M.; Yoshida, K.; Kurouchi, M.; Araki, T.; Nanishi, Y.; Sugano, H.; Ahiko, Y.; Nakamura, H. High-Sensitivity Ozone Sensing Using 280 nm Deep Ultraviolet Light-Emitting Diode for Detection of Natural Hazard Ozone. *J. Environ. Prot.* **2012**, *3*, 695–699. [CrossRef]

22. Degner, M.; Damaschke, N.; Ewald, H.; Lewis, E. Real time exhaust gas sensor with high resolution for onboard sensing of harmful components. *IEEE Sens.* **2008**, 973–976.

23. Bui, D.A.; Hauser, P.C. A deep-UV light-emitting diode-based absorption detector for benzene, toluene, ethylbenzene, and the xylene compounds. *Sens. Actuators B Chem.* **2016**, *235*, 622–626. [CrossRef]

24. Matsuura, Y.; Miyagi, M. Hollow optical fibers for ultraviolet and vacuum ultraviolet light. *IEEE J. Sel. Top. Quantum Electron.* **2004**, *10*, 1430–1434. [CrossRef]

25. JCGM, J. Evaluation of Measurement Data—Guide to the Expression of Uncertainty in Measurement (Évaluation des Données de Mesure—Guide pour L'expression de L'incertitude de Mesure.). *Int. Organ. Stand. Geneva* **2008**, *50*, 134.

26. Loock, H.P.; Wentzell, P.D. Detection limits of chemical sensors: Applications and misapplications. *Sens. Actuators B Chem.* **2012**, *173*, 157–163. [CrossRef]

27. Yin, M.; Gu, B.; An, Q.-F.; Yang, C.; Guan, Y.L.; Yong, K.-T. Recent development of fiber-optic chemical sensors and biosensors: Mechanisms, materials, micro/nano-fabrications and applications. *Coord. Chem. Rev.* **2018**, *376*, 348–392. [CrossRef]

28. Kacik, D.; Martincek, I. Toluene optical fibre sensor based on air microcavity in PDMS. *Opt. Fiber Technol.* **2017**, *34*, 70–73. [CrossRef]

micromachines

MDPI

Article

Micro Milled Microfluidic Photoionization Detector for Volatile Organic Compounds

Gustavo C. Rezende [1], **Stéphane Le Calvé** [2,3], **Jürgen J. Brandner** [4] **and David Newport** [1,*]

[1] Bernal Institute, School of Engineering, University of Limerick, V94 T9PX Limerick, Ireland;
     gustavo.coelho@ul.ie
[2] Université de Strasbourg, Centre national de la recherche scientifique (CNRS), ICPEES UMR 7515,
     F-67087 Strasbourg, France; slecalve@unistra.fr
[3] In'Air Solutions, 25 rue Becquerel, 67087 Strasbourg, France
[4] Institute of Microstructure Technology (IMT), Karlsruhe Institute of Technology,
     Hermann-von-Helmholtz-Platz 1, 76344 Eggenstein-Leopoldshafen, Germany; juergen.brandner@kit.edu
*   Correspondence: david.newport@ul.ie; Tel.: +353-61-202-849

Received: 8 February 2019; Accepted: 28 March 2019; Published: 30 March 2019

Abstract: Government regulations and environmental conditions are pushing the development of improved miniaturized gas analyzers for volatile organic compounds. One of the many detectors used for gas analysis is the photoionization detector (PID). This paper presents the design and characterization of a microfluidic photoionization detector (or µPID) fabricated using micro milling and electrical discharge machining techniques. This device has no glue and facilitates easy replacement of components. Two materials and fabrication techniques are proposed to produce a layer on the electrodes to protect from ultraviolet (UV) light and possible signal noise generation. Three different microchannels are tested experimentally and their results are compared. The channel with highest electrode area (31.17 mm^2) and higher volume (6.47 µL) produces the highest raw signal and the corresponding estimated detection limit is 0.6 ppm for toluene without any amplification unit.

Keywords: photoionization detector; microfluidics; microfabrication; volatile organic compound (VOC) detection; toluene

1. Introduction

Volatile organic compounds (VOCs) are a class of carbon-containing chemicals with a high vapor pressure at ambient temperature. Typical indoor sources of VOCs are varnishes, paints, solvents, cleaning materials, etc. In addition, outdoor sources, such as automobile and industrial waste, also contribute to both indoor and outdoor VOC pollution [1]. Many VOCs are harmful to humans, including benzene, which is carcinogenic and has no safe recommended level of exposure [2,3]. The European Commission established, with effect from 2010, a regulation for benzene exposure at a maximum limit of 5 µg/m^3 (1.6 ppb) [4]. In addition, within the European Union, regulations and guidelines are established to maintain a healthy indoor air quality [4,5].

Air must be monitored in order to identify and quantify pollutants so that appropriate action can be taken to clean the air. Gas chromatographs (GCs) equipped with detectors are suitable gas analyzers to accomplish this task because they can separate, identify and quantify different chemicals, including VOCs. GCs in the market with low detection limits are still heavy, bulky, slow and lab-based [6–8], while current regulations raise the demand for efficient gas analyzers with enhanced portability, reduced resource consumption, improved robustness, high analysis speed, low cost and reduced detection limits. One possible way to achieve these improvements is through miniaturization of all the gas analyzer components, including the detector itself.

Among the existing detectors and techniques to detect VOCs [9], the photoionization detector (PID) is commonly used with a GC and it is suitable for miniaturization [10]. The PID uses ionization of gaseous compounds by light as a working principle to quantify chemicals in the gas samples. This detector can be classified according to the ionization source. When the ionization source is unseparated from the ionization chamber by a window, the ionization source is usually a discharge in a noble gas, such as helium, and is referred to as a discharge photo ionization detector (D-PID); or when the ionization source has no fluidic connection to the ionization chamber, which is known as the lamp photo ionization detector (L-PID). In both devices, a gas sample containing the species to be detected flows through the ionization chamber, where photons emitted by the ionization source reach the sample molecules. As a general rule, if the ionization energy of the photon is greater than the ionization potential of the molecule, ionization occurs. The electrodes establish an electric field in the ionization chamber where the ionized molecules generate an ionization current proportional to their concentration. Based on an external calibration, the electrical signal can be expressed as a chemical compound concentration.

Commercial detectors either have large ionization chamber volumes around 50–100 µL [11,12] or membranes, which increase the ionization chamber fill time [8,13]. For photoionization detectors applied to GC, reducing the external size of the PID is not the main point to ensure high efficiency, because existing portable PIDs are already small (20 mm) and lightweight (8 g) [8,13]. The challenge, therefore, is to develop a micro PID (µPID) detector with a rapid ionization chamber fill time, low gas flow rate and high sensitivity that is compatible for the development of a portable µGC equipped with a small carrier gas cylinder. This can be achieved with a miniaturized flow-through PID ionization chamber.

Reducing the ionization chamber volume of the PID can play an important role in overall GC-PID miniaturization. A small ionization chamber can improve both the gas analyzer performance and the portability. Ionization chamber miniaturization results in a higher surface to volume ratio, which should translate into a more sensitive signal. A small ionization chamber should also result in a higher signal-to-noise ratio. In addition to that, the smaller the ionization chamber volume is, the smaller the gas sample volume can be (without depleting signal intensity), and for a fixed carrier gas flow rate, this means also a faster analysis time and lower consumption of the carrier gas. Low carrier gas consumption is important to enable reduction in carrier gas cylinder size (reducing gas analyzer weight) and maintain carrier gas cylinder autonomy, both important qualities for portable GC-PID.

Recent scientific publications reveal innovations in the miniaturization of photoionization detectors, also in reducing the size of the ionization chamber. Table 1 shows the main publications, the type of PID, materials, design features and ionization chamber volume. These improvements in the ionization chamber volume were made possible by using silicon or etching based microfabrication techniques. Such techniques are often complex and expensive. This work proposes a new µPID prototype that uses simpler fabrication techniques, allows easy mount-dismount of the device components, easy exchange of parts and small ionization chamber volume.

Table 1. Main works on microfluidic photoionization detector (µPIDs).

Reference	Ionization Source	Manufacturing Main Materials	Design Main Features and Dimensions	Ionization Chamber
[14]	UV Lamp, 10.6 eV	-	Introduced nozzle inside a conventional ionization chamber.	10 µL
[15]	UV Lamp, 10.6 eV	Highly doped p-type <100> single-sided polished conductive Si wafers with resistivity 0.001–0.005 Ω.cm and 380 µm thickness; 500 µm thick Pyrex glass wafers.	Ionization chamber is a microchannel with cross-section 150 µm (width), 380 µm (depth) and length 2.3 cm. Entire overall channel size is 15 mm × 15 mm. Microchannel area covered by lamp is 2.4 mm × 2.4 mm.	1.3 µL
[16]	UV Lamp, 10.6 eV	Conductive p-type <100> silicon wafer and glass.	Channel etched 380 µm (width) × 380 µm (depth) × 2 cm (length).	0.5 µL
[17]	Helium discharge	Silicon and glass architecture.	Micro separation column fabricated on the same chip. Overall size (1.5 cm × 3 cm)	Not mentioned
[18]	Helium discharge	500 µm thick p-type <100> double side polished Si wafer with 500 nm thick thermal oxide layers; 100 µm thick Borofloat 33 glass wafer; 500 µm thick Borofloat 33 glass wafer.	Microchannels formed by Si and glass. Three main channels: 1) Auxiliary helium; 2) Analytes; 3) Outlet channel. Cross-section 380 µm (width) and 500 µm (depth);	1.4 µL
[19]	Helium discharge	Two (bottom and top) Borosilicate glass wafers 700 µm thickness and 100 mm diameter used as substrate.	Channel etched 250 µm (depth).	Not mentioned
This work	UV Lamp, 10.6 eV	Micromilled PMMA and PVC. Copper plate.	Modular assembly of components. No use of glue. Microchannels width vary between 400 and 500 µm.	1.75–6.42 µL

2. Micro PID

This section describes the µPID design fabricated by micro milling and electrical discharge machining. The components are simply assembled together, dispensing any chemical or physical bonding process. Initially, the design, main components, fabrication and assembly are presented (divided in two sections: Core and shell), followed by a description of the electrode coating.

2.1. Design, Fabrication and Assembly

The detector has two main structures: The core, which is the heart of the device, responsible for the detection; and the shell, whose main functions are to clamp together all parts and enable the world-to-chip connections (electronic and fluidic).

2.1.1. Core

The components of the core when assembled yield the microchannel where the gas sample flows, which is the ionization chamber, where the detection of the sample occurs. The main parts of the core are numbered in Figure 1. The top and bottom poly(methyl methacrylate) (PMMA) are made using micro milling and the electrodes are fabricated with electrical discharge machining (both processes with tolerances ±0.02 mm). PMMA was chosen due to its transparency (to facilitate visual inspection of the assembly), its structural stability, chemical inertness to the VOC gas samples used and its low porosity (minimal sorption-desorption of the gasses passing through the ionization chamber). Copper was chosen for the electrodes owing to its high electrical conductivity and ready availability. The O-rings are made of Viton (ERIKS, Utrecht, The Netherlands), since its softness is required to avoid breaking the UV window which is fragile material (MgF$_2$). In addition, a commercial UV lamp (Baseline MOCON, Lyons, CO, USA) with 10.6 eV output was used.

The bottom layer of PMMA constitutes the bottom wall of the microchannel, which contains two micro-pores of 500 µm diameter to form the inlet/outlet of the channel. It also has two holes to fit the electronic connectors to the copper electrodes. The electrodes fit on top of the bottom PMMA to form the lateral wall of the microchannel. The elevation in the bottom PMMA serves both to fit the electrodes to the right position in order to constitute the channel and to electrically isolate the two sides of the electrodes. To enclose the microchannel (ionization chamber), the UV lamp is placed on top of the copper electrodes, two O-rings with 1 mm cross section fit in the pockets designed on the top

PMMA, which closes the channel and ensures that the lamp is fitted to the right position in the center of the fluidic path between the inlet/outlet pores. The top layer of PMMA is designed to fit tightly to the bottom PMMA layer to ensure alignment of the components and reduce the probability of leaks.

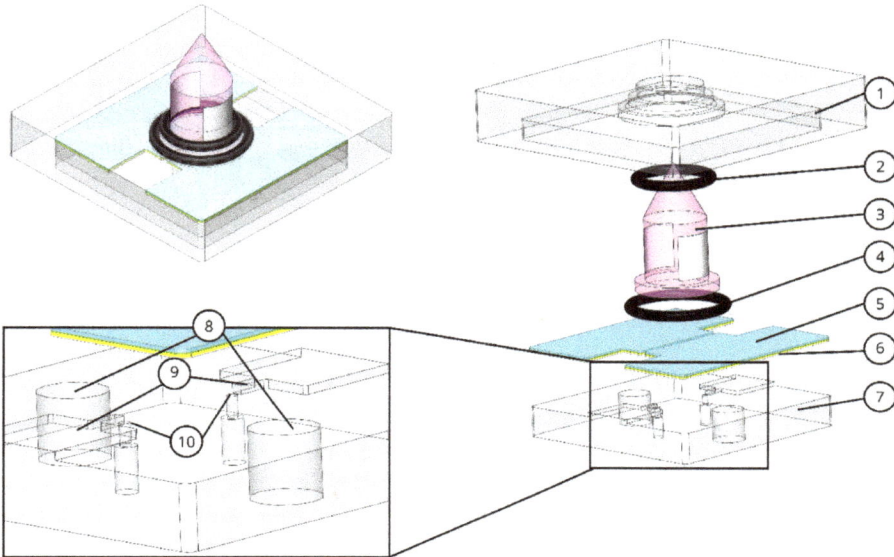

Figure 1. Assembled and exploded view of the detector core. (**1**) Top poly(methyl methacrylate) (PMMA), (**2**) O-ring 1, (**3**) UV lamp, (**4**) O-ring 2, (**5**) shield, (**6**) copper electrodes, (**7**) bottom PMMA, (**8**) electrode connection holes, (**9**) elevation structure, (**10**) micro-pores.

2.1.2. Shell

Another important structure is the shell of the device, which mechanically presses all the components together to minimize leakage and facilitates world-to-chip connectivity (fluidic and electronic). The electronic connections are to power the UV lamp, apply voltage on the electrodes and acquire the photoionization current. The fluidic connections connect the sample inlet and sample outlet. The main components of the shell are indicated in Figure 2. The top and bottom shells were made of PVC and fabricated using micro milling. The bottom shell has a groove to fit the spring and copper tape, which makes the electronic connection to the copper electrodes. A wire is soldered to the copper tape to connect the power supply and photoionization current signal acquisition. The bottom shell also has O-ring pockets (O-rings dimensions are: 3 mm outside diameter, 1 mm cross section and 1 mm internal diameter) on its top surface to connect the inlet/outlet gas sample to the ionization chamber. The O-rings are also made of Viton. The bottom surface of the bottom shell contains M3 threads to assemble the FESTO connector type QSM-M3-2 (Festo AG & Co. KG, Esslingen am Neckar, Germany), which is a plug in connector to 2 mm outside diameter tubing.

The top shell fits the top PMMA and positions the core so that it is aligned with the fluidic connections. It also encloses the UV lamp for safety measures. Two lateral pins are designed to slide freely and fit an electronic spring pin that reaches the lateral electrodes of the UV lamp. The spring allows freedom of the pin movement after touching the lamp, which is important to avoid impact and load being transferred to the fragile glass lamp. When the pins reach the right position, two M3 screws on top of the pins aid in fixing the pin to the right position. The lateral pins also ensure complete enclosing of the UV lamp to avoid its light. The top and bottom shells have passing holes for M3 screws that press all the parts at appropriate pressure to ensure fluidic sealing and structural stability, meaning no part would move unnecessarily during operation of the device.

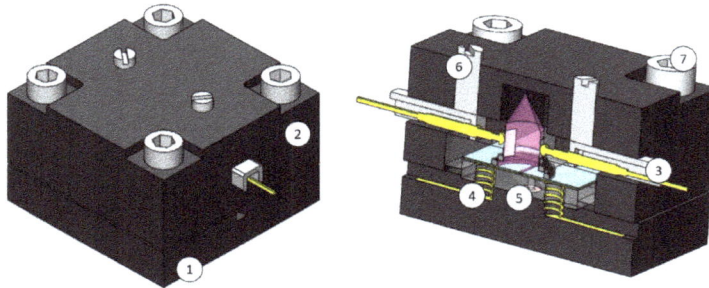

Figure 2. Isometric and cut view of the shell. (**1**) Bottom shell, (**2**) top shell, (**3**) lateral pin, (**4**) electrodes connection, (**5**) fluidic connections, (**6**) M3 pin fix screws, (**7**) M3 screw.

Using no bonding might give rise to undesirable leakage. However, in comparison with other μPID designs [15,18–22], the device presented here has the following advantages: (1) Easy prototyping, since the fabrication techniques are more straightforward than other approaches adopted in the literature; (2) Lower fabrication cost, this device does not require clean-room complex and high-cost facilities and processes. In addition, the materials used are cheaper compared to silicon wafers and it is easy to exchange different microchannel geometries, if needed; (3) Easy mount-dismount, which facilitates maintenance and ease of substitution of components such as the UV lamp (e.g., to use lamps with different energies).

2.2. Ionization Chamber

The ionization chamber of the μPID is the microchannel formed between the bottom PMMA, top PMMA, lamp and electrodes. This design proposed can produce an ionization chamber with volume up to 100 times lower compared to some commercial PIDs. The height of the channel, 500 μm, is the thickness of the copper plate and the width is initially chosen to keep the aspect ratio of the microchannel close to 1, since a wider channel could increase unnecessarily the chamber volume and a smaller width could reduce excessively the UV illumination area. The length and shape of the microchannel are designed to maximize the use of the illumination diameter of the commercial UV lamp (estimated illumination diameter ~6 mm).

Important channel properties are: (1) Width, representing the distance between the copper electrodes; (2) Height, directly proportional to the electrode area; (3) Electrode area, which depends on the shape and the height of the microchannel; (4) Lamp illumination area, surface of the channel illuminated by the UV lamp and (5) Volume of the channel from inlet to outlet, which corresponds to the ionization chamber volume. Figure 3 shows three different electrodes/channel designs and Table 2 presents their dimensions. The three shapes displayed have the objective to change gradually the area of the electrodes and ionization chamber volume.

Table 2. Ionization chamber properties.

Channel n	w [μm]	$A_{electrodes}$ [mm^2]	$V_{inlet/outlet}$ [μL]	A/A_1	V/V_1
1	500	5.98	1.75	1.0	1.0
2	500	12.67	2.75	2.1	1.6
3	400	31.17	6.42	5.2	3.7

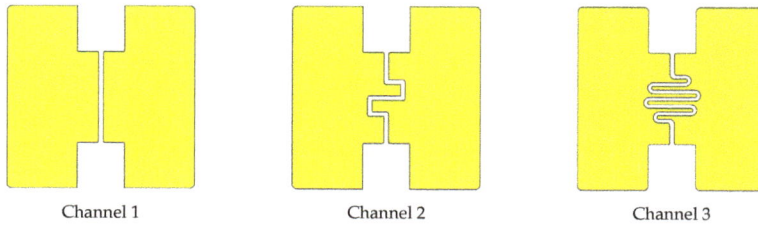

Channel 1   Channel 2   Channel 3

Figure 3. Microchannels for ionization chamber.

2.3. Coating Shield

When the copper electrodes are exposed to radiation at a frequency higher than the copper work function, electrons are ejected from the surface and cause signal noise. Therefore, it is desirable to protect the electrodes from the high-energy radiation coming from the UV lamp. A possible solution is to create an UV light coating shield on the copper surface. Ideally, a good shield will filter out all the radiation with an energy greater than the work function of the metal used as electrodes (in the case of copper, 4.48–4.94 eV [23]). In addition to that, it should be thin to ensure that the thickness of the coating does not increase the volume of the ionization chamber. Moreover, it must not cover the area of the electrode used for ion and electron collection. Other important characteristics are: Chemical compatibility with the electrode materials and gas samples, resistance to temperature oscillation, humidity and mechanical deformation. Two materials were considered in this paper for use as an electrode shield: Diamond-like carbon (DLC) and polymethyl methacrylate (PMMA).

2.3.1. DLC Coating

Diamond-like carbon is a metastable form of amorphous carbon that can be used as a protective coating. It can be applied as thin film with dimensions ranging from a few nm to μm [24]. Even quite thin films can block unwanted radiation, which renders it potentially suitable as an electrode coating in this application. A 0.1 μm thickness layer of DLC can have 5–18% UV transparency at 350–450 nm wavelength range and 0.2 μm DLC thickness yields 2–8 % UV transparency [24]. Also, DLC preferentially blocks short wavelengths [25,26].

A first sample of DLC coating was produced on one side of a 1 mm thick copper plate. Before the coating was applied, the surfaces were polished and etched (to remove impurities) for better adhesion of the DLC. The coating was made using plasma enhanced chemical vapor deposition and had 1 μm thickness. Figure 4 shows the difference between the coated and uncoated side. It can be seen that the DLC coating yields a grey surface on top of the copper. This sample started to peel off after a few weeks, likely due to exposure to sunlight and summer temperatures/humidity. Figure 4c shows the state of the peeled of sample. The scanning electron microscopy (SEM) images on Figure 5 shows that the DLC coating detached from copper like a peeled off thin foil, suggesting poor adhesion between the coating and substrate. A second sample was produced on the same type of substrate, this time incorporating a titanium adhesive layer between the DLC and copper to increase adhesion. The new sample had considerably better stability and did not fail (crack or break) during the test period.

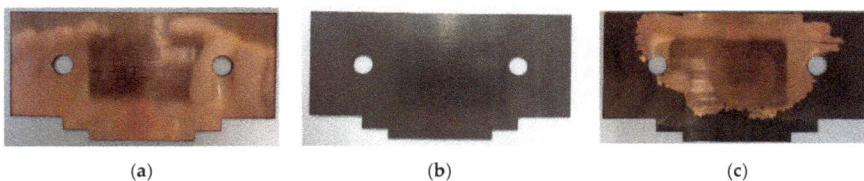

(a)   (b)   (c)

Figure 4. Diamond-like carbon (DLC) coating before and after failure. (a) Uncoated side before failure, (b) Coated side before failure, (c) Coated side after failure (4.5 weeks).

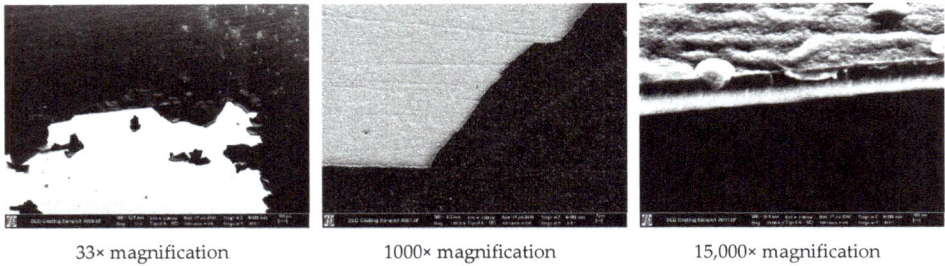

| 33× magnification | 1000× magnification | 15,000× magnification |

Figure 5. Scanning electron microscopy (SEM) images of the DLC coating failure.

2.3.2. PMMA Coating

The second type of electrode coating considered is PMMA because it does not transmit wavelengths below 360 nm [27], which should be sufficient to avoid photoelectric effects in the copper electrodes. PMMA is a relatively cheap material compared to DLC and can be deposited as a top layer on the electrodes by spin coating, a process that is not expensive and does not require complex machinery. Disadvantages to be considered are: (1) Difficulty to produce perfectly flat surfaces to avoid leakage; (2) difficulty to produce very thin layers (< 0.1 µm) and (3) during fabrication, the PMMA might run inside of the channel, isolating the electrodes from the ions, then reducing the active area inducing a lower signal.

3. Results

Toluene was chosen as the representative compound of the VOCs for all the following experiments aiming at characterizing the µPID signal where different parameters were changed, such as voltage applied to the electrodes or total gas flow rate.

3.1. Experimental Setup and Material

An experimental setup was built to evaluate the µPID prototype performance, represented in Figure 6. A gas sample supply, containing either pure nitrogen or toluene 100 ppm was connected to a mass flow controller (MFC), Bronkhorst El Flow Select F-201CV-050-AAD-33-V (Bronkhorst High-Tech BV., AK Ruurlo, The Netherlands), with a full scale of 20 mL/min and error of 0.5% of reading plus 0.1% of full scale. The MFC regulates the flow rate of gas sample into the µPID, and the flow rate leaving the device is read by a mass flow meter (MFM), Bronkhorst El Flow Select F-100D-AAD-33-V, full scale 20 mL/min and same error as the MFC. To power up the UV lamp, the lateral electrodes of the lamp are plugged inside the lamp drive circuit of the commercial Baseline MOCON PID. This circuit is powered by a power supply at 4V and yields a current of about 35 mA. The power supply was also used to provide the voltage on the electrodes in the ionization chamber and the digital electrometer Keithley 616A (Keithley Instruments, Solon, OH, USA) was used to measure the current signal.

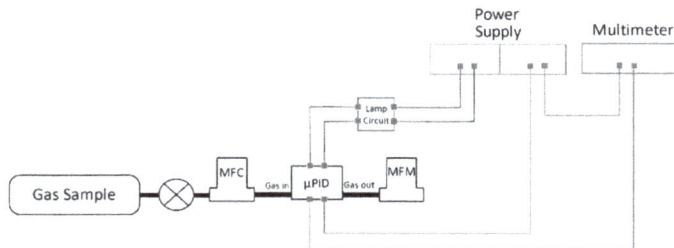

Figure 6. Experimental setup used for signal measurements from µPID prototype.

3.2. Optical Microscope Images

In order to investigate the effects of the fabrication and design, digital optical microscope images were taken from the microchannel formed by the electrodes and the bottom PMMA. Figure 7a shows the image of channel 1 previously used in experiments with the UV lamp. It is possible to notice round circles marked by the use of the lamp, the central circle suggest that UV light is more intense within a diameter of approximately 4.5 mm. The image zoomed on the fitting of PMMA and electrodes (Figure 7b) reveal imperfections of the fitting between them, which can lead to gaps ranging from 15–60 μm. The consequence of those gaps will be leakage and an undesirable variation in the distance between the electrodes. The fitting of electrodes presented in Figure 7a presented an average channel width of 504 μm (5.5% error), while the channel 2 shown in Figure 7c presented considerable deviations in the nominal value of channel width, varying from 454 μm to 548 μm.

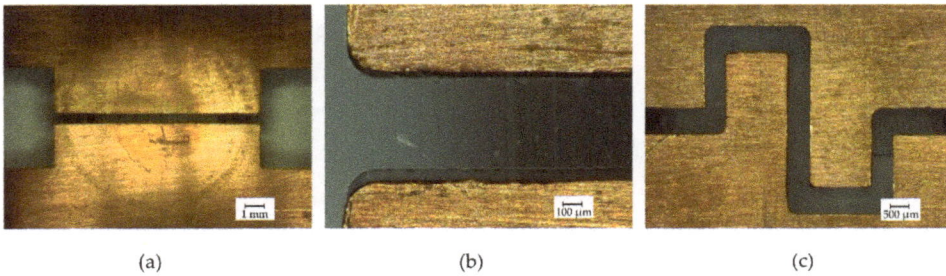

(a) (b) (c)

Figure 7. Digital optical microscope images of copper electrodes assembled on bottom polymethyl methacrylate (PMMA). (**a**) Channel 1; (**b**) Zoom at bottom PMMA and copper electrodes assembly; (**c**) Channel 2.

3.3. Experimental Results and Discussion

The signals from the μPID were obtained without an amplifier, which should be developed in the future in order to improve the signal/noise ratio and thus the sensitivity. The multimeter (Keithley 616A) acquired μPID signals ranging from about 70 pA to 3.5 nA. The current standard uncertainty is calculated considering two main factors: (i) Multimeter resolution (R), which varies from 0.1 pA to 10 pA depending on the current measurement range and (ii) measurement repeatability, thus the combined standard uncertainty is given by

$$u_C^2 = u_1^2 + u_2^2 \tag{1}$$

The standard uncertainty calculated due to resolution and repeatability are obtained with

$$u_1 = R/\sqrt{3} \tag{2}$$

and

$$u_2 = \sigma(X)/\sqrt{n} \tag{3}$$

where $\sigma(X)$ is the standard deviation of the sample and n are the number of samples. The expanded uncertainty (U) is calculated by

$$U = t \cdot u_C \tag{4}$$

where t is the Student's distribution factor for 95% confidence interval with v degrees of freedom calculated by the Welch-Satterthwaite Equation.

Figure 8 shows the μPID signal for channel 3 using toluene 100 ppm gas sample when voltage and flow rate are changed (leakage as a function of flow rate is presented in the supplementary file Figure S1). A range of voltages was applied from 0.1 V to 30 V and flow rates of 0.5, 10 and

20 mL/min were used. To obtain the data corresponding to the three flow rates, the μPID signal was stabilized, which lasted about 2h for 0.5 mL/min and 1h for 10 mL/min and 20 mL/min, this time being necessary to remove humidity traces which may perturb the signal. After stabilization, the voltage was varied from 0.1 to 30 V for each flow rate. Each point displayed on Figure 8 corresponds to an average of 10 signal measurements taken within 2 min interval at each flow rate and voltage. The repeatability is calculated individually for each point and corresponds to the standard deviation of the 10 measurements. The relative uncertainty of the μPID signal ($U/Signal$) varies from 7.5% to 0.1% at low and high voltages respectively.

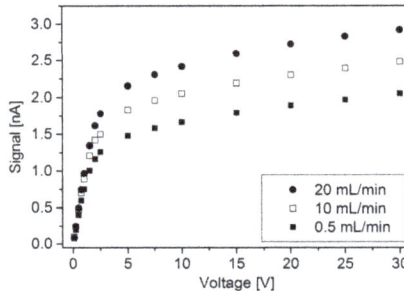

Figure 8. Channel 3 signal with variation of voltage on copper electrodes for various gas sample flow-rates. Uncertainty bars are not displayed in the graphic since their size is small compared to the symbols representing the points.

The three flow rate datasets (0.5, 10 and 20 mL/min) present similar curve shape and the signal increases with flow rate. Two linear regions of the data can be identified: (i) 0.1 V to 2.5 V and (ii) 5V to 30 V. The slope of the first region is ~ 25 higher than the second region. At 5 V, signals for 0.5, 10 and 20 mL/min are about 1.5, 1.8 and 2.2 nA, meaning that low flow rates might not be so detrimental for signal generation.

For gas analyzer portability, low flow rate and low voltage are desirable specifications, which means longer duration of portable carrier gas cylinder and longer battery life. As an initial analysis from Figure 8, a voltage of 5 V and flow rate 2 mL/min, might be a good choice to yield best cost benefit of portability vs. μPID signal level. However, since the currents generated by the device are very low, and those experiments were performed without a signal amplifier, the flow rate and voltages yielding the highest signal possible were chosen for the rest of the work and in particularly to obtain the results shown in Figures 9 and 10.

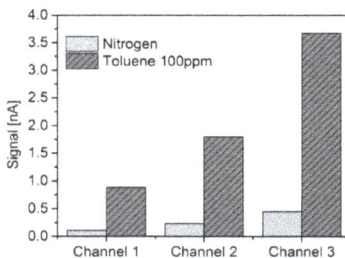

Design	Relative error	
	Nitrogen	Toluene 100 ppm
Channel 1	1.9 %	0.3 %
Channel 2	0.8 %	0.2 %
Channel 3	0.4 %	0.4 %

Figure 9. Signal for Channels 1, 2 and 3 using nitrogen and toluene 100 ppm as gas samples. Flow rate 20 mL/min and 30 V on electrodes. Relative uncertainty for signals of Channel 1 with nitrogen and toluene 100 ppm are 1.9% and 0.3%. Relative uncertainty for signals of Channel 2 with nitrogen and toluene 100 ppm are 0.8% and 0.2% respectively. Relative uncertainty for signals of Channel 3 with nitrogen and toluene 100 ppm are both 0.4%.

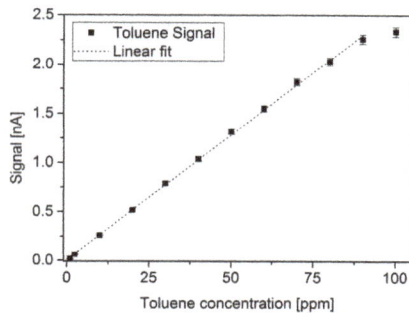

Figure 10. Channel 3 signal at 1–100 ppm toluene range. 20 mL/min and 30 V applied. Relative error is 2% for all data points.

In order to compare the performance of the three channels fabricated, Figure 9 presents the current signal for nitrogen and toluene 100 ppm for each of the three channels. The gas sample was injected at 20 mL/min and 30 V was applied on the electrodes. The current was measured after signal stabilization and five measurements were taken, the values displayed on Figure 9 are the average. The same procedure was adopted to calculate the uncertainties of this data. However, the repeatability corresponds to the average of the five measurements taken after stabilization. The values of uncertainties are given in the table of Figure 9 and the highest yielded value was 1.9%.

When comparing the signal for the three channels, it is evident that higher area yields higher signal. Also, the difference between the nitrogen and toluene 100 ppm signal increases, meaning higher sensitivity and possible better detection limit. Channels 1 and 2 have both 500 μm as nominal channel width, and it is interesting to notice that the proportion of area increase from channel 1 to channel 2 is similar to the signal increase observed, about 2 times higher. It is important to note that the signal of channel 3 for toluene 100 ppm was taken on a different day, yielding a level slightly higher than that observed in Figure 8.

Because the signal is higher with channel 3, the following measurements were performed in this configuration. Figure 10 shows signal of channel 3 for various toluene concentrations, ranging from 1 to 100 ppm. Again, the same conditions of voltage and flow were applied to get this data (30 V, 20 mL/min). Before and after measuring the signal of toluene, the μPID was purged with nitrogen. The data shown corresponds to the values of the signal for toluene at each concentration subtracted by the average of the purge signal (pure N_2) obtained before and after the toluene signal acquisition. It can be noticed that up to 90 ppm, the response of the μPID is linear and a linear fit can be obtained ($r^2 = 0.9996$), which is presented in the graphic. The relative error is calculated with same procedure as in Figures 8 and 9, yielding a maximum of error of 2%.

To estimate the detection limit, the maximum variation of the blank within a 2 min interval was obtained (after stabilizing) $\Delta S_{max} = 0.005$ nA. The detection limit for toluene is the concentration corresponding to $3 \times \Delta S_{max}$ (0.015 nA), which is estimated from a rule of three with the lowest concentration investigated, i.e., 2.5 ppm and 0.063 nA from the graph. The resulting estimated detection limit is therefore 0.6 ppm.

4. Conclusions

This paper presents the design, fabrication and characterization of micro milled μPID. The design's simple construction allows easy mount-dismount for components change, cleanse or replacement. The ionization chamber volume ranges from 1.75 to 6.42 μL, depending on the electrodes. Two materials have been proposed for electrode shielding. DLC needs to have a titanium adhesive layer for improved stabilization on the surface of copper electrodes. PMMA structures on copper electrodes are stable. However, the fabrication technique should be improved for adhesion only on top surface of the electrodes.

Optical microscope images of microchannels shows that electrodes might not fit well on bottom PMMA, possibly leading to distortion in the width of the microchannel and moderate leakage. μPID signal level increases with flow rate, voltage and electrode area. Channel 3 has highest signal and electrode area. However, it has the highest ionization chamber volume. The device presents an estimated detection limit of 0.6 ppm for toluene. Future design improvements should be done to mitigate leakage, improve electrodes fitting on bottom PMMA. Future evaluations will be done on response time, residence time, possible adsorption of analytes on the interior surfaces, influence of pressure and humidity on the signal. The device will also be compared to commercial detectors and coupled to a separation column and pre-concentrator.

Supplementary Materials: The following is available online at http://www.mdpi.com/2072-666X/10/4/228/s1, Figure S1: Leakage for three channel designs. Uncertainty varies from 1% at high flow rates to 5% at low flow-rates.

Author Contributions: Conceptualization, G.C.R., D.N., S.L.C. and J.J.B.; Data curation, G.C.R., D.N., S.L.C. and J.J.B.; Formal analysis, G.C.R., D.N. and S.L.C.; Funding acquisition, D.N., S.L.C. and J.J.B.; Investigation, G.C.R.; Methodology, G.C.R., D.N., S.L.C. and J.J.B.; Project administration, D.N., S.L.C. and J.J.B.; Resources, D.N., S.L.C. and J.J.B.; Supervision, D.N., S.L.C. and J.J.B.; Visualization, G.C.R.; Writing—original draft, G.C.R.; Writing—review & editing, G.C.R., D.N., S.L.C. and J.J.B..

Funding: This study was developed during the ITN Research Project, MIGRATE, supported by European Community H2020 Framework under the Grant Agreement No. 643095.

Acknowledgments: The authors are grateful to the Bernal Institute, the University of Limerick, Karlsruhe Institute of Technology, Centre national de la recherche scientifique (CNRS), the University of Strasbourg and In'Air Solutions.

Conflicts of Interest: The authors declare no conflict of interest.

References

1. Wallace, L.A.; Pellizzari, E.; Leaderer, B.; Zelon, H.; Sheldon, L. Emissions of Volatile Organic Compounds from Building Materials and Consumer Products. *Atmos. Environ.* **1987**, *21*, 385–393. [CrossRef]
2. WHO. *Air Quality Guidelines for Europe*, 2nd ed.; WHO Regional Office for Europe: Copenhagen, Denmark, 2000.
3. WHO. *WHO Guidelines for Indoor Air Quality: Selected Pollutants*; WHO Regional Office for Europe: Copenhagen, Denmark, 2010.
4. Council of the European Union; European Parliament. DIRECTIVE 2000/69/EC OF THE EUROPEAN PARLIAMENT AND OF THE COUNCIL of 16 November 2000 Relating to Limit Values for Benzene and Carbon Monoxide in Ambient Air. *Off. J. Eur. Communities* **2000**, *313*, 12–21.
5. Ministère de l'Ecologie. Décret No 2011-1727 Du 2 Décembre 2011 Relatif Aux Valeurs-Guides Pour l'air Intérieur Pour Le Formaldéhyde et Le Benzène. *J. Off. de la République Française* **2011**.
6. BTEX Analyzer with PID Detector: ChromaPID. Available online: http://www.chromatotec.com/BTEX,analyzer,with,PID,detector,chromaPID-Article-138-ChromaGC-Product-14.html (accessed on 19 July 2018).
7. BTEX (Model GC955-600) Benzene/Toluene/Xylene Analyser. Available online: http://www.et.co.uk/products/air-quality-monitoring/continuous-gas-analysers/synspec-gc955-601-btx/ (accessed on 19 July 2018).
8. Baseline MOCON. Available online: http://www.baseline-mocon.com (accessed on 6 July 2018).
9. Khan, S.; Newport, D.; Le Calvé, S. Development of a Toluene Detector Based on Deep UV Absorption Spectrophotometry Using Glass and Aluminum Capillary Tube Gas Cells with a LED Source. *Micromachines* **2019**, *10*, 193. [CrossRef] [PubMed]
10. Rezende, G.C.; Le Calvé, S.; Brandner, J.J.; Newport, D. Micro photoionization detectors. *Sens. Actuators B Chem.* **2019**, *287*, 86–94. [CrossRef]
11. Model 4430, IO Analytical. Available online: http://aimanalytical.com/Manuals/PIDFIDmanual.pdf (accessed on 20 July 2018).
12. Photoionization Detector, SRI Instruments. Available online: https://www.srigc.com/home/product_detail/pid---photo-ionization-detector (accessed on 4 October 2018).

Micromachines **2019**, *10*, 228

13. Ion Science. Available online: https://www.ionscience.com/ (accessed on 6 July 2018).
14. Sun, J.; Guan, F.; Cui, D.; Chen, X.; Zhang, L.; Chen, J. An Improved photoionization detector with a micro gas chromatography column for portable rapid gas chromatography system. *Sens. Actuators B Chem.* **2013**, *188*, 513–518. [CrossRef]
15. Zhu, H.; Nidetz, R.; Zhou, M.; Lee, J.; Buggaveeti, S.; Kurabayashi, K.; Fan, X. Flow-through microfluidic photoionization detectors for rapid and highly sensitive vapor detection. *Lab Chip* **2015**, *15*, 3021–3029. [CrossRef] [PubMed]
16. Lee, J.; Zhou, M.; Zhu, H.; Nidetz, R.; Kurabayashi, K.; Fan, X. Fully Automated Portable Comprehensive 2-Dimensional Gas Chromatography Device. *Anal. Chem.* **2016**, *88*, 10266–10274. [CrossRef] [PubMed]
17. Akbar, M.; Shakeel, H.; Agah, M. GC-on-Chip: Integrated column and photoionization detector. *Lab Chip* **2015**, *15*, 1748–1758. [CrossRef] [PubMed]
18. Zhu, H.; Zhou, M.; Lee, J.; Nidetz, R.; Kurabayashi, K.; Fan, X. Low-power miniaturized helium dielectric barrier discharge photoionization detectors for highly sensitive vapor detection. *Anal. Chem.* **2016**, *88*, 8780–8786. [CrossRef] [PubMed]
19. Narayanan, S.; Rice, G.; Agah, M. A Micro-discharge photoionization detector for micro-gas chromatography. *Microchim. Acta* **2014**, *181*, 493–499. [CrossRef]
20. Narayanan, S.; Rice, G.; Agah, M. Characterization of a micro-helium discharge detector for gas chromatography. *Sens. Actuators B Chem.* **2015**, *206*, 190–197. [CrossRef]
21. Akbar, M.; Restaino, M.; Agah, M. Chip-scale gas chromatography: from injection through detection. *Microsyst. Nanoeng.* **2015**, *1*, 15039. [CrossRef]
22. Zhou, M.; Lee, J.; Zhu, H.; Nidetz, R.; Kurabayashi, K.; Fan, X. A fully automated portable gas chromatography system for sensitive and rapid quantification of volatile organic compounds in water. *RSC Adv.* **2016**, *6*, 49416–49424. [CrossRef]
23. Haynes, W.M. *CRC Handbook of Chemistry and Physics*, 97th ed.; CRC Press: Boca Raton, FL, USA, 2016.
24. Rahman, F. *Nanostructures in Electronics and Photonics*, 1st ed.; CRC Press: Boca Raton, FL, USA, 2008.
25. Mednikarov, B.; Spasov, G.; Pirov, J.; Sahatchieva, M.; Popov, C.; Kulischa, W. Optical properties of diamond-like carbon and nanocrystalline diamond films. *J. Optoelectron. Adv. Mater.* **2005**, *7*, 1407–1413.
26. Lin, C.R.; Wei, D.H.; Chang, C.K.; Liao, W.H. Optical properties of diamond-like carbon films for antireflection coating by RF magnetron sputtering method. *Phys. Procedia.* **2011**, *18*, 46–50. [CrossRef]
27. Joram, C. *Transmission Curves of Plexiglass (PMMA) and Optical Grease*; PH-EP-Tech-Note-2009-003; CERN Document Server: Meyrin, Switzerland, 2009.

micromachines

MDPI

Article

Sub-ppb Level Detection of BTEX Gaseous Mixtures with a Compact Prototype GC Equipped with a Preconcentration Unit

Irene Lara-Ibeas [1,2], Alberto Rodríguez-Cuevas [3], Christina Andrikopoulou [1], Vincent Person [3], Lucien Baldas [2], Stéphane Colin [2] and Stéphane Le Calvé [1,3,*]

[1] ICPEES UMR 7515, Université de Strasbourg/CNRS, F-67000 Strasbourg, France;
   ilaraibeas@unistra.fr (I.L.-l.); candrikopoulou@unistra.fr (C.A.)
[2] Institut Clément Ader (ICA), Université de Toulouse/CNRS, INSA, ISAE-SUPAERO, Mines-Albi, UPS,
   31400 Toulouse, France; baldas@insa-toulouse.fr (L.B.); stephane.colin@insa-toulouse.fr (S.C.)
[3] In'Air Solutions, 25 rue Becquerel, 67087 Strasbourg, France; arodriguez@inairsolutions.com (A.R.-C.);
   vperson@inairsolutions.fr (V.P.)
*   Correspondence: slecalve@unistra.fr; Tel.: +33-3-6885-0368

Received: 8 February 2019; Accepted: 7 March 2019; Published: 13 March 2019

Abstract: In this work, a compact gas chromatograph prototype for near real-time benzene, toluene, ethylbenzene and xylenes (BTEX) detection at sub-ppb levels has been developed. The system is composed of an aluminium preconcentrator (PC) filled with Basolite C300, a 20 m long Rxi-624 capillary column and a photoionization detector. The performance of the device has been evaluated in terms of adsorption capacity, linearity and sensitivity. Initially, PC breakthrough time for an equimolar 1 ppm BTEX mixture has been determined showing a remarkable capacity of the adsorbent to quantitatively trap BTEX even at high concentrations. Then, a highly linear relationship between sample volume and peak area has been obtained for all compounds by injecting 100-ppb samples with volumes ranging from 5–80 mL. Linear plots were also observed when calibration was conducted in the range 0–100 ppb using a 20 mL sampling volume implying a total analysis time of 19 min. Corresponding detection limits of 0.20, 0.26, 0.49, 0.80 and 1.70 ppb have been determined for benzene, toluene, ethylbenzene, m/p-xylenes and o-xylene, respectively. These experimental results highlight the potential applications of our device to monitor indoor or outdoor air quality.

Keywords: preconcentrator; microfluidics; miniaturized gas chromatograph; BTEX; PID detector

---

1. Introduction

In recent years, there has been an increasing interest in air pollution since numerous studies have demonstrated its impact on human health [1–3]. Among the broad variety of identified air contaminants, benzene, toluene, ethylbenzene and xylenes (BTEX) require particular attention not only for their toxic, mutagenic, and/or carcinogenic effects but also for their key role in photochemical reactions. These compounds can be emitted by several indoor as well as outdoor sources. In outdoor air, the main sources are related to fuel storage or combustion processes such as road traffic, petrol stations or industrial activities [4–7]. In closed environments, most of BTEX are emitted by cleaning products, furnishings as well as building materials like flooring materials, wall coverings, adhesives, varnishes or paints [8–10]. Smoking and cooking are also considered as other major sources of BTEX in indoor air [11,12]. Numerous investigations have demonstrated the harmful effects of BTEX on human health even at low concentrations [13]. Fatigue, loss of coordination, memory problems, headache, skin and eyes irritation as well as other more severe effects like asthma, kidney damage or neurological problems have been linked to BTEX exposure [14,15]. In addition, benzene is a known carcinogen compound, with long-term exposure being associated with the development of leukaemia [16]. Thus, in 2013, a threshold limit value of 5 μg m^{-3} (1.6 ppb) for benzene was set by the European Union in public buildings. In France, this limit was decreased to 2 μg m^{-3} (0.6 ppb) in 2016. Therefore, to check if indoor air quality (IAQ) is in accordance with the new legislation, on-site rapid and sensitive analysis is required.

So far, methods based on gas chromatography have typically been the techniques employed for BTEX analysis in indoor and outdoor air. Benchtop chromatographs are sensitive and accurate instruments enabling BTEX detection in the order of parts per trillion (ppt). However, their large size, heavy weight and high energy consumption limit their use for in-situ measurements. Consequently, many air quality measurement campaigns have been conducted using commercial sampling cartridges [8,17–21] that were subsequently analyzed in a laboratory increasing not only the total analysis time but also the risk of sample degradation during storage or transport. Furthermore, these off-line analyses do not provide concentration-time profiles since each measurement represents an average value of pollutant concentration over the selected sampling time.

During the past two decades, great efforts have been made to develop real-time sensitive miniaturized gas chromatographs. Portable instruments containing micro-electro-mechanical systems (MEMS)-based components have grown in popularity due to their small dimensions, low energy consumption and short time of analysis. Table 1 summarizes the most remarkable commercially available and laboratory prototypes of miniaturized gas chromatographs (GC) developed in the last decade. All devices presented are highly portable (weight < 5 kg) enabling on-site measurements. Most of these instruments perform BTEX analysis in less than 15 min providing high-resolution concentration-time profiles. Nevertheless, these analyzers have a detection limit of a few ppb, which is not enough to comply with the requirements of the new regulation. So far, only the GC-metal oxide semiconductor (MOS) reported by Zampolli et al. [22] and the GC- photoionization detector (PID) developed by Skog et al. [23] were able to detect concentrations below 1 ppb. High sensitivities are usually correlated with long sampling times. In the first case, 55 min of sampling and around 12 min for separation were needed to achieve sub-ppb detection of BTEX [22]. Shorter sampling (20 min) and similar separation (15 min) times were required by Skog et al. [23] to reach limits of detection (LOD) in the order of ppt. Both devices imply a considerable long total analysis time, hindering the possibility to provide near real-time BTEX concentrations. Regarding the MEMS-based instruments available in the literature, a sensitive and rapid BTEX analysis remains a challenging issue.

The present work reports the development of a compact GC prototype for near real-time BTEX detection at sub-ppb levels and its validation under controlled laboratory conditions. The device is based on the miniaturized GC developed by Nasreddine et al. [24] where a preconcentration unit has been added in order to significantly improve its sensitivity. An improvement ratio between 1.55 and 4.46 in detection limit was achieved for the different compounds compared to the previous version. To our knowledge, this is the first work to report a GC prototype able to perform sub-ppb BTEX levels detection in such a short analysis time.

The paper is organized as follows: Section 2 describes the GC working principle and presents the experimental setup for BTEX generation employed in this project. Section 3 presents the performances of the GC prototype in reference to adsorption capacity, linearity and sensitivity. Section 4 summarizes the conclusions and proposes several improvements for the future work.

Table 1. Most representative compact gas chromatographs (GC) for BTEX analysis over last decade.

Ref.	Size (cm)	Weight (kg)	Sampling Time (min)	Analysis Time (min)	Preconcentrator l (mm) × d (µm) × w (mm)	Adsorbent	GC Column l (m) × d (µm) × w (µm)	Stationary Phase and Thickness	Carrier Gas Flow Rate (mL/min)	Detector	LOD (ppb)
This work	32 × 29 × 14	~5	4	15	Cavity 4.6 × 350 × 7.4	Basolite C300 5.8 mg	Capillary 20 × 180 (i.d.)	Rxi 624 Sil MS 1 µm	N_2 2.5	PID	0.1–1.6 (BTEX)
GC-PID [23]	31 × 30 × 20	32	20	~15	Tube 0.165 cm i.d.	ResSil-B 75 mg	Capillary 15 × 530 (i.d.)	MXT-1 3 µm	A.A. 0.8–2.2	PID	0.002–0.011 (BTEX)
GC-MOS [25]	n. d.	n. d.	5	4	Cavity with micro-pillars 10 × 400 × 5	Zeolite DaY ~ 13 µm	Circular spiral 5 × 100 × 100	PDMS 100 nm	7	MOS	24 (toluene) 5 (o-xylene)
GC-PID [26]	n. d.	n. d.	1	5 (5 comp.)	4 Parallel channels n.d × 400 × 0.6	SWNTs 0.15 mg	Serpentine with micropillars 4 × 350 × 320	OV-101 0.2 µm	A.A. 5	PID	<1 (benzene)
GC-MOS [27]	n. d.	n. d.	n. d.	60 (3 comp.)	n. p.	n. p.	Serpentine 1.6 × 1200 × 600	Porapak Q	A.A.	MOS	5 (benzene)
PEMM-1 [28]	19 × 30 × 14	3.5	1	4 (17 comp.)	2 Cavities (V ~ 9.4 µL) 380 (d)	C-B 2.0 mg C-X 2.3 mg	Square spiral 3.1 × 240 × 150	PDMS 0.20 µm	He 3	5 µCR	420–890 (BTEX)
Frog 4000 [29]	25 × 19 × 37	<2.2	0.5	5	n. d.	Silica gel aerogel	4.8 *	PDMS 0.8 µm	A.A.	PID	~ppb
GC-PID [9]	n. d.	n. d.	50	13	n. d.	EtOxBox 10 mg	n. p.	n. p.	A.A. 30	PID	1.25 (benzene)
GC-PID [31]	60 × 50 × 10	<5	2	14.2 (50 comp.)	Cavity 8.15 × 250 × 2.9	C-B 1.135 mg	1D: 10 × 250 (i.d.) 2D: 3 × 250 (i.d.)	1D: Rtx-5MS 2D: Rtx-200 0.25 µm	He 2	µPID	n. d.
GC-CR [32]	20 × 15 × 9	2.1		2.5 min (9 comp.)	2 Cavities (V ~ 9.4 µL)	C-B 2.0 mg C-X 2.3 mg	6*	PDMS 0.2 µm	n. d.	µCR	n. d.
GC-CMOS [33]	16 × 11 × 11	n. d.	n. d.	n. d.	Cavity with micro-pillars 10 × 250 × 2	Carbon film	Square spiral 3 × 250 × 100	DB-1	n. d.	CMOS	15 (1,3,5-TMB)
iGC32 [34]	8 × 10	n. d.	120	10	U shape n.d. × 300 × 1350	C-B + C-X	2 Serpentines 0.30 × 230 (i.d.)	OV-1 0.2 µm	A.A. 0.2	2 CD	10-2 (BTEX)
Zebra GC [30]	15 × 30 × 10	~1.8	10	<2	Cavity with micro-pillars 13 × 240 × 13	Tenax TA ~ 200 nm	Serpentine 2 × 70 × 240	OV-1 ~250 nm	He 1	TCD	~25 (TEX)
GC-PID [24]	32 × 29 × 14	~4	1	10	n. p.	n. p.	Capillary 20 × 180 (i.d.)	Rxi 624 Sil MS 1 µm	N_2 2.5	PID	0.8–3.2 (BTEX)
GC-MOX [22]	n. d.	n. d.	55	~12	Ten parallel channels 800 µm depth	QxCav	Square spiral 0.5 × 900 × 900	Carbograph 2 0.2% Carbowax	A.A. 15	MOS	0.1 (benzene)

i.d.: internal diameter, comp.: compounds. n. d.: not defined, n. p.: not present, * only column length is reported. A.A: ambient air. MOS: metal oxide semiconductor. CR: chemiresistor detector. CD: capacitive detector. CMOS: complementary metal oxide sensor. MOX: metal oxide sensor. QxCav: quinoxaline bridged cavitand.

2. Materials and Methods

2.1. Prototype of Micro Gas Chromatograph (GC)

The GC laboratory prototype employed in this study is presented in Figure 1. The system is based on the GC laboratory prototype developed by Nasreddine et al. [24]. In the present work, a preconcentration module has been integrated into the system enabling an improvement in sensitivity. The detailed operating principle of the GC is described in [24] so that only a brief description of the overall system is given below while the preconcentration module developed in the present work is highly detailed in the next section. The system operates according to four steps: sampling, preconcentration, separation and detection. Samples are collected and introduced in the preconcentrator (PC) by means of a SP 570 EC-BL micropump (Schwarzer Precision, Germany) and a EL-FLOW flow controller (Bronkhorst, Ruurlo, Netherlands). Separation step is carried out using a commercial 20-m long capillary column (i.d. 0.18 mm, RXi-624 stationary phase, 1 μm film thickness, Restek, Bellefonte, PA, USA). Polydimethylsiloxane (PDMS) columns are typically used for BTEX separation because as they are non-polar, they provide high resolution but long separation times as well. In environmental monitoring, a compromise between short analysis time and reasonable resolution is required, therefore, a slightly more polar column (Rxi-624Sil MS) was selected. Detection is conducted employing an eVx Blue mini photoionization detector (PID) (Baseline MOCON, Lyons, CO, USA) equipped with a 10.6 eV ultra-violet lamp. This GC laboratory prototype was controlled by a computer using a homemade software.

In a previous study [24], injection time was set to 20 s to transfer the sampling loop content to the column. In this new version, injection time of 80 s has been fixed according to the temperature ramp of the heating system, where 150 °C were reached in 60 s and were maintained for 20 s. After desorption, the six port-valve returned to the initial position to start the next sampling.

In the aforementioned work [24], flow rate of 2.5 mL min^{-1} and 80 °C were selected as the optimal conditions for BTEX separation. Using these conditions, BTEX analysis was performed in 10 min and detection limits between 1–3 ppb were found for the different compounds. Analytical performances of this device were validated under controlled laboratory conditions and in real environments [36,37]. In the present version, as the enriched peaks are larger, temperature of the column had to be reduced of 10 °C to avoid coelution between ethylbenzene and m/p-xylenes peaks; therefore, the total analysis time was increased to 15 min.

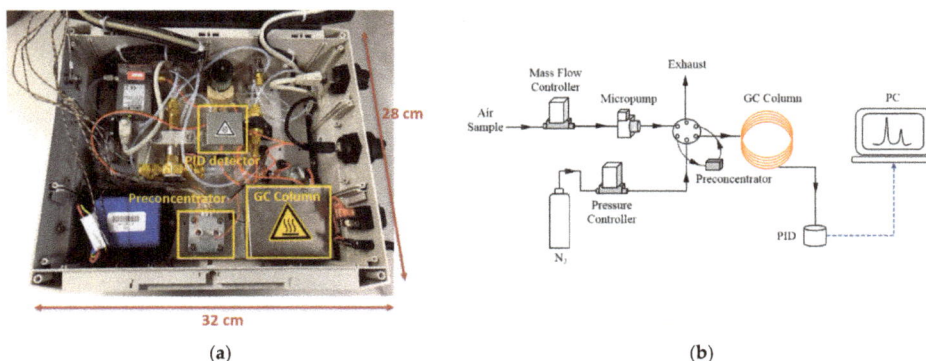

(a)   (b)

Figure 1. (**a**) Photograph of the compact gas chromatographs (GC) prototype and (**b**) schematic view of the device updated with a preconcentration unit.

2.2. Preconcentration Module

An aluminium PC with dimensions of 40 mm × 40 mm × 12.3 mm and weight of 54.9 g was manufactured and integrated into the GC laboratory prototype. The device is presented in Figure 2.

The design is based on the preconcentrator proposed by Camara et al. [38] in which a symmetrical manifold fluidic system was fabricated at the inlet and outlet of the adsorbent cavity to promote a uniform flow distribution. However, there are some differences between both designs. In this PC, two metal porous filters (GKN Sinter Metals, Bonn, Germany) are located between the channel system and the microfluidic cavity to ensure that the adsorbent remains in the cavity and to prevent clogging of the microchannels. The manifold consists of one inlet channel split in two channels of 350 μm which are also split to obtain finally four channels of 300 μm connected to a microfluidic cavity of 4.6 mm × 7.4 mm (see Figure 2b) where the adsorbent is placed. In this cavity, 5.8 mg of Basolite was packed manually. Heating system consists of three heating cartridges (Watlow, St. Louis, MO, USA) of 70 W each. The system allows to reach a temperature ramp of about 150 °C/min. In the experiments, desorption was performed at 150 °C and this temperature was maintained for 20 s. Temperature is measured with a type K thermocouple (RS Components SAS, Beauvais, France). Prior to the beginning of the experiments, the adsorbent was conditioned at 180°C under a nitrogen flow of 10 mL/min during 2 h.

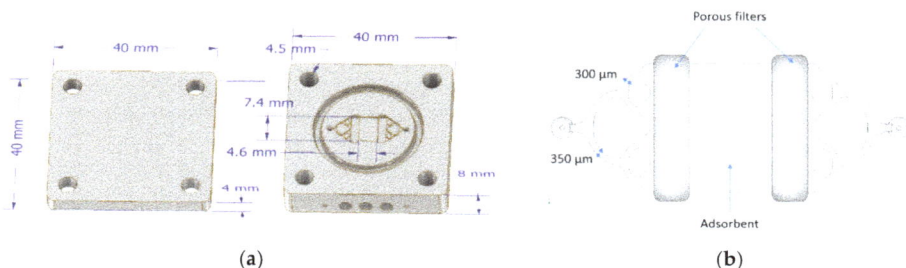

Figure 2. (a) Design of the preconcentrator and (b) zoom on the microfluidic system and the adsorbent cavity.

2.3. Experimental Setup for BTEX Generation

Different BTEX concentrations were generated using the experimental device shown in Figure 3. A standard mixture of BTEX purchased from Messer (Folschviller, France) was diluted with nitrogen (99.999% purity) using mass flow controllers 1 and 2. The initial concentration of every compound was equal to 100 ppb with a 10% uncertainty. This setup allows generating different concentrations in the range 2–100 ppb. Mass flow controller 3 was used to select the sampling flow rate which was set to 5 mL/min for all the experiments.

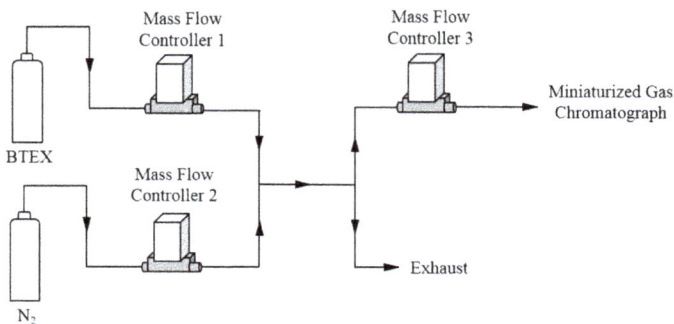

Figure 3. Experimental setup for BTEX generation.

2.4. Breakthrough Test

The above described preconcentrator was mounted on the set-up showed in Figure 4. The 5 mL/min of an equimolar 1 ppm BTEX mixture diluted in nitrogen (Air Products SAS, Aubervilliers, France) flowed through the preconcentration unit. This concentration was too high to be analyzed by a standard gas chromatograph (μBTEX-1 In'Air Solutions, Strasbourg, France) equipped with a 200 μL sampling loop. Therefore, the gas stream from the preconcentrator outlet was diluted prior to the analysis using an additional flow of 95 mL/min of nitrogen (99.999% purity, Messer, Folschviller, France) to avoid saturation of the gas chromatograph detector. The whole diluted gas stream was connected directly to the chromatograph sampling loop which was constantly renewed. Every 11 min, the content of the sampling loop was injected into the separation column and analyzed.

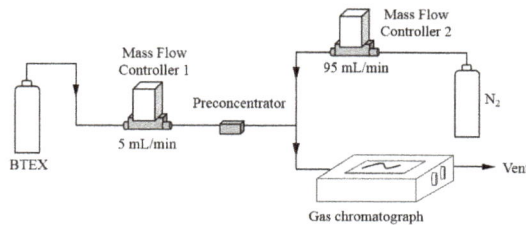

Figure 4. Schematic drawing of set-up used for breakthrough experiments.

3. Results and Discussion

Numerous experiments were conducted to evaluate the performances of the GC prototype: (1) A breakthrough experiment was carried out with a high BTEX concentration of 1 ppm in order to determine the adsorption capacity of the preconcentrator in the studied conditions; (2) repeatability was investigated; (3) the sample volume varied in the range 0–80 mL for a fixed gaseous concentration of BTEX; and (4) the gaseous BTEX concentration varied between 0–100 ppb for a given sample volume.

3.1. Adsorption Capacity

The adsorption capacity is dependent on the preconcentrator geometry, the adsorbent itself and the gas flow rate. The breakthrough time is defined as the time at which 5% of the molecules are leaving the adsorbent bed [39] so that it indicates the maximum sampling time to conduct a quantitative analysis at this concentration. Consequently, the breakthrough point is reached when 5% of the injected concentration (C_0) is passing through the adsorbent and, thus, the measured concentration at the outlet (C) is equal to 0.05 C_0. It should be noted that the breakthrough time will depend on the injected gas concentration, the flow rate and on the molecule in the case of a competitive adsorption.

Basolite C300 was selected as adsorbent as it has been already employed for benzene preconcentration in other analytical devices [40], demonstrating better preconcentration performance than Tenax TA, one of the most common materials for BTEX preconcentration.

In order to assess the adsorption capacity of the preconcentrator, a breakthrough experiment was performed. In this experiment, an equimolar 1 ppm (C_0) mixture of BTEX flowed through the preconcentrator at a flow rate of 5 mL/min. Usually, BTEX concentrations in indoor air are lower than 10 ppb and do not exceed 100 ppb, so that 1 ppm BTEX mixture simulates a highly polluted environment. The effluent concentration (C) at the preconcentrator outlet was continuously analyzed by a BTEX analyzer provided by In'Air Solutions (μBTEX-1 In'Air Solutions, Strasbourg, France) and the relative BTEX concentrations (C/C_0) were plotted versus time (see Figure 5a).

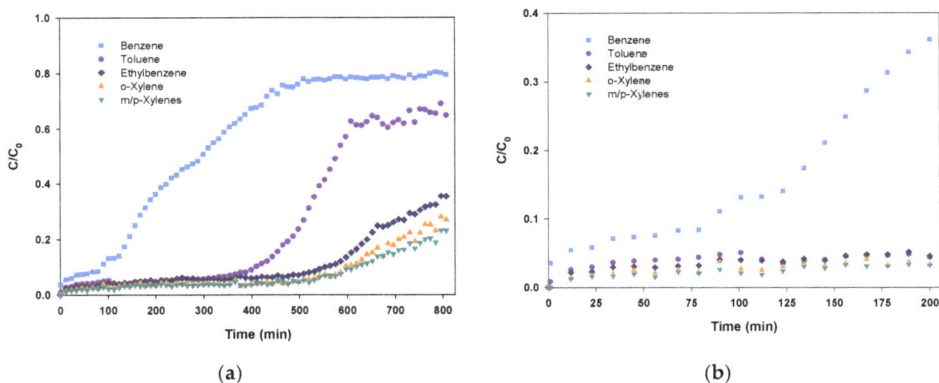

Figure 5. (**a**) C/C_0 vs. time during a breakthrough experiment performed with an equimolar 1 ppm mixture of BTEX over Basolite-filled preconcentrator and (**b**) enlarged view of the first 200 min of the experiment.

As it can be observed in Figure 5a, breakthrough curves have two different parts. The first part is almost a flat line indicating that the concentration at the outlet is close to zero and, thus, all the molecules are being adsorbed. Once the breakthrough time is reached ($C/C_0 = 0.05$), the second step begins. In this interval, the adsorbent starts to be overloaded and some of the molecules are not being adsorbed, leading to an increase of concentration at the PC outlet. This concentration rise is sharp at the beginning and starts to be less pronounced as the preconcentrator is close to the total saturation. In our experiment, only benzene and toluene have reached this point with relative concentrations of 0.7–0.8. As the objective of this test was to determine the breakthrough time, the experiment was stopped before achieving the total saturation of the adsorbent ($C/C_0 = 1$).

It is visible in Figure 5 that the adsorption capacity is remarkably different depending on the compound. Breakthrough times were determined as follows: 12 min (benzene), 156 min (toluene and ethylbenzene), 454 min (m/p-xylenes) and 457 min (o-xylene). The remarkable adsorption obtained for most of the compounds can be attributed to the large micropore volume of Basolite C300 as reported in other studies [41]. This noticeable adsorption capacity enables the collection of large sample volumes.

To illustrate the differences between benzene and the other compounds, Figure 5b presents an enlarged view of the first 200 min of the experiment where benzene clearly started to leave the adsorbent bed before the others. However, our results suggest that a sampling time of 12 min at 5 mL/min, corresponding to a sample volume of 60 mL, can be performed in a considerably polluted environment without any sample loss due to breakthrough, which would lead to an underestimation of the real gas concentration in air.

3.2. Repeatability

To evaluate the repeatability of the device, ten 20-mL samples containing 100 ppb of BTEX were consecutively injected. Between two consecutive injections, a cleaning step was conducted. During the separation step and once the sample was injected, the preconcentrator was maintained at 180 °C for 10 min to ensure there were no residues of the previous injection. After each cleaning step, a blank was conducted to verify that all BTEX were desorbed before starting the next injection. Peak area and retention time were determined for every replicate. Relative standard deviations (RSD) of peak area and retention time were calculated to verify the stability of the measurements. The corresponding RSD for the peak areas were 3.8, 6.3, 11.0, 14.7 and 13.4% for benzene, toluene, ethylbenzene, m/p-xylenes and o-xylene, respectively. The greater dispersion of ethylbenzene and m/p xylenes peak areas is probably due to the peak integration itself; as these species are coeluted, the error in the integration could be higher than in the case of benzene and toluene. Furthermore, in the current prototype, sample injection and desorption as well as the cleaning step were conducted manually because these functions were not yet integrated in the software although they can obviously impact the repeatability of the measurements. Indeed, it is expected to substantially improve the repeatability of the measurements when all the steps will be operated automatically by the software. Despite the lack of automatization, retention times were remarkably stable with RSD of 0.5, 0.7, 1.7, 1.4 and 1.4% for benzene, toluene, ethylbenzene, m/p-xylenes and o-xylene, respectively. Peak area of each blank was also determined, showing that after the cleaning step, the residues are less than 3% even after the injection of 100 ppb samples. It must be noted that a concentration of 100 ppb BTEX mimics a highly polluted environment where a preconcentrator is usually not required for analysis. Even if these results are far from ideal, the repeatability tests demonstrated that desorption is repeatable and, consequently, the aforementioned results show that the integration of the preconcentration unit into the previous GC version has not significantly influenced the repeatability of the device.

3.3. GC Signal Versus Sample Volume

After validation of the high adsorption capacity of BTEX on Basolite at high concentrations of 1 ppm and the stability of the measurements, the influence of sample volume passing through the preconcentrator was evaluated. For this, different volumes varying from 5–80 mL of a 100-ppb standard gaseous mixture of BTEX were injected in duplicate at a fixed flow rate of 5 mL/min. As it is described above, a cleaning step was performed after each analysis to ensure all BTEX have been desorbed. As illustrated in Figure 6, peak areas calculated from the chromatograms increase proportionally with the corresponding sampling volumes with correlation coefficients (R^2) of 0.9946, 0.9892, 0.9824, 0.9608 and 0.9838 for benzene, toluene, ethylbenzene, m/p-xylenes and o-xylene, respectively. Therefore, different sample volumes can be used for the analysis in the studied range without modifying the performances of the experimental device. However, longer sampling improves the sensitivity but decreases the temporal resolution, i.e., the time needed for sampling and analysis increases. Therefore, a sample volume of 20 mL was selected as a compromise between sampling volume and time resolution, the flow rate being fixed at 5 mL/min to avoid any breakthrough.

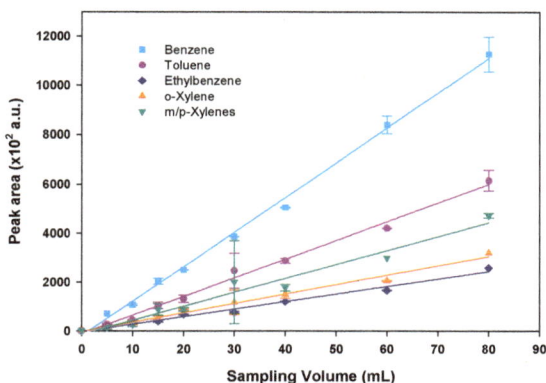

Figure 6. Peak area variation with sampling volume (BTEX concentration of 100 ppb). The vertical error bars show the standard deviation for duplicate injections.

3.4. Calibration Curves and Detection Limit

Once adsorption capacity validated and sample volume fixed to 20 mL at a flow rate of 5 mL/min, a calibration was performed using different gaseous concentrations of the targeted compounds ranging from 2.5–100 ppb. Each concentration was injected in duplicate. As mentioned before, a cleaning step after each analysis was performed. The mean peak area was then calculated and plotted versus the injected concentration for each compound. Note that m- and p-xylenes were not separated. The obtained calibration curves are displayed in Figure 7 for each species.

The peak area increases linearly with the injected concentrations, demonstrating that the preconcentrator operates in a very satisfactory way as confirmed by the obtained correlation coefficients in the range 0.9777–0.9959 (see Table 2). The calibration slopes decrease with the molecular weight of the compound, being steeper for benzene (C_6H_6) and toluene (C_7H_8) than for the other compounds (C_8H_{10}) and, thus, leading to a greater sensitivity for the weakest compounds as already observed by Nasreddine et al. [24].

Detection limits were calculated from a signal-to-noise ratio of 3 for the two lowest injected concentrations (2.5 and 5 ppb). Using a sample volume of 20 mL, detection limits of the order of a few hundred ppt were obtained for all compounds, excepting o-xylene (see Table 2).

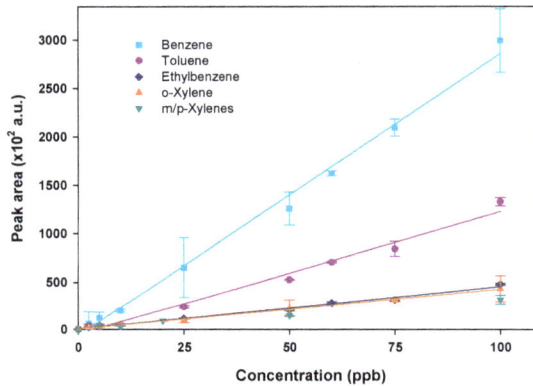

Figure 7. Calibration curves of BTEX performed at 5 mL/min and with a sampling volume of 20 mL. The vertical error bars show the standard deviation for duplicate injections.

Table 2. Calibration equations and limits of detection (LOD) obtained for BTEX with and without the preconcentration module (this work, Nasreddine et al. [24]).

Compound	Calibration Equation	R^2	LOD 1 (ppb) * (This Work)	LOD 2 (ppb) Nasreddine et al. [24]	Ratio LOD 2/LOD 1
Benzene	y = 2828.2 x	0.9913	0.20	0.72	3.6
Toluene	y = 1206.3 x	0.9777	0.26	1.16	4.46
Ethylbenzene	y = 454.2 x	0.9895	0.49	2.10	4.40
m/p-Xylenes	y = 311.9 x	0.9959	0.80	1.40	1.75
o-Xylene	y = 427.0 x	0.9949	1.70	2.63	1.55

* LOD (ppb) = (3 × lowest injected concentration)/(S/N of the lowest injected concentration).

These results show a considerable sensitivity enhancement compared to the previous development made by Nasreddine et al. [24] with an improvement ratio ranging between 1.55–4.46. To illustrate this improvement, a chromatogram of 100 ppb BTEX with and without the preconcentration stage is displayed in Figure 8. As evidenced by the figure, peak intensities are significantly greater in the new version, with the noise also increasing, but to a lesser extent.

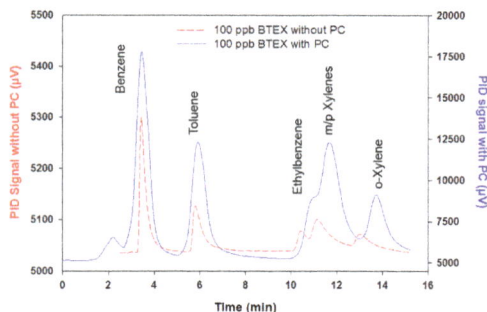

Figure 8. Red dashed line: Chromatogram of 100 ppb BTEX without preconcentration step (sample volume of 200 μL). Blue solid line: Chromatogram of 100 ppb BTEX with preconcentration step (sample volume of 20 mL).

It should be noted that lower detection limits can be achieved with this new prototype by increasing the sample volume passing through the preconcentration unit. In addition, the benzene limit of detection of 0.2 ppb obtained with the novel miniaturized GC integrating a preconcentration module is now consistent with the threshold limit value of 0.6 ppb imposed by the French regulation.

Apart from peak intensities, peak areas have been significantly enhanced. To quantify this improvement, a preconcentration factor (PF) has been defined as the ratio between peak areas with and without the preconcentration step. This factor is expected to be 100 as the sampling volume is increased by a factor of 100 (20 mL/200 μL) in the new prototype version. For the chromatogram presented in Figure 8, PF were 63.2, 82.4, 81.3, 89.3 and 77.9 for benzene, toluene, ethylbenzene, m/p xylenes and o-xylene, respectively. The difference between the obtained PF and the expected PF of 100 could be explained by two facts. Firstly, as it was demonstrated by the blanks performed in the repeatability test, some BTEX residues were not desorbed during the first desorption and, secondly, as the microfluidic cavity was manually packed with the adsorbent, it is probable that it was not completely filled and, thus, some BTEX molecules can pass through the cavity without being in contact with the adsorbent. For example, as illustrated in Figure 5b for the breakthrough experiments, some benzene molecules left the cavity without being adsorbed, even from the beginning of the experiments. This lower benzene trapping yield of 96% can explain the lower benzene PF of 63 observed. Nevertheless, satisfactory and repeatable results were obtained when varying the injected volume or concentration, showing that the trapping yield of BTEX and the preconcentrator factor are reproducible and have little significance in the accuracy of the analysis once the device is properly calibrated.

By sampling a volume of 20 mL, one full analysis cycle is performed in 19 min. Since preconcentrator cleaning and cooling down operated during the separation step, the resulting total analysis time was 19 min, which is acceptable to establish concentration–time profiles. Even if these results are successful, there are still significant scopes for improvements in terms of analysis time and peak resolution. It is expected to decrease the time of analysis by replacing the current separation column by a MEMS-based column enabling BTEX separation in 5 min, as it was reported in other studies [26,29]. Peak resolution can be enhanced by improving the performance of the PC heating system and/or reducing the detector response time. More compact devices with an integrated heating system can achieve a faster temperature ramp, which would result in shorter injection time and narrower desorption peaks. Additionally, a most powerful heating system could achieve higher temperatures in shorter times decreasing the amount of non-desorbed BTEX and, therefore, increasing the PF. Further work is planned to investigate alternative adsorbents that could potentially improve the trapping yield for benzene. On the other hand, a μPID detector with small detection chamber could also improve the peak resolution since the sample renewal inside the chamber would be faster and, thus, peak broadening would be reduced.

4. Conclusions

We reported the development of a compact GC prototype for near-real time BTEX analysis in sub-ppb range. Using a 20 mL sampling volume, BTEX analysis was conducted in 19 min and corresponding detection limits of 0.20, 0.26, 0.49, 0.80 and 1.70 ppb were calculated for benzene, toluene, ethylbenzene, m/p-xylenes and o-xylene, respectively. Considering the extremely low detection limit achieved by this prototype, it becomes possible to extend its use to other fields of application such as the food industry, early cancer diagnosis or explosives detection, by measuring other VOCs families and/or by changing the nature of adsorbent.

Several improvements can be proposed for future versions of the compact GC. First, the preconcentrator mass should be reduced in order to achieve faster heating, which would lead to more rapid temperature transfer to the adsorbent and, thus, faster desorption and then to thinner chromatographic peaks. Secondly, a shorter separation column based on MEMS technology could potentially reduce the total analysis time. In addition, the miniaturization of both components will reduce the energy consumption of the prototype, increasing its autonomy and improving its portability.

Finally, the automation of the device should integrate the cleaning step of adsorbent during pollutants chromatographic separation between two adsorption/desorption cycles, in order to improve the analysis quality.

Author Contributions: Conceptualization, V.P.; Methodology, I.L.-l., A.R.-C. and S.L.C.; Investigation, I.L.-l., A.R.C. and C.A; writing—original draft preparation, I.L.-l.; writing—review and editing, L.B., S.C. and S.L.C.; supervision, S.L.C., S.C. and L.B.

Funding: This work has received funding from the Clean Sky 2 Joint Undertaking under the European Union's Horizon 2020 research and innovation program under grant agreement No 687014. This project was also supported through the ELCOD project which was implemented as a part of the INTERREG V Oberrhein/Rhin Supérieur program and was supported by the European Regional Development Fund (ERDF) and the co-financed project partners Region Grand Est in France and the countries of Baden-Württemberg and Rhineland-Palatinate. This project has also received funding from the European Union's Framework Programme for Research and Innovation Horizon 2020 (2014–2020) under the Marie Skłodowska-Curie Grant Agreement No. 643095. This ITN Research Project is supported by European Community H2020 Framework under the Grant Agreement No. 643095 (H2020-MSCA-INT-2014).

Acknowledgments: This study was supported by the founders aforementioned cited and by CNRS and University of Strasbourg.

Conflicts of Interest: The authors declare no conflict of interest.

References

1. Bruce, N.; Perez-Padilla, R.; Albalak, R.; Organization, W.H. The health effects of indoor air pollution exposure in developing countries. *World Health Organ.* **2002**, *78*, 1078–1092.
2. Buka, I.; Koranteng, S.; Osornio-Vargas, A.R. The effects of air pollution on the health of children. *Paediatr. Child Health* **2006**, *11*, 513–516.
3. Kampa, M.; Castanas, E. Human health effects of air pollution. *Environ. Pollut.* **2008**, *151*, 362–367. [CrossRef]
4. Esmaelnejad, F.; Hajizadeh, Y.; Pourzamani, H.; Amin, M. Monitoring of benzene, toluene, ethyl benzene, and xylene isomers emission from Shahreza gas stations in 2013. *Int. J. Environ. Health Eng.* **2015**, *4*, 17.
5. Li, J.; Xie, S.D.; Zeng, L.M.; Li, L.Y.; Li, Y.Q.; Wu, R.R. Characterization of ambient volatile organic compounds and their sources in Beijing, before, during, and after Asia-Pacific Economic Cooperation China 2014. *Atmos. Chem. Phys.* **2015**, *15*, 7945–7959. [CrossRef]
6. Madhoun, W.A.A.; Ramli, N.A.; Yahaya, A.S.; Yusuf, N.F.F.M.; Ghazali, N.A.; Sansuddin, N. Levels of benzene concentrations emitted from motor vehicles in various sites in Nibong Tebal, Malaysia. *Air Qual. Atmos. Health* **2011**, *4*, 103–109. [CrossRef]
7. Miri, M.; Rostami Aghdam Shendi, M.; Ghaffari, H.R.; Ebrahimi Aval, H.; Ahmadi, E.; Taban, E.; Gholizadeh, A.; Yazdani Aval, M.; Mohammadi, A.; Azari, A. Investigation of outdoor BTEX: Concentration, variations, sources, spatial distribution, and risk assessment. *Chemosphere* **2016**, *163*, 601–609. [CrossRef]

8. Campagnolo, D.; Saraga, D.E.; Cattaneo, A.; Spinazzè, A.; Mandin, C.; Mabilia, R.; Perreca, E.; Sakellaris, I.; Canha, N.; Mihucz, V.G.; et al. VOCs and aldehydes source identification in European office buildings—The OFFICAIR study. *Build. Environ.* **2017**, *115*, 18–24. [CrossRef]

9. Cheng, Y.-H.; Lin, C.-C.; Hsu, S.-C. Comparison of conventional and green building materials in respect of VOC emissions and ozone impact on secondary carbonyl emissions. *Build. Environ.* **2015**, *87*, 274–282. [CrossRef]

10. Martins, E.M.; de Sá Borba, P.F.; dos Santos, N.E.; dos Reis, P.T.B.; Silveira, R.S.; Corrêa, S.M. The relationship between solvent use and BTEX concentrations in occupational environments. *Environ. Monit. Assess.* **2016**, *188*, 608. [CrossRef]

11. Bari, M.A.; Kindzierski, W.B.; Wheeler, A.J.; Héroux, M.-È.; Wallace, L.A. Source apportionment of indoor and outdoor volatile organic compounds at homes in Edmonton, Canada. *Build. Environ.* **2015**, *90*, 114–124. [CrossRef]

12. Romagnoli, P.; Balducci, C.; Perilli, M.; Vichi, F.; Imperiali, A.; Cecinato, A. Indoor air quality at life and work environments in Rome, Italy. *Environ. Sci. Pollut. Res. Int.* **2016**, *23*, 3503–3516. [CrossRef]

13. Ueno, Y.; Horiuchi, T.; Morimoto, T.; Niwa, O. Microfluidic Device for Airborne BTEX Detection. *Anal. Chem.* **2001**, *73*, 4688–4693. [CrossRef]

14. Hazrati, S.; Rostami, R.; Farjaminezhad, M.; Fazlzadeh, M. Preliminary assessment of BTEX concentrations in indoor air of residential buildings and atmospheric ambient air in Ardabil, Iran. *Atmos. Environ.* **2016**, *132*, 91–97. [CrossRef]

15. Kandyala, R.; Raghavendra, S.P.C.; Rajasekharan, S.T. Xylene: An overview of its health hazards and preventive measures. *J. Oral Maxillofac. Pathol. JOMFP* **2010**, *14*, 1–5. [CrossRef]

16. Rinsky, R.A.; Smith, A.B.; Hornung, R.; Filloon, T.G.; Young, R.J.; Okun, A.H.; Landrigan, P.J. Benzene and Leukemia. *N. Engl. J. Med.* **1987**, *316*, 1044–1050. [CrossRef]

17. Mandin, C.; Trantallidi, M.; Cattaneo, A.; Canha, N.; Mihucz, V.G.; Szigeti, T.; Mabilia, R.; Perreca, E.; Spinazzè, A.; Fossati, S.; et al. Assessment of indoor air quality in office buildings across Europe—The OFFICAIR study. *Sci. Total Environ.* **2017**, *579*, 169–178. [CrossRef]

18. Norbäck, D.; Hashim, J.H.; Hashim, Z.; Ali, F. Volatile organic compounds (VOC), formaldehyde and nitrogen dioxide (NO$_2$) in schools in Johor Bahru, Malaysia: Associations with rhinitis, ocular, throat and dermal symptoms, headache and fatigue. *Sci. Total Environ.* **2017**, *592*, 153–160. [CrossRef]

19. Xu, J.; Szyszkowicz, M.; Jovic, B.; Cakmak, S.; Austin, C.C.; Zhu, J. Estimation of indoor and outdoor ratios of selected volatile organic compounds in Canada. *Atmos. Environ.* **2016**, *141*, 523–531. [CrossRef]

20. Yurdakul, S.; Civan, M.; Özden, Ö.; Gaga, E.; Döğeroğlu, T.; Tuncel, G. Spatial variation of VOCs and inorganic pollutants in a university building. *Atmos. Pollut. Res.* **2017**, *8*, 1–12. [CrossRef]

21. Zhong, L.; Su, F.-C.; Batterman, S. Volatile Organic Compounds (VOCs) in Conventional and High Performance School Buildings in the U.S. *Int. J. Environ. Res. Public Health* **2017**, *14*, 100. [CrossRef]

22. Zampolli, S.; Elmi, I.; Mancarella, F.; Betti, P.; Dalcanale, E.; Cardinali, G.C.; Severi, M. Real-time monitoring of sub-ppb concentrations of aromatic volatiles with a MEMS-enabled miniaturized gas-chromatograph. *Sens. Actuators B Chem.* **2009**, *141*, 322–328. [CrossRef]

23. Skog, K.M.; Xiong, F.; Kawashima, H.; Doyle, E.; Soto, R.; Gentner, D.R. Compact, Automated, Inexpensive, and Field-Deployable Vacuum-Outlet Gas Chromatograph for Trace-Concentration Gas-Phase Organic Compounds. *Anal. Chem.* **2019**, *91*, 1318–1327. [CrossRef]

24. Nasreddine, R.; Person, V.; Serra, C.A.; Le Calvé, S. Development of a novel portable miniaturized GC for near real-time low level detection of BTEX. *Sens. Actuators B Chem.* **2016**, *224*, 159–169. [CrossRef]

25. Gregis, G.; Sanchez, J.-B.; Bezverkhyy, I.; Guy, W.; Berger, F.; Fierro, V.; Bellat, J.-P.; Celzard, A. Detection and quantification of lung cancer biomarkers by a micro-analytical device using a single metal oxide-based gas sensor. *Sens. Actuators B Chem.* **2018**, *255*, 391–400. [CrossRef]

26. Sun, J.; Xue, N.; Wang, W.; Wang, H.; Liu, C.; Ma, T.; Li, T. Compact Prototype GC-PID System integrated with Micro-PC and Micro GC Column. *J. Micromech. Microeng.* **2019**, *29*, 1–6. [CrossRef]

27. Sun, J.; Geng, Z.; Xue, N.; Liu, C.; Ma, T. A Mini-System Integrated with Metal-Oxide-Semiconductor Sensor and Micro-Packed Gas Chromatographic Column. *Micromachines* **2018**, *9*, 408. [CrossRef]

28. Wang, J.; Bryant-Genevier, J.; Nuñovero, N.; Zhang, C.; Kraay, B.; Zhan, C.; Scholten, K.; Nidetz, R.; Buggaveeti, S.; Zellers, E.T. Compact prototype microfabricated gas chromatographic analyzer for autonomous determinations of VOC mixtures at typical workplace concentrations. *Microsyst. Nanoeng.* **2018**, *4*, 17101. [CrossRef]

29. Soo, J.-C.; Lee, E.G.; LeBouf, R.F.; Kashon, M.L.; Chisholm, W.; Harper, M. Evaluation of a portable gas chromatograph with photoionization detector under variations of VOC concentration, temperature, and relative humidity. *J. Occup. Environ. Hyg.* **2018**, *15*, 351–360. [CrossRef]

30. Trzciński, J.W.; Pinalli, R.; Riboni, N.; Pedrini, A.; Bianchi, F.; Zampolli, S.; Elmi, I.; Massera, C.; Ugozzoli, F.; Dalcanale, E. In Search of the Ultimate Benzene Sensor: The EtQxBox Solution. *ACS Sens.* **2017**, *2*, 590–598. [CrossRef]

31. Lee, J.; Zhou, M.; Zhu, H.; Nidetz, R.; Kurabayashi, K.; Fan, X. Fully Automated Portable Comprehensive 2-Dimensional Gas Chromatography Device. Available online: http://pubs.acs.org/doi/abs/10.1021/acs.analchem.6b03000 (accessed on 14 January 2019).

32. Wang, J.; Nuñovero, N.; Lin, Z.; Nidetz, R.; Buggaveeti, S.; Zhan, C.; Kurabayashi, K.; Steinecker, W.H.; Zellers, E.T. A Wearable MEMS Gas Chromatograph for Multi-Vapor Determinations. *Procedia Eng.* **2016**, *168*, 1398–1401. [CrossRef]

33. Tzeng, T.-H.; Kuo, C.-Y.; Wang, S.-Y.; Huang, P.-K.; Huang, Y.-M.; Huang, Y.-M.; Hsieh, W.-C.; Huang, Y.-J.; Kuo, P.-H.; Yu, S.-A.; et al. A Portable Micro Gas Chromatography System for Lung Cancer Associated Volatile Organic Compound Detection. *IEEE J. Solid-State Circuits* **2016**, *51*, 259–272.

34. Qin, Y.; Gianchandani, Y.B. A fully electronic microfabricated gas chromatograph with complementary capacitive detectors for indoor pollutants. *Microsyst. Nanoeng.* **2016**, *2*, 15049. [CrossRef]

35. Garg, A.; Akbar, M.; Vejerano, E.; Narayanan, S.; Nazhandali, L.; Marr, L.C.; Agah, M. Zebra GC: A mini gas chromatography system for trace-level determination of hazardous air pollutants. *Sens. Actuators B Chem.* **2015**, *212*, 145–154. [CrossRef]

36. Nasreddine, R.; Person, V.; Serra, C.A.; Schoemaecker, C.; Le Calvé, S. Portable novel micro-device for BTEX real-time monitoring: Assessment during a field campaign in a low consumption energy junior high school classroom. *Atmos. Environ.* **2016**, *126*, 211–217. [CrossRef]

37. Lara-lbeas, I.; Trocquet, C.; Nasreddine, R.; Andrikopoulou, C.; Person, V.; Cormerais, B.; Englaro, S.; Le Calvé, S. BTEX near real-time monitoring in two primary schools in La Rochelle, France. *Air Qual. Atmos. Health* **2018**, *11*, 1091–1107. [CrossRef]

38. Camara, E.H.M.; Breuil, P.; Briand, D.; de Rooij, N.F.; Pijolat, C. A micro gas preconcentrator with improved performance for pollution monitoring and explosives detection. *Anal. Chim. Acta* **2011**, *688*, 175–182. [CrossRef]

39. Dettmer, K.; Engewald, W. Adsorbent materials commonly used in air analysis for adsorptive enrichment and thermal desorption of volatile organic compounds. *Anal. Bioanal. Chem.* **2002**, *373*, 490–500. [CrossRef]

40. Leidinger, M.; Rieger, M.; Sauerwald, T.; Alépée, C.; Schütze, A. Integrated pre-concentrator gas sensor microsystem for ppb level benzene detection. *Sens. Actuators B Chem.* **2016**, *236*, 988–996. [CrossRef]

41. Vellingiri, K.; Kumar, P.; Deep, A.; Kim, K.-H. Metal-organic frameworks for the adsorption of gaseous toluene under ambient temperature and pressure. *Chem. Eng. J.* **2017**, *307*, 1116–1126. [CrossRef]

MDPI
St. Alban-Anlage 66
4052 Basel
Switzerland
Tel. +41 61 683 77 34
Fax +41 61 302 89 18
www.mdpi.com

Micromachines Editorial Office
E-mail: micromachines@mdpi.com
www.mdpi.com/journal/micromachines

www.ingramcontent.com/pod-product-compliance
Lightning Source LLC
Chambersburg PA
CBHW051842210326
41597CB00033B/5750